THREE NEGRO CLASSICS

THREE NEGRO CLASSICS

UP FROM SLAVERY
Booker T Washington

THE SOULS OF
BLACK FOLK
W E B DuBois

THE AUTOBIOGRAPHY
OF AN
EX-COLORED MAN
James Weldon Johnson

INTRODUCTION BY JOHN HOPE FRANKLIN

AVON BOOKS

An Imprint of HarperCollins*Publishers*

AVON BOOKS
An Imprint of HarperCollins*Publishers*
10 East 53rd Street
New York, New York 10022-5299

Introduction copyright © 1965 by Avon Book Division, The Hearst
Corporation
Published by arrangement with the author
ISBN: 0-380-01581-1
www.avonbooks.com

First Bard Printing: February 1999
First Avon Books Printing: February 1965

Avon Trademark Reg. U.S. Pat. Off. and in Other Countries, Marca
Registrada, Hecho en U.S.A.
HarperCollins® is a trademark of HarperCollins Publishers Inc.

Printed in the U.S.A.

59   58   57   56   55   54   53   52   51

CONTENTS

INTRODUCTION

The last quarter of the nineteenth century witnessed the steady deterioration of the position of Negroes in the United States. The end of Reconstruction had left them without any protection from the merciless attacks of the Klan and other terrorist groups who continued to use the mythical threat of "Negro rule" as their excuse for lawlessness. A few Negroes still voted, but they did so at great peril; and the number declined with every passing year. Educational opportunities for Negroes were quite limited, with segregated schools in the South suffering from wanton discrimination in the allocation of public funds and with struggling denominational schools inadequately supported by their Northern benefactors. Negroes had little economic security, moreover. In the South the new industries were closed to them; and most were forced to subsist as small farmers or hapless sharecroppers. In the North new labor unions barred them from membership, while "new" immigrants showed their hostility in a variety of ways.

As the new century opened Negroes had little reason to be optimistic. Some Southern states had effectively disfranchised them in their new constitutions, and others were preparing to do so. Segregation statutes were multiplying rapidly, and the doctrine of "separate but equal" had been written into the law of the land by the Supreme Court decision in *Plessy v. Ferguson* in 1896. As Negroes moved from the farm to the city they experienced new and frightening forms of hostility and resistance. And among the new forms, none was more disturbing than the race riot. There were major riots in Wilmington, North Carolina, Atlanta, Georgia, and Brownsville, Texas, as well as in Springfield, Ohio, and Springfield, Illinois. And there were minor riots in dozens of towns and cities, North and South.

Among the numerous other means employed by white

Americans to keep the Negro in a subordinate position—
the argument that the Negro was innately inferior—was
widely and effectively advanced. Between 1890 and 1910
leading magazines and newspapers depicted Negroes as
ignorant, lazy, improvident, clownish, irresponsible, childish, and criminal. Social Darwinists, taking their cue from
the laws promulgated for the animal world in general, declared that if Negroes did not survive it was because they
were not "the fittest" and that no law promulgated by
the state could change "the natural order of things." The
local, state, and federal governments refused to argue with
these conclusions; Negroes—with the support of a few
white friends—were forced to fend for themselves.

Articulate Negroes were not unaccustomed to stating
their own case, even if they did not agree on what should
be done to improve the situation. As early as 1828 a free
Negro second-hand clothes dealer in Boston, David
Walker, lashed out against race prejudice and injustice
and called on Negroes in slavery to rise up against their
masters. A few years later Negroes began holding annual
conventions to consider ways and means of improving their
lot. At the end of the Civil War, Negroes, meeting in conventions in many parts of the United States—both North
and South—called on the state and federal governments
to protect them from the vengeful attacks by the former
Confederates. In pessimism and despair, men like Martin
Delany and Henry McNeal Turner advocated the return
of Negroes to Africa. But Frederick Douglass, T. Thomas
Fortune, the Afro-American League, and the Equal Rights
League insisted that Negroes must fight for equality *in*
the United States.

Other articulate Negroes, moreover, faced head-on the
contention that they were inferior and not entitled to an
equal place in American life. George Washington Williams
published his *History of the Negro Race in America* in
1883, in which he sought to "give the world more correct
ideas" about Negroes. Within a few years Williams and
several other Negroes who had fought in the Civil War
published histories of the Negro's part in that conflict.
Toward the end of the century Negroes sought to defend
themselves against assertions that they had made no contributions to the development of the United States. W. H.
Crogman wrote *The Progress of a Race; the Remarkable
Advancement of the Afro-American* in 1902. "The un-

precedented leap the Negro made when freed from the oppressing withes of bondage," Crogman said, "is more than deserving of a high place in history. . . . The world needs to know of what mettle these people are built." Perhaps William T. Alexander had put it more bluntly in 1888 when he wrote in the introduction to his *History of the Colored Race in America*, "so far as noble characteristics are concerned, the Colored Race possess those traits to fully as great a degree as do the white."

It seemed that the harder they tried, the more unsuccessful Negroes were in convincing white America that they deserved more than second-class citizenship and the general derogation to which they were subjected. Lynchings increased in number and barbarity; riots spread tragically in the early years of the century; and segregation was the order of the day in every part of the country. Small wonder that William E.B. DuBois declared that "The problem of the twentieth century is the problem of the color line," or that Booker T. Washington asserted that race prejudice was eating away at the vitals of the South, or that James Weldon Johnson cried out, "This land is ours by right of birth."

*   *   *

The world into which Washington, DuBois, and Johnson were born was a peculiarly difficult one for intelligent and sensitive young men. It was a world of conflict, industrial expansion, and material progress. It was a world in which cynics thrived and realists adjusted without difficulty; but one in which idealists found the going rather tough. For young Negroes who were intelligent, sensitive, and ambitious, the going was particularly rough. They did not conform to the stereotype, and if they tried to overcome the stereotype, the likelihood was that they would not only be rejected but would experience hurt as well as hardship. Washington, DuBois, and Johnson were three such men; and their lives, spanning the better portion of the last half of the nineteenth century and a goodly portion of the present century, tell a great deal about the problems of adjustment such Negroes faced as they attempted to live as normal, intelligent, well-educated Americans.

Writing in 1903 DuBois said, "Easily the most striking

thing in the history of the American Negro since 1876 is the ascendancy of Mr. Booker T. Washington." Washington's rise to a position of power and leadership had indeed been remarkable. It began with his birth in a slave cabin in Virginia in 1857 or 1858, the son of a Negro slave woman and a white father. As a very young child he had a searching desire to learn, and the sight of the school which his young mistress attended was a veritable paradise to him He soon memorized some numbers while sitting on his stepfather's barrel in the salt mines of West Virginia; upon acquiring a "blue-back" speller, he was on the way to realizing his ambition for an education.

The story of Washington's early life—how he gained a few hours of schooling each day in Malden, West Virginia, how he made his way to Hampton Institute, and how he made an enviable record as a student and worker—is told by Washington in *Up From Slavery* better than it could possibly be recounted here. Perhaps more important than the formal education he received was the encouragement that the Hampton principal, Samuel C. Armstrong, and others gave him to work unremittingly and to develop habits of dedicated service and attitudes of good will and cooperation toward his white neighbors. Soon, in his home town, where he taught for several years, and back at Hampton, to which he had been invited by the principal, he was able to test these habits and attitudes with gratifying success. Had he not impressed Samuel C. Armstrong in this regard, his career might well have ended as a teacher at Hampton.

When the white citizens of Alabama, charged with the responsibility of selecting a person to head a school for Negroes, approached the principal of Hampton Institute, he had one name to recommend: Booker T. Washington. There was opposition to such a school, but Washington eliminated most of the opposition by assuring the whites that the education he was offering would not lure Negroes from the farms or spoil them for service in the white community. He soon had their support; having made peace with the whites of Alabama, he was able to tap the resources of Northern philanthropy. Tuskegee Institute, founded in 1881, grew rapidly, acquiring several thousand acres of land, constructing many buildings, and attracting students from many parts of the United States and several foreign countries.

By the middle of the nineties Washington's name and influence extended far beyond the Tuskegee community. Whites in all parts of the country came to believe that Washington's program of education and self-help and his policy of cooperating with whites in "all things essential to mutual progress" would lead to a final solution of the race problem. In his celebrated address at the Cotton States Exposition in Atlanta in 1895, he placated white supremacists by renouncing social equality and by urging Negroes to make friends with the whites and to pursue careers in "agriculture, mechanics, in commerce, in domestic service, and in the professions." From time to time he counseled Negroes to develop habits and skills that would win for them respected places in the Southern community. He called for the intelligent management of farms, ownership of land, habits of thrift, patience, and perseverance, and the cultivation of high morals and good manners.

By the beginning of the new century, Washington was one of the most powerful men in the United States. Great philanthropists and industrialists such as Andrew Carnegie and John D. Rockefeller listened to him courteously and were influenced by his advice. Presidents such as Theodore Roosevelt and William Howard Taft depended on him for suggestions regarding the resolution of problems involving race. Southern whites in high places knew that a good word in their behalf by Washington would open doors previously closed to them. If most whites and many Negroes regarded Washington as one of the great and gifted leaders of their time, some Negroes entertained doubts about the validity of his position and the effectiveness of his leadership. William Monroe Trotter, fiery editor of the *Boston Guardian*, regarded Washington as a traitor to his race. W. E. B. DuBois recognized his strength but seriously questioned his wisdom and even his integrity. Perhaps the thing that disturbed DuBois most was what he later called "The Tuskegee Machine," manifested in the power that Washington enjoyed in determining the future course of Negroes in general as well as the careers of individual Negroes.

The criticism of Washington by DuBois, Trotter, and others like them did little to dislodge Washington from his position as spokesman for Negroes and as arbiter in Negro-white relations. The fact that race relations did not substantially improve during the period of his leadership, that lynchings increased yearly, and that the modern race riot was born when he was at the height of his power and influence convinced DuBois that Washington did not have the answers. But these events did little to shake his position with his powerful white friends. When he died in 1915, he was still firmly entrenched in the position of leading educator, leading racial adviser at many levels, and leading advocate of what DuBois called the "gospel of work and money."

Among Negro Americans, there could hardly be a greater contrast than the careers of Booker T. Washington and W. E. B. DuBois. Born in Great Barrington, Massachusetts, in the year that Ulysses S. Grant was elected President of the United States, DuBois was relatively late in discovering that his color placed him at a disadvantage. "I remember well when the shadow swept across me," he later wrote. "I was a little thing, away up in the hills of New England, where the dark Housatonic winds between Hoosac and Taghkanie to the sea. In a wee wooden schoolhouse, something put it into the boys' and girls' heads to buy gorgeous visiting-cards—ten cents a package—and exchange. The exchange was merry, till one girl, a tall newcomer, refused my card—refused it peremptorily, with a glance. Then it dawned upon me with a certain suddenness that I was different from the others; or like, mayhap, in heart and life and longing, but shut out from their world by a vast veil. . . ." But DuBois was different in still another way. He excelled among his white classmates and went on to achieve high academic honors.

At Fisk University DuBois learned more about the world of color as well as the worlds of philosophy, history, and the classics. During the academic year he wrote essays for his courses and feature articles for the student publication, *The Herald*. During the summers he secured employment as a teacher of young Negroes in a school in the Tennessee hills, where he met "new and intricate and unconscious discrimination." From Fisk he proceeded to Harvard where he received another degree of Bachelor of Arts and, as one of the six commencement speakers, selected "Jefferson Davis" as the subject of his discourse. Within a few years DuBois not only had earned the degree of Doctor of Philosophy at Harvard but had also pursued post-graduate studies in Berlin. By this time he had reached the conclusion that the most important contribution he could make toward the improvement of the Negro's position in the United States was to undertake a series of studies of the Negro, "primarily scientific—a careful search for Truth conducted as thoroughly, broadly, and honestly as the material resources and mental equipment will allow."

To carry forward his aim, DuBois wrote *The Philadelphia Negro: A Social Study,* which was published in 1899 by the University of Pennsylvania. At Atlanta University, where he had accepted an invitation to teach sociology, he set up a program of studies of the American Negro and published works, among others, on the Negro common school, the Negro church, and the Negro artisan. These studies, that have been described as the "first real sociological research in the South," did not satisfy DuBois. Scientific inquiry seemed to prove nothing to the people of the United States regarding the Negro; and DuBois was becoming distressed over the manner in which the doctrines of Booker T. Washington were being hailed as indisputable truth. It was at this juncture, in 1903, that DuBois turned from his sociological and historical studies to write *The Souls of Black Folk.* Although he would write many scholarly works in the future, scientific scholarship would never again claim the undivided attention of DuBois. In 1905 he organized the Niagara Movement, a group of young Negro intellectuals who made it clear that they would not compromise in their quest for complete equality in American life. In 1909 he and most of the members of the Niagara Movement joined with other Negroes and their white supporters to found the National Association for the Advancement of Colored People, in which he became director of research and editor of the *Crisis.* In his new position DuBois pressed hard for legal and political recognition of the rights of the Negro. He supported Woodrow Wilson for President in 1912 under the mistaken impression that Wilson would take a stand in support of Negro rights. During World War I he urged Negroes to give full support to the Allied cause, in the hope that if the world was saved for democracy, Negroes would also be saved.

Disillusionment began to overcome DuBois as he began to see clearly "the connection of economics and politics."

If the Berlin Conference of 1885 began the final degradation of darker peoples, the Versailles Conference of
1919, DuBois feared, would complete it. To combat this
possibility and to demonstrate the common plight of Negroes at home and abroad, DuBois organized the Pan-
African Congress in Paris in February, 1919. By his own
admission the results of the meeting were small, but the
convening of fifty-seven Negro delegates from Africa, the
West Indies, and the United States contributed to the
growing awareness of the problem of color in the twentieth century. For the next decade the race problem in
the United States, with riots, an unsuccessful attempt to
secure federal legislation against lynching, and a vigorous
drive to widen the voting privileges of Negroes, kept Du-
Bois busy. But the gains were small, too small for Du-
Bois; and he seriously began to question the premises
and assumptions of the organization that was leading in
the fight for Negro equality.

By 1930 DuBois had become convinced that the basic
policies and ideals of the National Association for the
Advancement of Colored People must be modified and
changed; "that in a world where economic dislocation
had become so great as in ours, a mere appeal based on
the old liberalism, a mere appeal to justice and further
effort at legal decision, was missing the essential need;
that the essential need was to guard and better the chances
of Negroes, educated and ignorant, to earn a living, safeguard their income, and raise the level of their employment." DuBois looked to the increase of the power and
economic organization among Negroes, and he proposed
to support this policy through the pages of the *Crisis*.
When he did so and also began to criticize the organization, his position was regarded as no longer tenable, and
he left the NAACP, returned to Atlanta University and
scholarly pursuits.

It seems almost incredible that DuBois still had more
than thirty years in which to pursue several careers, including the founding of *Phylon*, the Atlanta University
Review of Race and Culture, the return to the NAACP
as Director of Special Research, the advocacy of closer
ties among the darker peoples of the world, the steady
movement toward socialism culminating in his membership
in the Communist Party of the United States, his re-
nunciation of his citizenship in the United States, and

his taking out citizenship in Ghana, where he died in August, 1963. Always the seeker, highly sensitive to the persistent flux in the social order, DuBois had come to feel before the end of his life that nothing short of a drastic reorientation of the economic and social order could bring even a semblance of justice to Negroes in the United States and to colonial peoples.

DuBois was only three years old when James Weldon Johnson, the youngest of the three men under consideration, was born in 1871. Like DuBois, Johnson had the good fortune to enjoy a stable family life; even in his native Jacksonville, Florida, he experienced no shattering experiences connected with race at an early age. Indeed, he later said that if anything he developed "an unconscious race-superiority complex" and his early ambitions to be an important person someday were accepted as perfectly normal. When he was twelve years old he made a trip to New York; and this experience awakened his love for cosmopolitanism and caused him to conclude that he was "born to be a New Yorker." It was about this time that a Spanish-speaking companion stimulated his interest and enthusiasm for foreign languages and cultures.

It was not until Johnson went to Atlanta University that he had his first Jim Crow experience, the result of the recent enactment of a Florida statute providing for the separation of the races in railroad cars. When Johnson and a Cuban friend were proceeding to Atlanta, the conductor on the train was about to move him to the "Negro car," when the conversation in Spanish by the two young men impressed the conductor into silence. This caused Johnson to remark later that "in such situations, any kind of Negro will do; provided he is not one who is an American citizen" or appears to be an American citizen. This and similar experiences led Johnson to the conclusion that the experience of being a Negro in America was so narrowing that he must constantly guard against the danger of being choked or suffocated by it.

There was little danger that Johnson would be choked or suffocated by being a Negro. As Rebecca Barton has observed, Johnson never allowed himself to slip into a rut vocationally; "alternating between the poor pay of idealistic work and lucrative employment, between serious-minded purpose and frivolous interlude, he continually replenished his life energy and renewed his vision." After

his studies at Atlanta University he became a teacher,
then a principal. Still dissatisfied with his career, he read
law and passed the Florida bar, although he never became a practicing attorney.

He had said that he was "born to be a New Yorker,"
and it was to New York that he turned, after resigning
his principalship unceremoniously and resolving to take
his chances in the world of entertainment. With his
brother J. Rosamond Johnson, a talented and well-trained
pianist, he began to write musical comedies for the Broadway stage. Their success was remarkable. James Weldon
wrote the lyrics for a series of hits, including *Under The
Bamboo Tree, My Castle on the Nile,* and *Nobody's
Lookin' But the Owl and the Moon.* On tour with their
partner, Bob Cole, the Johnsons heard their works performed in many parts of the United States and Europe.
In his autobiography, *Along This Way,* Johnson wrote in
1933, "From the day I set foot in France, I became
aware of the working of a miracle within me. . . . I
recaptured for the first time since childhood the sense
of being just a human being."

But the restless Johnson continued to seek new experiences and new ways to exploit his talents. At Columbia University he studied literature and drama and began
to write seriously. It was his Columbia professor, Brander
Matthews, who encouraged him to write his first novel,
The Autobiography of an Ex-Colored Man. Already he
had become United States Consul in Venezuela; and in
1909 he received a similar appointment for Nicaragua. In
1914 he returned to New York, and two years later he
accepted the position as Field Secretary of the NAACP.
Working feverishly to build up the most important organization fighting to advance the position of the Negro,
Johnson increased the number of branches of the NAACP
from sixty-eight to three hundred and ten. In 1920 he
became general secretary of the organization, a position
he was to hold for a decade.

Johnson's finest hour as a race leader came in his
effort to secure the passage of the Dyer Anti-lynching Bill
by the United States Congress. For two years, 1921-1922,
he lobbied in Washington to persuade members of Congress to vote for the bill. Its passage by the House of
Representatives was due in no small degree to his relentless efforts. When it failed in the Senate, never having

come to a vote, Johnson came as close as he ever came to being embittered. It was at this time that he mused over the "thoughtlessness of people who take it as a matter of course that American Negroes should love their country."

By 1930 Johnson once more felt the call of the contemplative life and left the NAACP to become Adam K. Spence Professor of Creative Literature at Fisk University. There he wrote his autobiography and other works and gave public lectures regularly on "Life, Art, and Letters." At Fisk he hoped to influence students to develop additional racial strength and fitness "and in shaping fresh forces against bigotry and racial wrong." One of America's most intelligent and creative forces was lost to Fisk students and the country when Johnson was killed in an automobile accident in Maine in 1938.

* * *

Washington, DuBois, and Johnson were not only quite active on a variety of fronts, but they were all extremely prolific writers. Washington's first writings were his lectures and public addresses, which enjoyed a wide readership in all parts of the country. Many of them were later published in book form. There was such great curiosity about Washington as a person that he was urged to publish a series of articles about himself in the *Outlook*. These articles formed the basis for *Up From Slavery*, published in 1901; and this was a much more successful work than his first attempt at autobiography, *The Story of My Life and Work*, which had been published the previous year. Two works that followed, *Working with the Hands* (1904) and *Tuskegee and Its People* (1905) were concerned mainly with the Tuskegee story; while *My Larger Education* (1911) told of his experiences since the publication of *Up From Slavery*. He also lent his name to works that, at best, he merely supervised. His biography of Frederick Douglass (1907) has been attributed largely to Washington's friend, S. Laing Williams, while Robert E. Park did most of the work on *The Story of the Negro* (1909) and *The Man Farthest Down* (1912). There were other articles and books on Negro business, the Negro and labor, character building, and education.

Washington's writings, like his appearances, were in great demand, and he was hard put to satisfy it.

Nothing Washington wrote approached in popularity or importance his *Up From Slavery*. It is a great success story, the kind that appealed to millions of Americans. The fact that its subject was a Negro increased its appeal, and his almost unbecoming modesty won Washington and Tuskegee many new friends. In the pages of this work, as much as in the Atlanta speech, are the assurances that white America sought that the Negro would not make a serious bid for political and social equality. "I believe it is the duty of the Negro," he writes, "to deport himself modestly in regard to political claims, depending upon the slow but sure influences that proceed from the possession of property, intelligence, and high character for the full recognition of his political rights." He would insist upon the same deportment on the part of whites, but while white Americans hailed his position on the political aspirations of Negroes, they paid scant attention to what he had to say about whites.

DuBois began writing when he was editor of the Fisk *The Herald* in 1887 and never stopped until he died in 1963. During those years he produced a vast number of books, articles, and pamphlets; a bibliography would fill many pages. Outstanding among his works are his *Suppression of the African Slave Trade*, his doctoral dissertation, that was published as the first volume of the Harvard Historical Studies in 1896. Then followed the study of Philadelphia, the monographs on various aspects of Negro life in the United States, and the collection of essays called *The Souls of Black Folk*, published in 1903. In the year that the NAACP was founded, 1909, DuBois published his biography of John Brown, which he regarded as one of the best things he had done. Then followed his first novel, in 1911, *The Quest of the Silver Fleece*, his first historical sketch of the Negro (1915) that appeared in a revised and enlarged form in 1939 under the title *Black Folk, Then and Now, Darkwater* in 1920, and the *Gift of Black Folk* in 1924. Four years later his second novel and his favorite book, *Dark Princess*, appeared.

The years of his second period at Atlanta University as well as the later years were most prolific. In 1935 his most significant historical work, *Black Reconstruction*, was published; and while it was an angry book, it demanded and secured a complete re-examination of the troubled years following the Civil War. Ten years later he turned his attention to the problem of colonies and world peace, publishing *Color and Democracy* in 1945 and *The World and Africa* in 1947. Toward his last years he devoted his efforts to completing an ambitious fictional trilogy, *The Ordeal of Mansard*, and doing preparatory work in Accra on the *Encyclopaedia Africana*. The latter effort was a resumption of a much earlier attempt—in the thirties—to bring out an Encyclopedia of the Negro under the auspices of the Phelps-Stokes Fund.

The Souls of Black Folk is easily the most widely read and the most successful of DuBois' works. Almost immediately it became a platform for young Negro intellectuals who found something painfully missing in the program of Booker T. Washington. It was several things at once: a deeply moving statement on the consciousness of color, a searching criticism of the philosophy of Washington, a quite scholarly examination of certain phases of the history of the Negro, and an evaluation of some of the mainsprings of the culture of the Negro American. There were those who shuddered at the impertinence of this "overly trained" Negro, but there were others who recognized this as an unusual and highly gifted talent.

One of the most striking things about the essays in *The Souls of Black Folk* is the control and restraint of the writing. Even as DuBois turned from purely scientific inquiry as a means of delivering the Negro from subordination, he wrote as an advocate who had been disciplined by years of scholarly study and writing. If this book was a fitting preparation for a man who was about to enter into a struggle in behalf of human dignity and equality that was to last for more than a half century, it was also a work whose twenty-odd editions attest the continuing interest and fascination on the part of a reading public in search for a succinct statement of the Negro's case for his right to equality. Small wonder that the hero in James Weldon Johnson's *Autobiography of an Ex-Colored Man* called it "remarkable."

Now and then DuBois tried his hand at verse, but it was never the preoccupation for him that it was for James Weldon Johnson. In addition to writing the lyrics for many of his brother's songs, Johnson wrote much serious verse. Three volumes, *Fifty Years and Other Poems* (1917), *God's Trombones* (1927) and *St. Peter Relates an Incident of the Resurrection Day* (1930) are representative of his best poetry. His *Black Manhattan* (1930) remains one of the best accounts ever written of the Negro in New York City and reveals not only his vast knowledge of the city but also his affection and enthusiasm for his favorite community. In 1933 he published his autobiography, *Along This Way,* which is as much a history of an important aspect of American life as it is the recounting of the personal experiences of one of the participants. That he continued to maintain a lively and serious interest in the problems of the Negro after he took up his professorship at Fisk University is revealed in his *Negro Americans, What Now?*, published in 1934. He made it clear that the struggle of the Negro for respect and recognition must be relentless and that even while making adjustments to new and difficult situations, he must never lose his own self-respect.

The Autobiography of an Ex-Colored Man, first published in 1912, is the only novel that James Weldon Johnson ever wrote. It deals with the problem of the light-skinned Negro whose situation is considerably aggravated by his frequently mistaken identity. The theme had been employed by earlier writers—by William Wells Brown in his *Clotel* in 1853 and by Charles W. Chesnutt in *The Wife of His Youth* in 1899—but Johnson's work is the first to deal primarily with the problem of "passing." It was not until 1927, when it was republished at the height of the Harlem Renaissance, that the work received wide critical attention. Its description of Bohemian life among Negroes in New York and its handling of many phases of Negro life at all levels attracted the attention of those who by that time were deeply interested in learning more about Negro aspirations and problems.

These three Negro classics are as different from each other as the three authors who wrote them. Yet, in its own special way, each reveals the deep apprehensions and the troubling dilemmas that virtually every sensitive Negro American has experienced. Each in its own way makes an attempt to overcome the crippling obstacles that every Negro in the United States has confronted. If every Negro cannot accommodate himself to the whims and foibles of white people or cannot throw himself into a passionate fight to destroy racial bigotry or "pass" into the society of his adversaries, he can at least maintain his equanimity and his dignity and his self-respect. And that is the message of each of these works, and why their publication together is an auspicious event.

John Hope Franklin

Up From Slavery

PREFACE

This volume is the outgrowth of a series of articles, dealing with incidents in my life, which were published consecutively in the *Outlook*. While they were appearing in that magazine I was constantly surprised at the number of requests which came to me from all parts of the country, asking that the articles be permanently preserved in book form. I am most grateful to the *Outlook* for permission to gratify these requests.

I have tried to tell a simple, straightforward story, with no attempt at embellishment. My regret is that what I have attempted to do has been done so imperfectly. The greater part of my time and strength is required for the executive work connected with the Tuskegee Normal and Industrial Institute, and in securing the money necessary for the support of the institution. Much of what I have said has been written on board trains, or at hotels or railroad stations while I have been waiting for trains, or during the moments that I could spare from my work while at Tuskegee. Without the painstaking and generous assistance of Mr. Max Bennett Thrasher I could not have succeeded in any satisfactory degree.

This volume is dedicated to my Wife
MARGARET JAMES WASHINGTON

And to my Brother
JOHN H. WASHINGTON

*Whose patience, fidelity, and hard work have gone far
to make the work at Tuskegee successful*

CHAPTER I

A Slave Among Slaves

I was born a slave on a plantation in Franklin County, Virginia. I am not quite sure of the exact place or exact date of my birth, but at any rate I suspect I must have been born somewhere and at some time. As nearly as I have been able to learn, I was born near a cross-roads post-office called Hale's Ford, and the year was 1858 or 1859. I do not know the month or the day. The earliest impressions I can now recall are of the plantation and the slave quarters—the latter being the part of the plantation where the slaves had their cabins.

My life had its beginning in the midst of the most miserable, desolate, and discouraging surroundings. This was so, however, not because my owners were especially cruel, for they were not, as compared with many others. I was born in a typical log cabin, about fourteen by sixteen feet square. In this cabin I lived with my mother and a brother and sister till after the Civil War, when we were all declared free.

Of my ancestry I know almost nothing. In the slave quarters, and even later, I heard whispered conversations among the coloured people of the tortures which the slaves, including, no doubt, my ancestors on my mother's side, suffered in the middle passage of the slave ship while being conveyed from Africa to America. I have been unsuccessful in securing any information that would throw any accurate light upon the history of my family beyond my mother. She, I remember, had a half-brother and a half-sister. In the days of slavery not very much attention was given to family history and family records—that is, black family records. My mother, I suppose, attracted the attention of a purchaser who was afterward my owner and hers. Her addition to the slave family attracted about as much attention as the purchase of a new horse or cow. Of my father I know even less than of my mother. I do not even know his name. I have heard reports to the ef-

fect that he was a white man who lived on one of the
near-by plantations. Whoever he was, I never heard of his
taking the least interest in me or providing in any way
for my rearing. But I do not find especial fault with him.
He was simply another unfortunate victim of the institution which the Nation unhappily had engrafted upon it
at that time.

The cabin was not only our living-place, but was also
used as the kitchen for the plantation. My mother was
the plantation cook. The cabin was without glass windows; it had only openings in the side which let in the
light, and also the cold, chilly air of winter. There was a
door to the cabin—that is, something that was called a
door—but the uncertain hinges by which it was hung, and
the large cracks in it, to say nothing of the fact that it was
too small, made the room a very uncomfortable one. In
addition to these openings there was, in the lower right-
hand corner of the room, the "cat-hole,"—a contrivance
which almost every mansion or cabin in Virginia possessed during the ante-bellum period. The "cat-hole" was
a square opening, about seven by eight inches, provided
for the purpose of letting the cat pass in and out of the
house at will during the night. In the case of our particular cabin I could never understand the necessity for this
convenience, since there were at least a half-dozen other
places in the cabin that would have accommodated the
cats. There was no wooden floor in our cabin, the naked
earth being used as a floor. In the centre of the earthen
floor there was a large, deep opening covered with
boards, which was used as a place in which to store sweet
potatoes during the winter. An impression of this potato-
hole is very distinctly engraved upon my memory, because
I recall that during the process of putting the potatoes in or
taking them out I would often come into possession of
one or two, which I roasted and thoroughly enjoyed.
There was no cooking-stove on our plantation, and all
the cooking for the whites and slaves my mother had to do
over an open fireplace, mostly in pots and "skillets." While
the poorly built cabin caused us to suffer with cold in
the winter, the heat from the open fireplace in summer
was equally trying.

The early years of my life, which were spent in the
little cabin, were not very different from those of thousands of other slaves. My mother, of course, had little

time in which to give attention to the training of her children during the day. She snatched a few moments for our care in the early morning before her work began, and at night after the day's work was done. One of my earliest recollections is that of my mother cooking a chicken late at night, and awakening her children for the purpose of feeding them. How or where she got it I do not know. I presume, however, it was procured from our owner's farm. Some people may call this theft. If such a thing were to happen now, I should condemn it as theft myself. But taking place at the time it did, and for the reason that it did, no one could ever make me believe that my mother was guilty of thieving. She was simply a victim of the system of slavery. I cannot remember having slept in a bed until after our family was declared free by the Emancipation Proclamation. Three children—John, my older brother, Amanda, my sister, and myself—had a pallet on the dirt floor, or, to be more correct, we slept in and on a bundle of filthy rags laid upon the dirt floor.

I was asked not long ago to tell something about the sports and pastimes that I engaged in during my youth. Until that question was asked it had never occurred to me that there was no period of my life that was devoted to play. From the time that I can remember anything, almost every day of my life has been occupied in some kind of labour; though I think I would now be a more useful man if I had had time for sports. During the period that I spent in slavery I was not large enough to be of much service, still I was occupied most of the time in cleaning the yards, carrying water to the men in the fields, or going to the mill, to which I used to take the corn, once a week, to be ground. The mill was about three miles from the plantation. This work I always dreaded. The heavy bag of corn would be thrown across the back of the horse, and the corn divided about evenly on each side; but in some way, almost without exception, on these trips, the corn would shift as to become unbalanced and would fall off the horse, and often I would fall with it. As I was not strong enough to reload the corn upon the horse, I would have to wait, sometimes for many hours, till a chance passer-by came along who would help me out of my trouble. The hours while waiting for some one were usually spent in crying. The time consumed in this way made me late in reaching the mill, and by the time I got my corn

ground and reached home it would be far into the night.
The road was a lonely one, and often led through dense
forests. I was always frightened. The woods were said to
be full of soldiers who had deserted from the army, and
I had been told that the first thing a deserter did to a
Negro boy when he found him alone was to cut off his
ears. Besides, when I was late in getting home I knew I
would always get a severe scolding or a flogging.

I had no schooling whatever while I was a slave, though
I remember on several occasions I went as far as the
schoolhouse door with one of my young mistresses to
carry her books. The picture of several dozen boys and
girls in a schoolroom engaged in study made a deep impression upon me, and I had the feeling that to get into
a schoolhouse and study in this way would be about the
same as getting into paradise.

So far as I can now recall, the first knowledge that I
got of the fact that we were slaves, and that freedom of
the slaves was being discussed, was early one morning
before day, when I was awakened by my mother kneeling
over her children and fervently praying that Lincoln and
his armies might be successful, and that one day she and
her children might be free. In this connection I have never
been able to understand how the slaves throughout the
South, completely ignorant as were the masses so far as
books or newspapers were concerned, were able to keep
themselves so accurately and completely informed about
the great National questions that were agitating the country. From the time that Garrison, Lovejoy, and others
began to agitate for freedom, the slaves throughout the
South kept in close touch with the progress of the movement. Though I was a mere child during the preparation
for the Civil War and during the war itself, I now recall
the many late-at-night whispered discussions that I heard
my mother and the other slaves on the plantation indulge
in. These discussions showed that they understood the
situation, and that they kept themselves informed of events
by what was termed the "grape-vine" telegraph.

During the campaign when Lincoln was first a candi-
date for the Presidency, the slaves on our far-off plantation, miles from any railroad or large city or daily newspaper, knew what the issues involved were. When war
was begun between the North and the South, every slave
on our plantation felt and knew that, though other issues

were discussed, the primal one was that of slavery. Even the most ignorant members of my race on the remote plantations felt in their hearts, with a certainty that admitted of no doubt, that the freedom of the slaves would be the one great result of the war, if the Northern armies conquered. Every success of the Federal armies and every defeat of the Confederate forces was watched with the keenest and most intense interest. Often the slaves got knowledge of the results of great battles before the white people received it. This news was usually gotten from the coloured man who was sent to the post-office for the mail. In our case the post-office was about three miles from the plantation and the mail came once or twice a week. The man who was sent to the office would linger about the place long enough to get the drift of the conversation from the group of white people who naturally congregated there, after receiving their mail, to discuss the latest news. The mail-carrier on his way back to our master's house would as naturally retail the news that he had secured among the slaves, and in this way they often heard of important events before the white people at the "big house," as the master's house was called.

I cannot remember a single instance during my childhood or early boyhood when our entire family sat down to the table together, and God's blessing was asked, and the family ate a meal in a civilized manner. On the plantation in Virginia, and even later, meals were gotten by the children very much as dumb animals get theirs. It was a piece of bread here and a scrap of meat there. It was a cup of milk at one time and some potatoes at another. Sometimes a portion of our family would eat out of the skillet or pot, while some one would eat from a tin plate held on the knees, and often using nothing but the hands with which to hold the food. When I had grown to sufficient size, I was required to go to the "big house" at meal-times to fan the flies from the table by means of a large set of paper fans operated by a pulley. Naturally much of the conversation of the white people turned upon the subject of freedom and the war, and I absorbed a good deal of it. I remember that at one time I saw two of my young mistresses and some lady visitors eating ginger-cakes, in the yard. At that time those cakes seemed to me to be absolutely the most tempting and desirable things that I had ever seen; and I then and

there resolved that, if I ever got free, the height of my
ambition would be reached if I could get to the point
where I could secure and eat ginger-cakes in the way that
I saw those ladies doing.

Of course as the war was prolonged the white people,
in many cases, often found it difficult to secure food for
themselves. I think the slaves felt the deprivation less than
the white, because the usual diet for the slaves was corn
bread and pork, and these could be raised on the plantation; but coffee, tea, sugar, and other articles which the
whites had been accustomed to use could not be raised
on the plantation, and the conditions brought about by
the war frequently made it impossible to secure these
things. The whites were often in great straits. Parched
corn was used for coffee, and a kind of black molasses
was used instead of sugar. Many times nothing was used
to sweeten the so-called tea and coffee.

The first pair of shoes that I recall wearing were wooden
ones. They had rough leather on the top, but the bottoms,
which were about an inch thick, were of wood. When I
walked they made a fearful noise, and besides this they
were very inconvenient, since there was no yielding to the
natural pressure of the foot. In wearing them one presented an exceedingly awkward appearance. The most trying ordeal that I was forced to endure as a slave boy,
however, was the wearing of a flax shirt. In the portion of
Virginia where I lived it was common to use flax as part
of the clothing for the slaves. That part of the flax from
which our clothing was made was largely the refuse, which
of course was the cheapest and roughest part. I can scarcely imagine any torture, except, perhaps, the pulling of a
tooth, that is equal to that caused by putting on a new
flax shirt for the first time. It is almost equal to the feeling that one would experience if he had a dozen or more
chestnut burrs, or a hundred small pinpoints, in contact
with his flesh. Even to this day I can recall accurately the
tortures that I underwent when putting on one of these
garments. The fact that my flesh was soft and tender
added to the pain. But I had no choice. I had to wear
the flax shirt or none; and had it been left to me to
choose, I should have chosen to wear no covering. In connection with the flax shirt, my brother John, who is several
years older than I am, performed one of the most generous
acts that I ever heard of one slave relative doing for an-

other. On several occasions when I was being forced to
wear a new flax shirt, he generously agreed to put it on in
my stead and wear it for several days, till it was "broken
in." Until I had grown to be quite a youth this single
garment was all that I wore.

One may get the idea from what I have said, that there
was bitter feeling toward the white people on the part
of my race, because of the fact that most of the white
population was away fighting in a war which would result in keeping the Negro in slavery if the South was successful. In the case of the slaves on our place this was not
true, and it was not true of any large portion of the slave
population in the South where the Negro was treated with
anything like decency. During the Civil War one of my
young masters was killed, and two were severely wounded.
I recall the feeling of sorrow which existed among the
slaves when they heard of the death of "Mars' Billy." It
was no sham sorrow but real. Some of the slaves had
nursed "Mars' Billy"; others had played with him when
he was a child. "Mars' Billy" had begged for mercy in the
case of others when the overseer or master was thrashing
them. The sorrow in the slave quarter was only second to
that in the "big house." When the two young masters
were brought home wounded, the sympathy of the slaves
was shown in many ways. They were just as anxious to
assist in the nursing as the family relatives of the wounded.
Some of the slaves would even beg for the privilege of sitting up at night to nurse their wounded masters. This
tenderness and sympathy on the part of those held in
bondage was a result of their kindly and generous nature.
In order to defend and protect the women and children
who were left on the plantations when the white males
went to war, the slaves would have laid down their lives.
The slave who was selected to sleep in the "big house"
during the absence of the males was considered to have
the place of honour. Any one attempting to harm "young
Mistress" or "old Mistress" during the night would have
had to cross the dead body of the slave to do so. I do
not know how many have noticed it, but I think that it
will be found to be true that there are few instances, either
in slavery or freedom, in which a member of my race
has been known to betray a specific trust.

As a rule, not only did the members of my race entertain no feelings of bitterness against the whites before

and during the war, but there are many instances of
Negroes tenderly caring for their former masters and
mistresses who for some reason have become poor and
dependent since the war. I know of instances where the
former masters of slaves have for years been supplied
with money by their former slaves to keep them from
suffering. I have known of still other cases in which the
former slaves have assisted in the education of the descendants of their former owners. I know of a case on a
large plantation in the South in which a young white
man, the son of the former owner of the estate, has become so reduced in purse and self-control by reason of
drink that he is a pitiable creature; and yet, notwithstanding the poverty of the coloured people themselves on this
plantation, they have for years supplied this young white
man with the necessities of life. One sends him a little
coffee or sugar, another a little meat, and so on. Nothing
that the coloured people possess is too good for the son
of "old Mars' Tom," who will perhaps never be permitted
to suffer while any remain on the place who knew directly or indirectly of "old Mars' Tom."

I have said that there are few instances of a member
of my race betraying a specific trust. One of the best
illustrations of this which I know of is in the case of an
ex-slave from Virginia whom I met not long ago in a
little town in the state of Ohio. I found that this man had
made a contract with his master, two or three years previous to the Emancipation Proclamation, to the effect that
the slave was to be permitted to buy himself, by paying so
much per year for his body; and while he was paying
for himself, he was to be permitted to labour where and
for whom he pleased. Finding that he could secure better
wages in Ohio, he went there. When freedom came, he
was still in debt to his master some three hundred dollars.
Notwithstanding that the Emancipation Proclamation
freed him from any obligation to his master, this black
man walked the greater portion of the distance back to
where his old master lived in Virginia, and placed the
last dollar, with interest, in his hands. In talking to me
about this, the man told me that he knew that he did
not have to pay the debt, but that he had given his word
to his master, and his word he had never broken. He felt
that he could not enjoy his freedom till he had fulfilled
his promise.

From some things that I have said one may get the idea that some of the slaves did not want freedom. This is not true. I have never seen one who did not want to be free, or one who would return to slavery.

I pity from the bottom of my heart any nation or body of people that is so unfortunate as to get entangled in the net of slavery. I have long since ceased to cherish any spirit of bitterness against the Southern white people on account of the enslavement of my race. No one section of our country was wholly responsible for its introduction, and, besides, it was recognized and protected for years by the General Government. Having once got its tentacles fastened on to the economic and social life of the Republic, it was no easy matter for the country to relieve itself of the institution. Then, when we rid ourselves of prejudice, or racial feeling, and look facts in the face, we must acknowledge that, notwithstanding the cruelty and moral wrong of slavery, the ten million Negroes inhabiting this country, who themselves or whose ancestors went through the school of American slavery, are in a stronger and more hopeful condition, materially, intellectually, morally, and religiously, than is true of an equal number of black people in any other portion of the globe. This is so to such an extent that Negroes in this country, who themselves or whose forefathers went through the school of slavery, are constantly returning to Africa as missionaries to enlighten those who remained in the fatherland. This I say, not to justify slavery—on the other hand, I condemn it as an institution, as we all know that in America it was established for selfish and financial reasons, and not from a missionary motive—but to call attention to a fact, and to show how Providence so often uses men and institutions to accomplish a purpose. When persons ask me in these days how, in the midst of what sometimes seem hopelessly discouraging conditions, I can have such faith in the future of my race in this country, I remind them of the wilderness through which and out of which, a good Providence has already led us.

Ever since I have been old enough to think for myself, I have entertained the idea that, notwithstanding the cruel wrongs inflicted upon us, the black man got nearly as much out of slavery as the white man did. The hurtful influences of the institution were not by any means con-

fined to the Negro. This was fully illustrated by the life
upon our own plantation. The whole machinery of slavery
was so constructed as to cause labour, as a rule, to be
looked upon as a badge of degradation, of inferiority.
Hence labour was something that both races on the slave
plantation sought to escape. The slave system on our place,
in a large measure, took the spirit of self-reliance and self-
help out of the white people. My old master had many
boys and girls, but not one, so far as I know, ever mastered
a single trade or special line of productive industry. The
girls were not taught to cook, sew or to take care of the
house. All of this was left to the slaves. The slaves, of
course, had little personal interest in the life of the plantation, and their ignorance prevented them from learning
how to do things in the most improved and thorough
manner. As a result of the system, fences were out of repair, gates were hanging half off the hinges, doors creaked,
window-panes were out, plastering had fallen but was not
replaced, weeds grew in the yard. As a rule, there was food
for whites and blacks, but inside the house, and on the
dining-room table, there was wanting that delicacy and re-
finement of touch and finish which can make a home the
most convenient, comfortable, and attractive place in the
world. Withal there was a waste of food and other ma-
terials which was sad. When freedom came, the slaves
were almost as well fitted to begin life anew as the master,
except in the matter of book-learning and ownership of
property. The slave owner and his sons had mastered no
special industry. They unconsciously had imbibed the feeling that manual labour was not the proper thing for them.
On the other hand, the slaves, in many cases, had mastered
some handicraft, and none were ashamed, and few unwilling, to labour.

Finally the war closed, and the day of freedom came.
It was a momentous and eventful day to all upon our
plantation. We had been expecting it. Freedom was in
the air, and had been for months. Deserting soldiers returning to their homes were to be seen every day. Others
who had been discharged, or whose regiments had been
paroled, were constantly passing near our place. The
"grape-vine telegraph" was kept busy night and day. The
news and mutterings of great events were swiftly carried
from one plantation to another. In the fear of "Yankee"
invasions, the silverware and other valuables were taken

from the "big house," buried in the woods, and guarded by trusted slaves. Woe be to any one who would have attempted to disturb the buried treasure. The slaves would give the Yankee soldiers food, drink, clothing—anything but that which had been specifically intrusted to their care and honour. As the great day drew nearer, there was more singing in the slave quarters than usual. It was bolder, had more ring, and lasted later into the night. Most of the verses of the plantation songs had some reference to freedom. True, they had sung those same verses before, but they had been careful to explain that the "freedom" in these songs referred to the next world, and had no connection with life in this world. Now they gradually threw off the mask: and were not afraid to let it be known that the "freedom" in their songs meant freedom of the body in this world. The night before the eventful day, word was sent to the slave quarters to the effect that something unusual was going to take place at the "big house" the next morning. There was little, if any, sleep that night. All was excitement and expectancy. Early the next morning word was sent to all the slaves, old and young, to gather at the house. In company with my mother, brother, and sister, and a large number of other slaves, I went to the master's house. All of our master's family were either standing or seated on the veranda of the house, where they could see what was to take place and hear what was said. There was a feeling of deep interest, or perhaps sadness, on their faces, but not bitterness. As I now recall the impression they made upon me, they did not at the moment seem to be sad because of the loss of property, but rather because of parting with those whom they had reared and who were in many ways very close to them. The most distinct thing that I now recall in connection with the scene was that some man who seemed to be a stranger (a United States officer, I presume) made a little speech and then read a rather long paper—the Emancipation Proclamation, I think. After the reading we were told that we were all free, and could go when and where we pleased. My mother, who was standing by my side, leaned over and kissed her children, while tears of joy ran down her cheeks. She explained to us what it all meant, that this was the day for which she had been so long praying, but fearing that she would never live to see.

For some minutes there was great rejoicing, and thanks-giving, and wild scenes of ecstasy. But there was no feel-ing of bitterness. In fact, there was pity among the slaves for our former owners. The wild rejoicing on the part of the emancipated coloured people lasted but for a brief period, for I noticed that by the time they returned to their cabins there was a change in their feelings. The great responsibility of being free, of having charge of them-selves, of having to think and plan for themselves and their children, seemed to take possession of them. It was very much like suddenly turning a youth of ten or twelve years out into the world to provide for himself. In a few hours the great questions with which the Anglo-Saxon race had been grappling for centuries had been thrown upon these people to be solved. These were the questions of a home, a living, the rearing of children, education, citizen-ship, and the establishment and support of churches. Was it any wonder that within a few hours the wild rejoicing ceased and a feeling of deep gloom seemed to pervade the slave quarters? To some it seemed that, now that they were in actual possession of it, freedom was a more seri-ous thing than they had expected to find it. Some of the slaves were seventy or eighty years old; their best days were gone. They had no strength with which to earn a living in a strange place and among strange people, even if they had been sure where to find a new place of abode. To this class the problem seemed especially hard. Be-sides, deep down in their hearts there was a strange and peculiar attachment to "old Marster" and "old Missus," and to their children, which they found it hard to think of breaking off. With these they had spent in some cases nearly a half-century, and it was no light thing to think of parting. Gradually, one by one, stealthily at first, the older slaves began to wander from the slave quarters back to the "big house" to have a whispered conversation with their former owners as to the future.

CHAPTER II

Boyhood Days

After the coming of freedom there were two points upon which practically all the people on our place were agreed, and I find that this was generally true throughout the South: that they must change their names, and that they must leave the old plantation for at least a few days or weeks in order that they might really feel sure that they were free.

In some way a feeling got among the coloured people that it was far from proper for them to bear the surname of their former owners, and a great many of them took other surnames. This was one of the first signs of freedom. When they were slaves, a coloured person was simply called "John" or "Susan." There was seldom occasion for more than the use of one name. If "John" or "Susan" belonged to a white man by the name of "Hatcher," sometimes he was called "John Hatcher," or as often "Hatcher's John." But there was a feeling that "John Hatcher" or "Hatcher's John" was not the proper title by which to denote a freeman; and so in many cases "John Hatcher" was changed to "John S. Lincoln" or "John S. Sherman," the initial "S" standing for no name, it being simply a part of what the coloured man proudly called his "entitles."

As I have stated, most of the coloured people left the old plantation for a short while at least, so as to be sure, it seemed, that they could leave and try their freedom on to see how it felt. After they had remained away for a time, many of the older slaves, especially, returned to their old homes and made some kind of contract with their former owners by which they remained on the estate.

My mother's husband, who was the stepfather of my brother John and myself, did not belong to the same owners as did my mother. In fact, he seldom came to our plantation. I remember seeing him there perhaps once a year, that being about Christmas time. In some way, during the war, by running away and following the Federal soldiers, it seems, he found his way into the new state of

West Virginia. As soon as freedom was declared, he sent
for my mother to come to the Kanawha Valley, in West
Virginia. At that time a journey from Virginia over the
mountains to West Virginia was rather a tedious and in
some cases a painful undertaking. What little clothing and
few household goods we had were placed in a cart, but
the children walked the greater portion of the distance,
which was several hundred miles.

I do not think any of us ever had been very far from the
plantation, and the taking of a long journey into another
state was quite an event. The parting from our former
owners and the members of our own race on the plantation was a serious occasion. From the time of our parting
till their death we kept up a correspondence with the
older members of the family, and in later years we have
kept in touch with those who were the younger members.
We were several weeks making the trip, and most of the
time we slept in the open air and did our cooking over a
log fire out of doors. One night I recall that we camped
near an abandoned log cabin, and my mother decided to
build a fire in that for cooking, and afterward to make a
"pallet" on the floor for our sleeping. Just as the fire had
gotten well started a large black snake fully a yard and a
half long dropped down the chimney and ran out on the
floor. Of course we at once abandoned that cabin. Finally we reached our destination—a little town called
Malden, which is about five miles from Charleston, the
present capital of the state.

At that time salt-mining was the great industry in that
part of West Virginia, and the little town of Malden
was right in the midst of the salt-furnaces. My stepfather
had already secured a job at a salt-furnace, and he had
also secured a little cabin for us to live in. Our new house
was no better than the one we had left on the old plantation in Virginia. In fact, in one respect it was worse. Notwithstanding the poor condition of our plantation cabin,
we were at all times sure of pure air. Our new home was in
the midst of a cluster of cabins crowded closely together,
and as there were no sanitary regulations, the filth about
the cabins was often intolerable. Some of our neighbours
were coloured people, and some were the poorest and
most ignorant and degraded white people. It was a motley
mixture. Drinking, gambling, quarrels, fights, and shock-
ingly immoral practices were frequent. All who lived in

the little town were in one way or another connected with the salt business. Though I was a mere child, my step-father put me and my brother at work in one of the furnaces. Often I began work as early as four o'clock in the morning.

The first thing I ever learned in the way of book knowledge was while working in this salt-furnace. Each salt-packer had his barrels marked with a certain number. The number allotted to my stepfather was "18." At the close of the day's work the boss of the packers would come around and put "18" on each of our barrels, and I soon learned to recognize that figure wherever I saw it, and after a while got to the point where I could make that figure, though I knew nothing about any other figures or letters.

From the time that I can remember having any thoughts about anything, I recall that I had an intense longing to learn to read. I determined, when quite a small child, that, if I accomplished nothing else in life, I would in some way get enough education to enable me to read common books and newspapers. Soon after we got settled in some manner in our new cabin in West Virginia, I induced my mother to get hold of a book for me. How or where she got it I do not know, but in some way she procured an old copy of Webster's "blue-back" spelling-book, which contained the alphabet, followed by such meaningless words as "ab," "ba," "ca," "da." I began at once to devour this book, and I think that it was the first one I ever had in my hands. I had learned from somebody that the way to begin to read was to learn the alphabet, so I tried in all the ways I could think of to learn it,—all of course without a teacher, for I could find no one to teach me. At that time there was not a single member of my race anywhere near us who could read, and I was too timid to approach any of the white people. In some way, within a few weeks, I mastered the greater portion of the alphabet. In all my efforts to learn to read my mother shared fully my ambition, and sympathized with me and aided me in every way that she could. Though she was totally ignorant, so far as mere book knowledge was concerned, she had high ambitions for her children, and a large fund of good, hard, common sense which seemed to enable her to meet and master every situation. If I have done anything in life worth attention, I feel sure that I inherited the disposition from my mother.

In the midst of my struggles and longing for an education, a young coloured boy who had learned to read in the state of Ohio came to Malden. As soon as the coloured people found out that he could read, a newspaper was secured, and at the close of nearly every day's work this young man would be surrounded by a group of men and women who were anxious to hear him read the news contained in the papers. How I used to envy this man! He seemed to me to be the one young man in all the world who ought to be satisfied with his attainments.

About this time the question of having some kind of a school opened for the coloured children in the village began to be discussed by members of the race. As it would be the first school for Negro children that had ever been opened in that part of Virginia, it was, of course, to be a great event, and the discussion excited the widest interest. The most perplexing question was where to find a teacher. The young man from Ohio who had learned to read the papers was considered, but his age was against him. In the midst of the discussion about a teacher, another young coloured man from Ohio, who had been a soldier, in some way found his way into town. It was soon learned that he possessed considerable education, and he was engaged by the coloured people to teach their first school. As yet no free schools had been started for coloured people in that section, hence each family agreed to pay a certain amount per month, with the understanding that the teacher was to "board 'round"—that is, spend a day with each family. This was not bad for the teacher, for each family tried to provide the very best on the day the teacher was to be its guest. I recall that I looked forward with an anxious appetite to the "teacher's day" at our little cabin.

This experience of a whole race beginning to go to school for the first time, presents one of the most interesting studies that has ever occurred in connection with the development of any race. Few people who were not right in the midst of the scenes can form any exact idea of the intense desire which the people of my race showed for an education. As I have stated, it was a whole race trying to go to school. Few were too young, and none too old, to make the attempt to learn. As fast as any kind of teachers could be secured, not only were day-schools filled, but night-schools as well. The great ambition of the older people was to try to learn to read the Bible before they died. With this end in view, men and women who were fifty or seventy-five years old would often be found in the night-school. Sunday-schools were formed soon after freedom, but the principal book studied in the Sunday-school was the spelling-book. Day-school, night-school, Sunday-school, were always crowded, and often many had to be turned away for want of room.

The opening of the school in the Kanawha Valley, however, brought to me one of the keenest disappointments that I ever experienced. I had been working in a salt-furnace for several months, and my stepfather had discovered that I had a financial value, and so, when the school opened, he decided that he could not spare me from my work. This decision seemed to cloud my every ambition. The disappointment was made all the more severe by reason of the fact that my place of work was where I could see the happy children passing to and from school, mornings and afternoons. Despite this disappointment, however, I determined that I would learn something, anyway. I applied myself with greater earnestness than ever to the mastering of what was in the "blue-back" speller.

My mother sympathized with me in my disappointment, and sought to comfort me in all the ways she could, and to help me find a way to learn. After a while I succeeded in making arrangements with the teacher to give me some lessons at night, after the day's work was done. These night lessons were so welcome that I think I learned more at night than the other children did during the day. My own experiences in the night-school gave me faith in the night-school idea, with which, in after years, I had to do both at Hampton and Tuskegee. But my boyish heart was still set upon going to the day-school, and I let no opportunity slip to push my case. Finally I won, and was permitted to go to the school in the day for a few months, with the understanding that I was to rise early in the morning and work in the furnace till nine o'clock, and return immediately after school closed in the afternoon for at least two more hours of work.

The schoolhouse was some distance from the furnace, and as I had to work till nine o'clock, and the school opened at nine, I found myself in a difficulty. School would always be begun before I reached it, and sometimes my class had recited. To get around this difficulty I yielded to a temptation for which most people, I suppose, will condemn me; but since it is a fact, I might as well state it. I have great faith in the power and influence of facts. It is seldom that anything is permanently gained by holding back a fact. There was a large clock in a little office in the furnace. This clock, of course, all the hundred or more workmen depended upon to regulate their hours of beginning and ending the day's work. I got the idea that the way for me to reach school on time was to move the clock hands from half-past eight up to the nine o'clock mark. This I found myself doing morning after morning, till the furnace "boss" discovered that something was wrong, and locked the clock in a case. I did not mean to inconvenience any body. I simply meant to reach that schoolhouse in time.

When, however, I found myself at the school for the first time, I also found myself confronted with two other difficulties. In the first place, I found that all of the other children wore hats or caps on their heads, and I had neither hat nor cap. In fact, I do not remember that up to the time of going to school I had ever worn any kind of covering upon my head, nor do I recall that either I or anybody else had even thought anything about the need of covering for my head. But, of course, when I saw how all the other boys were dressed, I began to feel quite uncomfortable. As usual, I put the case before my mother, and she explained to me that she had no money with which to buy a "store hat," which was a rather new institution at that time among the members of my race and was considered quite the thing for young and old to own, but that she would find a way to help me out of the difficulty. She accordingly got two pieces of "homespun" (jeans) and sewed them together, and I was soon the proud possessor of my first cap.

The lesson that my mother taught me in this has always remained with me, and I have tried as best I could to teach it to others. I have always felt proud, whenever I think of the incident, that my mother had strength of character enough not to be led into the temptation of seeming to be that which she was not—of trying to impress my schoolmates and others with the fact that she was able to buy me a "store hat" when she was not. I have always felt proud that she refused to go into debt for that which she did not have the money to pay for. Since that time I have owned many kinds of caps and hats, but never one of which I have felt so proud as of the cap made of the two pieces of cloth sewed together by my mother. I have noted the fact, but without satisfaction, I need not add, that several of the boys who began their careers with "store hats" and who were my schoolmates and used to join in the sport that was made of me because I had only a "homespun" cap, have ended their careers in the penitentiary, while others are not able now to buy any kind of hat.

My second difficulty was with regard to my name, or rather *a* name. From the time when I could remember anything, I had been called simply "Booker." Before going to school it had never occurred to me that it was needful or appropriate to have an additional name. When I heard the school-roll called, I noticed that all of the children had at least two names, and some of them indulged in what seemed to me the extravagance of having three. I was in deep perplexity, because I knew that the teacher would demand of me at least two names, and I had only one. By the time the occasion came for the enrolling of my name, an idea occurred to me which I thought would make me equal to the situation; and so, when the teacher asked me what my full name was, I calmly told him "Booker Washington," as if I had been called by that name all my life; and by that name I have since been known. Later in life I found that my mother had given me the name of "Booker Taliaferro" soon after I was born, but in some way that part of my name seemed to disappear, and for a long while was forgotten, but as soon as I found out about it I revived it, and made my full name "Booker Taliaferro Washington." I think there are not many men in our country who have had the privilege of naming themselves in the way that I have.

More than once I have tried to picture myself in the position of a boy or man with an honoured and distinguished ancestry which I could trace back through a period of hundreds of years, and who had not only inherited a name, but fortune and a proud family homestead; and yet I have sometimes had the feeling that if I had inherited these, and had been a member of a more popular race, I should have been inclined to yield to the temptation of depending upon my ancestry and my colour to do that for me which I should do for myself. Years ago I resolved that because I had no ancestry myself I would leave a record of which my children would be proud, and which might encourage them to still higher effort.

The world should not pass judgment upon the Negro, and especially the Negro youth, too quickly or too harshly. The Negro boy has obstacles, discouragements, and temptations to battle with that are little known to those not situated as he is. When a white boy undertakes a task, it is taken for granted that he will succeed. On the other hand, people are usually surprised if the Negro boy does not fail. In a word, the Negro youth starts out with the presumption against him.

The influence of ancestry, however, is important in helping forward any individual or race, if too much reliance is not placed upon it. Those who constantly direct attention to the Negro youth's moral weaknesses, and compare his advancement with that of white youths, do not consider the influence of the memories which cling about the old family homesteads. I have no idea, as I have stated elsewhere, who my grandmother was. I have, or have had, uncles and aunts and cousins, but I have no knowledge as to what most of them are. My case will illustrate that of hundreds of thousands of black people in every part of our country. The very fact that the white boy is conscious that, if he fails in life, he will disgrace the whole family record, extending back through many generations, is of tremendous value in helping him to resist temptations. The fact that the individual has behind and surrounding him proud family history and connection serves as a stimulus to help him to overcome obstacles when striving for success.

The time that I was permitted to attend school during the day was short, and my attendance was irregular. It was not long before I had to stop attending day-school altogether, and devote all of my time again to work. I resorted to the night-school again. In fact, the greater part of the education I secured in my boyhood was gathered through the night-school after my day's work was done. I had difficulty often in securing a satisfactory teacher. Sometimes, after I had secured some one to teach me at night, I would find, much to my disappointment, that the teacher knew but little more than I did. Often I would have to walk several miles at night in order to recite my night-school lessons. There was never a time in my youth, no matter how dark and discouraging the days might be, when one resolve did not continually remain with me, and that was a determination to secure an education at any cost.

Soon after we moved to West Virginia, my mother adopted into our family, notwithstanding our poverty, an orphan boy, to whom afterward we gave the name of James B. Washington. He has ever since remained a member of the family.

After I had worked in the salt-furnace for some time, work was secured for me in a coal-mine which was operated mainly for the purpose of securing fuel for the salt-furnace. Work in the coal-mine I always dreaded. One reason for this was that any one who worked in a coal-mine was always unclean, at least while at work, and it was a very hard job to get one's skin clean after the day's work was over. Then it was fully a mile from the opening of the coal-mine to the face of the coal, and all, of course, was in the blackest darkness. I do not believe that one ever experiences anywhere else such darkness as he does in a coal-mine. The mine was divided into a large number of different "rooms" or departments, and, as I never was able to learn the location of all these "rooms," I many times found myself lost in the mine. To add to the horror of being lost, sometimes my light would go out, and then, if I did not happen to have a match, I would wander about in the darkness until by chance I found some one to give me a light. The work was not only hard, but it was dangerous. There was always the danger of being blown to pieces by a premature explosion of powder, or of being crushed by falling slate. Accidents from one or the other of these causes were frequently occurring, and this kept me in constant fear. Many children of the tenderest years were compelled then, as is now true I fear, in most coal-mining districts, to spend a large part of their lives in these coal-mines, with little opportunity to get an education; and, what is worse, I have often noted that, as a rule, young boys who begin life in a coal-mine are often physically and mentally dwarfed. They soon lose ambition to do anything else than to continue as a coal-miner.

In those days, and later as a young man, I used to try to picture in my imagination the feelings and ambitions of a white boy with absolutely no limit placed upon his aspirations and activities. I used to envy the white boy who had no obstacles placed in the way of his becoming a Congressman, Governor, Bishop, or President by reason of the accident of his birth or race. I used to picture the way that I would act under such circumstances; how I would begin at the bottom and keep rising until I reached the highest round of success.

In later years, I confess that I do not envy the white boy as I once did. I have learned that success is to be measured not so much by the position that one has reached in life as by the obstacles which he has overcome while trying to succeed. Looked at from this standpoint, I almost reach the conclusion that often the Negro boy's birth and connection with an unpopular race is an advantage, so far as real life is concerned. With few exceptions, the Negro youth must work harder and must perform his task even better than a white youth in order to secure recognition. But out of the hard and unusual struggle which he is compelled to pass, he gets a strength, a confidence, that one misses whose pathway is comparatively smooth by reason of birth and race.

From any point of view, I had rather be what I am, a member of the Negro race, than be able to claim membership with the most favoured of any other race. I have always been made sad when I have heard members of any race claiming rights and privileges, or certain badges of distinction, on the ground simply that they were members of this or that race, regardless of their own individual worth or attainments. I have been made to feel sad for such persons because I am conscious of the fact that mere connection with what is known as a superior race will not permanently carry an individual forward unless he has individual worth, and mere connection with what is regarded as an inferior race will not finally hold an individual back if he possesses intrinsic, individual merit. Every persecuted individual and race should get much consolation out of the great human law, which is universal and eternal, that merit, no matter under what skin found, is in the long run, recognized and rewarded. This I have said here, not to call attention to myself as an individual, but to the race to which I am proud to belong.

CHAPTER III

The Struggle for an Education

One day, while at work in the coal-mine, I happened to overhear two miners talking about a great school for coloured people somewhere in Virginia. This was the first time that I had ever heard anything about any kind of school or college that was more pretentious than the little coloured school in our town.

In the darkness of the mine I noiselessly crept as close as I could to the two men who were talking. I heard one tell the other that not only was the school established for the members of my race, but that opportunities were provided by which poor but worthy students could work out all or part of the cost of board, and at the same time be taught some trade or industry.

As they went on describing the school, it seemed to me that it must be the greatest place on earth, and not even Heaven presented more attractions for me at that time than did the Hampton Normal and Agricultural Institute in Virginia, about which these men were talking. I resolved at once to go to that school, although I had no idea where it was, or how many miles away, or how I was going to reach it; I remembered only that I was on fire constantly with one ambition, and that was to go to Hampton. This thought was with me day and night.

After hearing of the Hampton Institute, I continued to work for a few months longer in the coal-mine. While at work there, I heard of a vacant position in the household of General Lewis Ruffner, the owner of the salt-furnace and coal-mine. Mrs. Viola Ruffner, the wife of General Ruffner, was a "Yankee" woman from Vermont. Mrs. Ruffner had a reputation all through the vicinity for being very strict with her servants, and especially with the boys who tried to serve her. Few of them had remained with her more than two or three weeks. They all left with the same excuse: she was too strict. I decided, however, that I would rather try Mrs. Ruffner's house than remain in the coal-mine, and so my mother applied to her for the vacant position. I was hired at a salary of $5 per month.

I had heard so much about Mrs. Ruffner's severity that I was almost afraid to see her, and trembled when I went into her presence. I had not lived with her many weeks, however, before I began to understand her. I soon began to learn that, first of all, she wanted everything kept clean about her, that she wanted things done promptly and systematically, and that at the bottom of everything she wanted absolute honesty and frankness. Nothing must be sloven or slipshod; every door, every fence, must be kept in repair.

I cannot now recall how long I lived with Mrs. Ruffner before going to Hampton, but I think it must have been a year and a half. At any rate, I here repeat what I have said more than once before, that the lessons that I learned in the home of Mrs. Ruffner were as valuable to me as any education I have ever gotten anywhere since. Even to this day I never see bits of paper scattered around a house or in the street that I do not want to pick them up at once. I never see a filthy yard that I do not want to clean it, a paling off of a fence that I do not want to put it on, an unpainted or unwhitewashed house that I do not want to paint or whitewash it, or a button off one's clothes, or a grease-spot on them or on a floor, that I do not want to call attention to it.

From fearing Mrs. Ruffner I soon learned to look upon her as one of my best friends. When she found that she could trust me she did so implicitly. During the one or two winters that I was with her she gave me an opportunity to go to school for an hour in the day during a portion of the winter months, but most of my studying was done at night, sometimes alone, sometimes under some one whom I could hire to teach me. Mrs. Ruffner always encouraged and sympathized with me in all my efforts to get an education. It was while living with her that I began to get together my first library. I secured a dry-goods box, knocked out one side of it, put some shelves in it, and began putting into it every kind of book that I could get my hands upon, and called it my "library."

Notwithstanding my success at Mrs. Ruffner's I did not give up the idea of going to the Hampton Institute. In the fall of 1872 I determined to make an effort to get there, although, as I have stated, I had no idea of the direction in which Hampton was, or what it would cost to go there. I do not think that any one thoroughly sympathized with me in my ambition to go to Hampton unless it was my mother, and she was troubled with a grave fear that I was starting out on a "wild-goose chase." At any rate, I got only a half-hearted consent from her that I might start. The small amount of money that I had earned had been consumed by my stepfather and the remainder of the family, with the exception of a very few dollars, and so I had very little with which to buy clothes and pay travelling expenses. My brother John helped me all that he could, but of course that was not a great deal, for his work was in the coal-mine, where he did not earn much, and most of what he did earn went in the direction of paying the household expenses.

Perhaps the thing that touched and pleased me most in connection with my starting for Hampton was the interest that many of the older coloured people took in the matter. They had spent the best days of their lives in slavery, and hardly expected to live to see the time when they would see a member of their race leave home to attend a boarding-school. Some of these older people would give me a nickel, others a quarter, or a handkerchief.

Finally the great day came, and I started for Hampton. I had only a small, cheap satchel that contained what few articles of clothing I could get. My mother at the time was rather weak and broken in health. I hardly expected to see her again, and thus our parting was all the more sad. She, however, was very brave through it all. At that time there were no through trains connecting that part of West Virginia with eastern Virginia. Trains ran only a portion of the way, and the remainder of the distance was travelled by stage-coaches.

The distance from Malden to Hampton is about five hundred miles. I had not been away from home many hours before it began to grow painfully evident that I did not have enough money to pay my fare to Hampton. One experience I shall long remember. I had been travelling over the mountains most of the afternoon in an old-fashioned stage-coach, when, late in the evening, the coach stopped for the night at a common unpainted house called a hotel. All the other passengers except myself were whites. In my ignorance I supposed that the little hotel existed for the purpose of accommodating the passengers who travelled on the stage-coach. The difference that the colour of one's skin would make I had not thought anything about. After all the other passengers had been shown rooms and were getting ready for supper, I shyly presented myself before the man at the desk. It is true I had practically no money in my pocket with which to pay for bed or food, but I had hoped in some way to beg my way into the good graces of the landlord, for at that season in the mountains of Virginia the weather was cold, and I wanted to get indoors for the night. Without asking as to whether I had any money, the man at the desk firmly refused to even consider the matter of providing me with food or lodging. This was my first experience in finding out what the colour of my skin meant. In some way I managed to keep warm by walking about, and so got through the night. My whole soul was so bent upon reaching Hampton that I did not have time to cherish any bitterness toward the hotel-keeper.

By walking, begging rides both in wagons and in the cars, in some way, after a number of days, I reached the city of Richmond, Virginia, about eighty-two miles from Hampton. When I reached there, tired, hungry, and dirty, it was late in the night. I had never been in a large city, and this rather added to my misery. When I reached Richmond, I was completely out of money. I had not a single acquaintance in the place, and, being unused to city ways, I did not know where to go. I applied at several places for lodging, but they all wanted money, and that was what I did not have. Knowing nothing else better to do, I walked the streets. In doing this I passed by many foodstands where fried chicken and half-moon apple pies were piled high and made to present a most tempting appearance. At that time it seemed to me that I would have promised all that I expected to possess in the future to have gotten hold of one of those chicken legs or one of those pies. But I could not get either of these, nor anything else to eat.

I must have walked the streets till after midnight. At last I became so exhausted that I could walk no longer. I was tired, I was hungry, I was everything but discouraged. Just about the time when I reached extreme physical exhaustion, I came upon a portion of a street where the board sidewalk was considerably elevated. I waited for a few minutes, till I was sure that no passers-by could see me, and then crept under the sidewalk and lay for the night upon the ground, with my satchel of clothing for a pillow. Nearly all night I could hear the tramp of feet over my head. The next morning I found myself refreshed. but I was extremely hungry. because it had been a long time since I had had sufficient food. As soon as it became light enough for me to see my surroundings I noticed that I was near a large ship, and that this ship seemed to be unloading a cargo of pig iron. I went at once to the vessel and asked the captain to permit me to help unload the vessel in order to get money for food. The captain, a white man, who seemed to be kind-hearted, consented. I worked long enough to earn money for my breakfast, and it seems to me, as I remember it now, to have been about the best breakfast that I have ever eaten.

My work pleased the captain so well that he told me if I desired I could continue working for a small amount per day. This I was very glad to do. I continued working on this vessel for a number of days. After buying food with the small wages I received there was not much left to add to the amount I must get to pay my way to Hampton. In order to economize in every way possible, so as to be sure to reach Hampton in a reasonable time, I continued to sleep under the same sidewalk that gave me shelter the first night I was in Richmond. Many years after that the coloured citizens of Richmond very kindly tendered me a reception at which there must have been two thousand people present. This reception was held not far from the spot where I slept the first night I spent in that city, and I must confess that my mind was more upon the sidewalk that first gave me shelter than upon the reception, agreeable and cordial as it was.

When I had saved what I considered enough money with which to reach Hampton, I thanked the captain of the vessel for his kindness, and started again. Without any unusual occurrence I reached Hampton, with a surplus of exactly fifty cents with which to begin my education. To me it had been a long, eventful journey; but the first sight of the large, three-story, brick school building seemed to have rewarded me for all that I had undergone in order to reach the place. If the people who gave the money to provide that building could appreciate the influence the sight of it had upon me, as well as upon thousands of other youths, they would feel all the more encouraged to make such gifts. It seemed to me to be the largest and most beautiful building I had ever seen. The sight of it seemed to give me new life. I felt that a new kind of existence had now begun—that life would now have a new meaning. I felt that I had reached the promised land, and I resolved to let no obstacle prevent me from putting forth the highest effort to fit myself to accomplish the most good in the world.

As soon as possible after reaching the grounds of the Hampton Institute, I presented myself before the head teacher for assignment to a class. Having been so long without proper food, a bath and change of clothing, I did not, of course, make a very favourable impression upon her, and I could see at once that there were doubts in her mind about the wisdom of admitting me as a student. I felt that I could hardly blame her if she got the idea that I was a worthless loafer or tramp. For some time she did not refuse to admit me, neither did she decide in my favour, and I continued to linger about her, and to impress her in all the ways I could with my worthiness. In the meantime I saw her admitting other students, and that added greatly to my discomfort, for I felt, deep down in my heart, that I could do as well as they, if I could only get a chance to show what was in me.

After some hours had passed, the head teacher said to me: "The adjoining recitation-room needs sweeping. Take the broom and sweep it."

It occurred to me at once that here was my chance. Never did I receive an order with more delight. I knew that I could sweep, for Mrs. Ruffner had thoroughly taught me how to do that when I lived with her.

I swept the recitation-room three times. Then I got a dusting-cloth and I dusted it four times. All the woodwork around the walls, every bench, table, and desk, I went over four times with my dusting-cloth. Besides, every piece of furniture had been moved and every closet and corner in the room had been thoroughly cleaned. I had the feeling that in a large measure my future depended upon the impression I made upon the teacher in the cleaning of that room. When I was through, I reported to the head teacher. She was a "Yankee" woman who knew just where to look for dirt. She went into the room and inspected the floor and closets; then she took her handkerchief and rubbed it on the woodwork about the walls, and over the table and benches. When she was unable to find one bit of dirt on the floor, or a particle of dust on any of the furniture, she quietly remarked, "I guess you will do to enter this institution."

I was one of the happiest souls on earth. The sweeping of that room was my college examination, and never did any youth pass an examination for entrance into Harvard or Yale that gave him more genuine satisfaction. I have passed several examinations since then, but I have always felt that this was the best one I ever passed.

I have spoken of my own experience in entering the Hampton Institute. Perhaps few, if any, had anything like the same experience that I had, but about that same period there were hundreds who found their way to Hampton and other institutions after experiencing something of the same difficulties that I went through. The young men and women were determined to secure an education at any cost.

The sweeping of the recitation-room in the manner that I did it seems to have paved the way for me to get through Hampton. Miss Mary F. Mackie, the head teacher, offered me a position as janitor. This, of course, I gladly accepted, because it was a place where I could work out nearly all the cost of my board. The work was hard and taxing, but I stuck to it. I had a large number of rooms to care for, and had to work late into the night, while at the same time I had to rise by four o'clock in the morning, in order to build the fires and have a little time in which to prepare my lessons. In all my career at Hampton, and ever since I have been out in the world, Miss Mary F. Mackie, the head teacher to whom I have referred, proved one of my strongest and most helpful friends. Her advice and encouragement were always helpful and strengthening to me in the darkest hour.

I have spoken of the impression that was made upon me by the buildings and general appearance of the Hampton Institute, but I have not spoken of that which made the greatest and most lasting impression upon me, and that was a great man—the noblest, rarest human being that it has ever been my privilege to meet. I refer to the late General Samuel C. Armstrong.

It has been my fortune to meet personally many of

what are called great characters, both in Europe and
America, but I do not hesitate to say that I never met
any man who, in my estimation, was the equal of General Armstrong. Fresh from the degrading influences of
the slave plantation and the coal-mines, it was a rare
privilege for me to be permitted to come into direct contact with such a character as General Armstrong. I shall
always remember that the first time I went into his presence he made the impression upon me of being a perfect
man: I was made to feel that there was something about
him that was superhuman. It was my privilege to know
the General personally from the time I entered Hampton
till he died, and the more I saw of him the greater he
grew in my estimation. One might have removed from
Hampton all the buildings, class-rooms, teachers, and industries, and given the men and women there the opportunity of coming into daily contact with General Armstrong, and that alone would have been a liberal education. The older I grow, the more I am convinced that there
is no education which one can get from books and costly
apparatus that is equal to that which can be gotten from
contact with great men and women. Instead of studying
books so constantly, how I wish that our schools and
colleges might learn to study men and things!

General Armstrong spent two of the last six months of
his life in my home at Tuskegee. At that time he was
paralyzed to the extent that he had lost control of his body
and voice in a very large degree. Notwithstanding his
affliction, he worked almost constantly night and day
for the cause to which he had given his life. I never saw a
man who so completely lost sight of himself. I do not believe he ever had a selfish thought. He was just as happy
in trying to assist some other institution in the South as
he was when working for Hampton. Although he fought
the Southern white man in the Civil War, I never heard
him utter a bitter word against him afterward. On the
other hand, he was constantly seeking to find ways by
which he could be of service to the Southern whites.

It would be difficult to describe the hold that he had
upon the students at Hampton, or the faith they had in
him. In fact, he was worshipped by his students. It never
occurred to me that General Armstrong could fail in anything that he undertook. There is almost no request
that he could have made that would not have been com-

plied with. When he was a guest at my home in Alabama, and was so badly paralyzed that he had to be wheeled about in an invalid's chair, I recall that one of the General's former students had occasion to push his chair up a long, steep hill that taxed his strength to the utmost. When the top of the hill was reached, the former pupil, with a glow of happiness on his face, exclaimed, "I am so glad that I have been permitted to do something that was real hard for the General before he dies!" While I was a student at Hampton, the dormitories became so crowded that it was impossible to find room for all who wanted to be admitted. In order to help remedy the difficulty the General conceived the plan of putting up tents to be used as rooms. As soon as it became known that General Armstrong would be pleased if some of the older students would live in the tents during the winter, nearly every student in school volunteered to go.

I was one of the volunteers. The winter that we spent in those tents was an intensely cold one, and we suffered severely—how much I am sure General Armstrong never knew, because we made no complaints. It was enough for us to know that we were pleasing General Armstrong, and that we were making it possible for an additional number of students to secure an education. More than once, during a cold night, when a stiff gale would be blowing, our tent was lifted bodily, and we would find ourselves in the open air. The General would usually pay a visit to the tents early in the morning, and his earnest, cheerful, encouraging voice would dispel any feeling of despondency.

I have spoken of my admiration for General Armstrong, and yet he was but a type of that Christlike body of men and women who went into the Negro schools at the close of the war by the hundreds to assist in lifting up my race. The history of the world fails to show a higher, purer, and more unselfish class of men and women than those who found their way into those Negro schools.

Life at Hampton was a constant revelation to me; was constantly taking me into a new world. The matter of having meals at regular hours, of eating on a tablecloth, using a napkin, the use of the bathtub and of the toothbrush, as well as the use of sheets upon the bed, were all new to me.

I sometimes feel that almost the most valuable lesson I

got at the Hampton Institute was in the use and value of
the bath. I learned there for the first time some of its
value, not only in keeping the body healthy, but in inspiring self-respect and promoting virtue. In all my travels
in the South and elsewhere since leaving Hampton I have
always in some way sought my daily bath. To get it sometimes when I have been the guest of my own people in a
single-roomed cabin has not always been easy to do, except by slipping away to some stream in the woods. I
have always tried to teach my people that some provision
for bathing should be a part of every house.

For some time, while a student at Hampton, I possessed
but a single pair of socks, but when I had worn these till
they became soiled, I would wash them at night and
hang them by the fire to dry, so that I might wear them
again the next morning.

The charge for my board at Hampton was ten dollars
per month. I was expected to pay a part of this in cash
and to work out the remainder. To meet this cash payment, as I have stated, I had just fifty cents when I
reached the institution. Aside from a very few dollars that
my brother John was able to send me once in a while, I
had no money with which to pay my board. I was determined from the first to make my work as janitor so valuable that my services would be indispensable. This I succeeded in doing to such an extent that I was soon informed that I would be allowed the full cost of my board
in return for my work. The cost of tuition was seventy
dollars a year. This, of course, was wholly beyond my
ability to provide. If I had been compelled to pay the
seventy dollars for tuition, in addition to providing for
my board, I would have been compelled to leave the
Hampton school. General Armstrong, however, very
kindly got Mr. S. Griffitts Morgan, of New Bedford, Mass.,
to defray the cost of my tuition during the whole time that
I was at Hampton. After I finished the course at Hampton
and had entered upon my lifework at Tuskegee, I had the
pleasure of visiting Mr. Morgan several times.

After having been for a while at Hampton, I found
myself in difficulty because I did not have books and
clothing. Usually, however, I got around the trouble about
books by borrowing from those who were more fortunate
than myself. As to clothes, when I reached Hampton I
had practically nothing. Everything that I possessed was

in a small hand satchel. My anxiety about clothing was increased because of the fact that General Armstrong made a personal inspection of the young men in ranks, to see that their clothes were clean. Shoes had to be polished, there must be no buttons off the clothing, and no grease-spots. To wear one suit of clothes continually, while at work and in the schoolroom, and at the same time keep it clean, was rather a hard problem for me to solve. In some way I managed to get on till the teachers learned that I was in earnest and meant to succeed, and then some of them were kind enough to see that I was partly supplied with second-hand clothing that had been sent in barrels from the North. These barrels proved a blessing to hundreds of poor but deserving students. Without them I question whether I should ever have gotten through Hampton.

When I first went to Hampton I do not recall that I had ever slept in a bed that had two sheets on it. In those days there were not many buildings there, and room was very precious. There were seven other boys in the same room with me; most of them, however, students who had been there for some time. The sheets were quite a puzzle to me. The first night I slept under both of them, and the second night I slept on top of both of them; but by watching the other boys I learned my lesson in this, and have been trying to follow it ever since and to teach it to others.

I was among the youngest of the students who were in Hampton at that time. Most of the students were men and women—some as old as forty years of age. As I now recall the scene of my first year, I do not believe that one often has the opportunity of coming into contact with three or four hundred men and women who were so tremendously in earnest as these men and women were. Every hour was occupied in study or work. Nearly all had had enough actual contact with the world to teach them the need for education. Many of the older ones were, of course, too old to master the text-books very thoroughly, and it was often sad to watch their struggles; but they made up in earnestness much of what they lacked in books. Many of them were as poor as I was, and, be-sides having to wrestle with their books, they had to strug-gle with a poverty which prevented their having the neces-sities of life. Many of them had aged parents who were

dependent upon them, and some of them were men who
had wives whose support in some way they had to provide
for.

The great and prevailing idea that seemed to take possession of every one was to prepare himself to lift up the
people at his home. No one seemed to think of himself.
And the officers and teachers, what a rare set of human
beings they were! They worked for the students night and
day, in season and out of season. They seemed happy only
when they were helping the students in some manner.
Whenever it is written—and I hope it will be—the part
that the Yankee teachers played in the education of the
Negroes immediately after the war will make one of the
most thrilling parts of the history of this country. The
time is not far distant when the whole South will appreciate this service in a way that it has not yet been able
to do.

CHAPTER IV

Helping Others

At the end of my first year at Hampton I was confronted with another difficulty. Most of the students went
home to spend their vacation. I had no money with which
to go home, but I had to go somewhere. In those days
very few students were permitted to remain at the school
during vacation. It made me feel very sad and homesick to
see the other students preparing to leave and starting for
home. I not only had no money with which to go home,
but I had none with which to go anywhere.

In some way, however, I had gotten hold of an extra,
second-hand coat which I thought was a pretty valuable
coat. This I decided to sell, in order to get a little money
for travelling expenses. I had a good deal of boyish pride,
and I tried to hide, as far as I could, from the other students the fact that I had no money and nowhere to go.
I made it known to a few people in the town of Hampton
that I had this coat to sell, and, after a good deal of per-
suading, one coloured man promised to come to my room
to look the coat over and consider the matter of buying
it. This cheered my drooping spirits considerably. Early

the next morning my prospective customer appeared. After looking the garment over carefully, he asked me how much I wanted for it. I told him I thought it was worth three dollars. He seemed to agree with me as to price, but remarked in the most matter-of-fact way: "I tell you what I will do; I will take the coat, and I will pay you five cents, cash down, and pay you the rest of the money just as soon as I can get it." It is not hard to imagine what my feelings were at the time.

With this disappointment I gave up all hope of getting out of the town of Hampton for my vacation work. I wanted very much to go where I might secure work that would at least pay me enough to purchase some much-needed clothing and other necessities. In a few days practically all the students and teachers had left for their homes, and this served to depress my spirits even more.

After trying for several days in and near the town of Hampton, I finally secured work in a restaurant at Fortress Monroe. The wages, however, were very little more than my board. At night, and between meals, I found considerable time for study and reading; and in this direction I improved myself very much during the summer.

When I left school at the end of my first year, I owed the institution sixteen dollars that I had not been able to work out. It was my greatest ambition during the summer to save money enough with which to pay this debt. I felt that this was a debt of honour, and that I could hardly bring myself to the point of even trying to enter school again till it was paid. I economized in every way that I could think of—did my own washing, and went without necessary garments—but still I found my summer vacation ending and I did not have the sixteen dollars.

One day, during the last week of my stay in the restaurant, I found under one of the tables a crisp, new ten-dollar bill. I could hardly contain myself, I was so happy. As it was not my place of business I felt it to be the proper thing to show the money to the proprietor. This I did. He seemed as glad as I was, but he coolly explained to me that, as it was his place of business, he had a right to keep the money, and he proceeded to do so. This, I confess, was another pretty hard blow to me. I will not say that I became discouraged, for as I now look back over my life I do not recall that I ever became discouraged over anything that I set out to accomplish. I have begun

everything with the idea that I could succeed, and I never
had much patience with the multitudes of people who are
always ready to explain why one cannot succeed. I have
always had a high regard for the man who could tell me
how to succeed. I determined to face the situation just as
it was. At the end of the week I went to the treasurer
of the Hampton Institute, General J. F. B. Marshall, and
told him frankly my condition. To my gratification he
told me that I could reënter the institution, and that he
would trust me to pay the debt when I could. During
the second year I continued to work as a janitor.

The education that I received at Hampton out of the
text-books was but a small part of what I learned there.
One of the things that impressed itself upon me deeply,
the second year, was the unselfishness of the teachers. It
was hard for me to understand how any individuals could
bring themselves to the point where they could be so
happy in working for others. Before the end of the year,
I think I began learning that those who are happiest are
those who do the most for others. This lesson I have tried
to carry with me ever since.

I also learned a valuable lesson at Hampton by coming
into contact with the best breeds of live stock and fowls.
No student, I think, who has had the opportunity of
doing this could go out into the world and content himself with the poorest grades.

Perhaps the most valuable thing that I got out of my
second year was an understanding of the use and value of
the Bible. Miss Nathalie Lord, one of the teachers, from
Portland, Me., taught me how to use and love the Bible.
Before this I had never cared a great deal about it, but
now I learned to love to read the Bible, not only for the
spiritual help which it gives, but on account of it as literature. The lessons taught me in this respect took such a
hold upon me that at the present time, when I am at home,
no matter how busy I am, I always make it a rule to read
a chapter or a portion of a chapter in the morning, before beginning the work of the day.

Whatever ability I may have as a public speaker I owe
in a measure to Miss Lord. When she found out that I
had some inclination in this direction, she gave me private lessons in the matter of breathing, emphasis, and
articulation. Simply to be able to talk in public for the
sake of talking has never had the least attraction for me.

In fact, I consider that there is nothing so empty and unsatisfactory as mere abstract public speaking; but from my early childhood I have had a desire to do something to make the world better, and then to be able to speak to the world about that thing.

The debating societies at Hampton were a constant source of delight to me. These were held on Saturday evening; and during my whole life at Hampton I do not recall that I missed a single meeting. I not only attended the weekly debating society, but was instrumental in organizing an additional society. I noticed that between the time when supper was over and the time to begin evening study there were about twenty minutes which the young men usually spent in idle gossip. About twenty of us formed a society for the purpose of utilizing this time in debate or in practice in public speaking. Few persons ever derived more happiness or benefit from the use of twenty minutes of time than we did in this way.

At the end of my second year at Hampton, by the help of some money sent me by my mother and brother John, supplemented by a small gift from one of the teachers at Hampton, I was enabled to return to my home in Malden, West Virginia, to spend my vacation. When I reached home I found that the salt-furnaces were not running, and that the coal-mine was not operating on account of the miners being out on a "strike." This was something which, it seemed, usually occurred whenever the men got two or three months ahead in their savings. During the strike, of course, they spent all that they had saved, and would often return to work in debt at the same wages, or would move to another mine at considerable expense. In either case, my observations convinced me that the miners were worse off at the end of a strike. Before the days of strikes in that section of the country, I knew miners who had considerable money in the bank, but as soon as the professional labour agitators got control, the savings of even the more thrifty ones began disappearing.

My mother and the other members of the family were, of course, much rejoiced to see me and to note the improvement that I had made during my two years' absence. The rejoicing on the part of all classes of the coloured people, and especially the older ones, over my return, was almost pathetic. I had to pay a visit to each family and take a meal with each, and at each place tell the story of

my experiences at Hampton. In addition to this I had to
speak before the church and Sunday-school, and at various
other places. The thing that I was most in search of,
though, work, I could not find. There was no work on account of the strike. I spent nearly the whole of the first
month of my vacation in an effort to find something to do
by which I could earn money to pay my way back to
Hampton and save a little money to use after reaching
there.

Toward the end of the first month, I went to a place
a considerable distance from my home, to try to find employment. I did not succeed, and it was night before I got
started on my return. When I had gotten within a mile
or so of my home I was so completely tired out that I
could not walk any farther, and I went into an old, abandoned house to spend the remainder of the night. About
three o'clock in the morning my brother John found me
asleep in this house, and broke to me, as gently as he
could, the sad news that our dear mother had died during
the night.

This seemed to me the saddest and blankest moment
in my life. For several years my mother had not been in
good health, but I had no idea, when I parted from her
the previous day, that I should never see her alive again.
Besides that, I had always had an intense desire to be with
her when she did pass away. One of the chief ambitions
which spurred me on at Hampton was that I might be
able to get to be in a position in which I could better
make my mother comfortable and happy. She had so often
expressed the wish that she might be permitted to live to
see her children educated and started out in the world.

In a very short time after the death of my mother our
little home was in confusion. My sister Amanda, although
she tried to do the best she could, was too young to know
anything about keeping house, and my stepfather was not
able to hire a housekeeper. Sometimes we had food cooked
for us, and sometimes we did not. I remember that more
than once a can of tomatoes and some crackers constituted
a meal. Our clothing went uncared for, and everything
about our home was soon in a tumble-down condition. It
seems to me that this was the most dismal period of my
life.

My good friend Mrs. Ruffner, to whom I have already
referred, always made me welcome at her home, and as-

sisted me in many ways during this trying period. Before the end of the vacation she gave me some work, and this, together with work in a coal-mine at some distance from my home, enabled me to earn a little money.

At one time it looked as if I would have to give up the idea of returning to Hampton, but my heart was so set on returning that I determined not to give up going back without a struggle. I was very anxious to secure some clothes for the winter, but in this I was disappointed, except for a few garments which my brother John secured for me. Notwithstanding my need of money and clothing, I was very happy in the fact that I had secured enough money to pay my travelling expenses back to Hampton. Once there, I knew that I could make myself so useful as a janitor that I could in some way get through the school year.

Three weeks before the time for the opening of the term at Hampton, I was pleasantly surprised to receive a letter from my good friend Miss Mary F. Mackie, the lady principal, asking me to return to Hampton two weeks before the opening of the school, in order that I might assist her in cleaning the buildings and getting things in order for the new school year. This was just the opportunity I wanted. It gave me a chance to secure a credit in the treasurer's office. I started for Hampton at once.

During these two weeks I was taught a lesson which I shall never forget. Miss Mackie was a member of one of the oldest and most cultured families of the North, and yet for two weeks she worked by my side cleaning windows, dusting rooms, putting beds in order, and what not. She felt that things would not be in condition for the opening of school unless every window-pane was perfectly clean, and she took the greatest satisfaction in helping to clean them herself. The work which I have described she did every year that I was at Hampton.

It was hard for me at this time to understand how a woman of her education and social standing could take such a delight in performing such service, in order to assist in the elevation of an unfortunate race. Ever since then I have had no patience with any school for my race in the South which did not teach its students the dignity of labour.

During my last year at Hampton every minute of my time that was not occupied with my duties as janitor was devoted to hard study. I was determined, if possible, to make such a record in my class as would cause me to be placed on the "honour roll" of Commencement speakers. This I was successful in doing. It was June of 1875 when I finished the regular course of study at Hampton. The greatest benefits that I got out of my life at the Hampton Institute, perhaps, may be classified under two heads:—

First was contact with a great man, General S. C. Armstrong, who, I repeat, was, in my opinion, the rarest, strongest, and most beautiful character that it has ever been my privilege to meet.

Second, at Hampton, for the first time, I learned what education was expected to do for an individual. Before going there I had a good deal of the then rather prevalent idea among our people that to secure an education meant to have a good, easy time, free from all necessity for manual labour. At Hampton I not only learned that it was not a disgrace to labour, but learned to love labour, not alone for its financial value, but for labour's own sake and for the independence and self-reliance which the ability to do something which the world wants done brings. At that institution I got my first taste of what it meant to live a life of unselfishness, my first knowledge of the fact that the happiest individuals are those who do the most to make others useful and happy.

I was completely out of money when I graduated. In company with other Hampton students, I secured a place as a table waiter in a summer hotel in Connecticut, and managed to borrow enough money with which to get there. I had not been in this hotel long before I found out that I knew practically nothing about waiting on a hotel table. The head waiter, however, supposed that I was an accomplished waiter. He soon gave me charge of a table at which there sat four or five wealthy and rather aristocratic people. My ignorance of how to wait upon them was so apparent that they scolded me in such a severe manner that I became frightened and left their table, leaving them sitting there without food. As a result of this I was reduced from the position of waiter to that of a dish-carrier.

But I determined to learn the business of waiting, and did so within a few weeks and was restored to my former position. I have had the satisfaction of being a guest in this hotel several times since I was a waiter there.

At the close of the hotel season I returned to my former

home in Malden, and was elected to teach the coloured school at that place. This was the beginning of one of the happiest periods of my life. I now felt that I had the opportunity to help the people of my home town to a higher life. I felt from the first that mere book education was not all that the young people of that town needed. I began my work at eight o'clock in the morning, and, as a rule, it did not end until ten o'clock at night. In addition to the usual routine of teaching, I taught the pupils to comb their hair, and to keep their hands and faces clean, as well as their clothing. I gave special attention to teaching them the proper use of the tooth-brush and the bath. In all my teaching I have watched carefully the influence of the tooth-brush, and I am convinced that there are few single agencies of civilization that are more far-reaching.

There were so many of the older boys and girls in the town, as well as men and women, who had to work in the daytime but still were craving an opportunity for some education, that I soon opened a night-school. From the first, this was crowded every night, being about as large as the school that I taught in the day. The efforts of some of the men and women, who in many cases were over fifty years of age, to learn, were in some cases very pathetic.

My day and night school work was not all that I undertook. I established a small reading-room and a debating society. On Sundays I taught two Sunday-schools, one in the town of Malden in the afternoon, and the other in the morning at a place three miles distant from Malden. In addition to this, I gave private lessons to several young men whom I was fitting to send to the Hampton Institute. Without regard to pay and with little thought of it, I taught any one who wanted to learn anything that I could teach him. I was supremely happy in the opportunity of being able to assist somebody else. I did receive, however, a small salary from the public fund, for my work as a public-school teacher.

During the time that I was a student at Hampton my older brother, John, not only assisted me all that he could, but worked all of the time in the coal-mines in order to support the family. He willingly neglected his own education that he might help me. It was my earnest wish to help him to prepare to enter Hampton, and to save enough money to assist him in his expenses there. Both of these objects I was successful in accomplishing. In three years my brother finished the course at Hampton, and he is now holding the important position of Superintendent of Industries at Tuskegee. When he returned from Hampton, we both combined our efforts and savings to send our adopted brother, James, through the Hampton Institute. This we succeeded in doing, and he is now the postmaster at the Tuskegee Institute. The year 1877, which was my second year of teaching in Malden, I spent very much as I did the first.

It was while my home was at Malden that what was known as the "Ku Klux Klan" was in the height of its activity. The "Ku Klux" were bands of men who had joined themselves together for the purpose of regulating the conduct of the coloured people, especially with the object of preventing the members of the race from exercising any influence in politics. They corresponded somewhat to the "patrollers" of whom I used to hear a great deal during the days of slavery, when I was a small boy. The "patrollers" were bands of white men—usually young men—who were organized largely for the purpose of regulating the conduct of the slaves at night in such matters as preventing the slaves from going from one plantation to another without passes, and for preventing them from holding any kind of meetings without permission and without the presence at these meetings of at least one white man.

Like the "patrollers" the "Ku Klux" operated almost wholly at night. They were, however, more cruel than the "patrollers." Their objects, in the main, were to crush out the political aspirations of the Negroes, but they did not confine themselves to this, because schoolhouses as well as churches were burned by them, and many innocent persons were made to suffer. During this period not a few coloured people lost their lives.

As a young man, the acts of these lawless bands made a great impression upon me. I saw one open battle take place at Malden between some of the coloured and white people. There must have been not from a hundred persons engaged on each side; many on both sides were seriously injured, among them being General Lewis Ruffner, the husband of my friend Mrs. Viola Ruffner. General Ruffner tried to defend the coloured people, and for this he was knocked down and so seriously wounded that he never completely recovered. It seemed to me as I watched this struggle between members of the two races, that there was no hope for our people in this country. The "Ku Klux" period was, I think, the darkest part of the Reconstruction days.

I have referred to this unpleasant part of the history of the South simply for the purpose of calling attention to the great change that has taken place since the days of the "Ku Klux." To-day there are no such organizations in the South, and the fact that such ever existed is almost forgotten by both races. There are few places in the South now where public sentiment would permit such organizations to exist.

CHAPTER V

The Reconstruction Period

The years from 1867 to 1878 I think may be called the period of Reconstruction. This included the time that I spent as a student at Hampton and as a teacher in West Virginia. During the whole of the Reconstruction period two ideas were constantly agitating the minds of the coloured people, or, at least, the minds of a large part of the race. One of these was the craze for Greek and Latin learning, and the other was a desire to hold office.

It could not have been expected that a people who had spent generations in slavery, and before that generations in the darkest heathenism, could at first form any proper conception of what an education meant. In every part of the South, during the Reconstruction period, schools, both day and night, were filled to overflowing with people of all ages and conditions, some being as far along in age as sixty and seventy years. The ambition to secure an education was most praiseworthy and encouraging. The idea, however, was too prevalent that, as soon as one secured a little education, in some unexplainable way he would be free from most of the hardships of the world, and, at any rate, could live without manual labour. There was a further feeling that a knowledge, however little, of the Greek and Latin languages would make one a

very superior human being, something bordering almost
on the supernatural. I remember that the first coloured
man whom I saw who knew something about foreign languages impressed me at that time as being a man of all
others to be envied.

Naturally, most of our people who received some little
education became teachers or preachers While among
these two classes there were many capable earnest godly
men and women still a large proportion took up teach
ing or preaching as an easy way to make a living Many
became teachers who could do little more than write their
names I remember there came into our neighbourhood
one of this class who was in search of a school to
teach, and the question arose while he was there as to
the shape of the earth and how he would teach the chil
dren concerning this subject He explained his position in
the matter by saying that he was prepared to teach that
the earth was either flat or round, according to the preference of a majority of his patrons.

The ministry was the profession that suffered most—
and still suffers, though there has been great improvement
—on account of not only ignorant but in many cases
immoral men who claimed that they were "called to
preach." In the earlier days of freedom almost every coloured man who learned to read would receive "a call to
preach" within a few days after he began reading At my
home in West Virginia the process of being called to the
ministry was a very interesting one Usually the "call"
came when the individual was sitting in church. Without
warning the one called would fall upon the floor as if
struck by a bullet, and would lie there for hours, speechless and motionless. Then the news would spread all
through the neighbourhood that this individual had received a "call." If he were inclined to resist the summons,
he would fall or be made to fall a second or third time.
In the end he always yielded to the call. While I wanted
an education badly, I confess that in my youth I had a
fear that when I had learned to read and write well I
would receive one of these "calls"; but, for some reason,
my call never came.

When we add the number of wholly ignorant men who
preached or "exhorted" to that of those who possessed
something of an education, it can be seen at a glance
that the supply of ministers was large. In fact, some time

ago I knew a certain church that had a total membership of about two hundred, and eighteen of that number
were ministers. But, I repeat, in many communities in the
South the character of the ministry is being improved,
and I believe that within the next two or three decades
a very large proportion of the unworthy ones will have
disappeared. The "calls" to preach, I am glad to say, are
not nearly so numerous now as they were formerly,
and the calls to some industrial occupation are growing
more numerous. The improvement that has taken place
in the character of the teachers is even more marked
than in the case of the ministers.

During the whole of the Reconstruction period our people throughout the South looked to the Federal Government for everything, very much as a child looks to its
mother This was not unnatural. The central government
gave them freedom, and the whole Nation had been en-
riched for more than two centuries by the labour of the
Negro Even as a youth, and later in manhood, I had the
feeling that it was cruelly wrong in the central government at the beginning of our freedom, to fail to make
some provision for the general education of our people in
addition to what the states might do, so that the people
would be the better prepared for the duties of citizenship.

It is easy to find fault, to remark what might have
been done, and perhaps, after all, and under all the circumstances, those in charge of the conduct of affairs did
the only thing that could be done at the time. Still, as I
look back now over the entire period of our freedom, I
cannot help feeling that it would have been wiser if some
plan could have been put in operation which would have
made the possession of a certain amount of education or
property, or both, a test for the exercise of the franchise,
and a way provided by which this test should be made
to apply honestly and squarely to both the white and
black races.

Though I was but little more than a youth during the
period of Reconstruction, I had the feeling that mistakes
were being made, and that things could not remain in
the condition that they were in then very long. I felt
that the Reconstruction policy, so far as it related to my
race, was in a large measure on a false foundation, was
artificial and forced. In many cases it seemed to me that
the ignorance of my race was being used as a tool with

which to help white men into office, and that there was
an element in the North which wanted to punish the
Southern white men by forcing the Negro into positions
over the heads of the Southern whites. I felt that the
Negro would be the one to suffer for this in the end.
Besides, the general political agitation drew the attention
of our people away from the more fundamental matters
of perfecting themselves in the industries at their doors
and in securing property.

The temptations to enter political life were so alluring
that I came very near yielding to them at one time, but
I was kept from doing so by the feeling that I would be
helping in a more substantial way by assisting in the laying of the foundation of the race through a generous
education of the hand, head, and heart. I saw coloured
men who were members of the state legislatures, and
county officers, who, in some cases, could not read or
write, and whose morals were as weak as their education.
Not long ago, when passing through the streets of a certain city in the South, I heard some brick-masons calling
out, from the top of a two-story brick building on which
they were working, for the "Governor" to "hurry up and
bring up some more bricks." Several times I heard the
command, "Hurry up, Governor!" "Hurry up, Governor!"
My curiosity was aroused to such an extent that I made
inquiry as to who the "Governor" was, and soon found
that he was a coloured man who at one time had held
the position of Lieutenant-Governor of his state.

But not all the coloured people who were in office during Reconstruction were unworthy of their positions, by
any means. Some of them, like the late Senator B. K.
Bruce, Governor Pinchback, and many others, were
strong, upright, useful men. Neither were all the class
designated as carpetbaggers dishonourable men. Some of
them, like ex-Governor Bullock, of Georgia, were men of
high character and usefulness.

Of course the coloured people, so largely without education, and wholly without experience in government, made
tremendous mistakes, just as any people similarly situated
would have done. Many of the Southern whites have a
feeling that, if the Negro is permitted to exercise his political rights now to any degree, the mistakes of the Reconstruction period will repeat themselves. I do not think
this would be true, because the Negro is a much stronger

and wiser man than he was thirty-five years ago, and he is fast learning the lesson that he cannot afford to act in a manner that will alienate his Southern white neighbours from him. More and more I am convinced that the final solution of the political end of our race problem will be for each state that finds it necessary to change the law bearing upon the franchise to make the law apply with absolute honesty, and without opportunity for double dealing or evasion, to both races alike. Any other course my daily observation in the South convinces me, will be unjust to the Negro, unjust to the white man, and unfair to the rest of the states in the Union, and will be, like slavery, a sin that at some time we shall have to pay for.

In the fall of 1878, after having taught school in Malden for two years, and after I had succeeded in preparing several of the young men and women, besides my two brothers, to enter the Hampton Institute, I decided to spend some months in study at Washington, D. C. I remained there for eight months. I derived a great deal of benefit from the studies which I pursued, and I came into contact with some strong men and women. At the institution I attended there was no industrial training given to the students, and I had an opportunity of comparing the influence of an institution with no industrial training with that of one like the Hampton Institute, that emphasized the industries. At this school I found the students, in most cases, had more money, were better dressed, wore the latest style of all manner of clothing, and in some cases were more brilliant mentally. At Hampton it was a standing rule that, while the institution would be responsible for securing some one to pay the tuition for the students, the men and women themselves must provide for their own board, books, clothing, and room wholly by work, or partly by work and partly in cash. At the institution at which I now was, I found that a large proportion of the students by some means had their own personal expenses paid for them. At Hampton the student was constantly making the effort through the industries to help himself, and that very effort was of immense value in character-building. The students at the other school seemed to be less self-dependent. They seemed to give more attention to more outward appearances. In a word, they did not appear to me to be beginning at the bottom, on a real, solid foundation, to the extent that they were at Hampton. They knew more about Latin and Greek when they left school, but they seemed to know less about life and its conditions as they would meet it at their homes. Having lived for a number of years in the midst of comfortable surroundings, they were not as much inclined as the Hampton students to go into the country districts of the South, where there was little of comfort, to take up work for our people, and they were more inclined to yield to the temptation to become hotel waiters and Pullman-car porters as their life-work.

During the time I was a student in Washington the city was crowded with coloured people, many of whom had recently come from the South. A large proportion of these people had been drawn to Washington because they felt that they could lead a life of ease there. Others had secured minor government positions, and still another large class was there in the hope of securing Federal positions. A number of coloured men—some of them very strong and brilliant—were in the House of Representatives at that time, and one, the Hon. B. K. Bruce, was in the Senate. All this tended to make Washington an attractive place for members of the coloured race. Then, too, they knew that at all times they could have the protection of the law in the District of Columbia. The public schools in Washington for coloured people were better then than they were elsewhere. I took great interest in studying the life of our people there closely at that time. I found that while among them there was a large element of substantial, worthy citizens, there was also a superficiality about the life of a large class that greatly alarmed me. I saw young coloured men who were not earning more than four dollars a week spend two dollars or more for a buggy on Sunday to ride up and down Pennsylvania Avenue in order that they might try to convince the world that they were worth thousands. I saw other young men who received seventy-five or one hundred dollars per month from the Government, who were in debt at the end of every month. I saw men who but a few months previous were members of Congress, then without employment and in poverty. Among a large class there seemed to be a dependence upon the Government for every conceivable thing. The members of this class had little ambition to create a position for themselves, but wanted the Federal officials to create one for them. How many times I wished then, and have often wished since, that by some power of magic I might remove the great bulk of these people into the country districts and plant them upon the soil, upon the solid and never deceptive foundation of Mother Nature, where all nations and races that have ever succeeded have gotten their start,—a start that at first may be slow and toilsome, but one that nevertheless is real.

In Washington I saw girls whose mothers were earning their living by laundrying. These girls were taught by their mothers, in rather a crude way it is true, the industry of laundrying. Later, these girls entered the public schools and remained there perhaps six or eight years. When the public-school course was finally finished, they wanted more costly dresses, more costly hats and shoes. In a word, while their wants had been increased, their ability to supply their wants had not been increased in the same degree. On the other hand, their six or eight years of book education had weaned them away from the occupation of their mothers. The result of this was in too many cases that the girls went to the bad. I often thought how much wiser it would have been to give these girls the same amount of mental training– and I favour any kind of training, whether in the languages or mathematics, that gives strength and culture to the mind—but at the same time to give them the most thorough training in the latest and best methods of laundrying and other kindred occupations.

CHAPTER VI

Black Race and Red Race

During the year that I spent in Washington, and for some little time before this there had been considerable agitation in the state of West Virginia over the question of moving the capital of the state from Wheeling to some other central point. As a result of this, the Legislature designated three cities to be voted upon by the citizens of the state as the permanent seat of government. Among these cities was Charleston, only five miles from Malden, my home. At the close of my school year in Washington

I was very pleasantly surprised to receive, from a committee of white people in Charleston, an invitation to canvass the state in the interests of that city. This invitation I accepted, and spent nearly three months in speaking in various parts of the state. Charleston was successful in winning the prize, and is now the permanent seat of government.

The reputation that I made as a speaker during this campaign induced a number of persons to make an earnest effort to get me to enter political life, but I refused, still believing that I could find other service which would prove of more permanent value to my race. Even then I had a strong feeling that what our people most needed was to get a foundation in education, industry, and property, and for this I felt that they could better afford to strive than for political preferment. As for my individual self, it appeared to me to be reasonably certain that I could succeed in political life, but I had a feeling that it would be a rather selfish kind of success—individual success at the cost of failing to do my duty in assisting in laying a foundation for the masses.

At this period in the progress of our race a very large proportion of the young men who went to school or to college did so with the expressed determination to prepare themselves to be great lawyers, or Congressmen, and many of the women planned to become music teachers; but I had a reasonably fixed idea, even at that early period in my life, that there was need for something to be done to prepare the way for successful lawyers, Congressmen, and music teachers.

I felt that the conditions were a good deal like those of an old coloured man, during the days of slavery, who wanted to learn how to play on the guitar. In his desire to take guitar lessons he applied to one of his young masters to teach him, but the young man, not having much faith in the ability of the slave to master the guitar at his age, sought to discourage him by telling him: "Uncle Jake, I will give you guitar lessons; but, Jake, I will have to charge you three dollars for the first lesson, two dollars for the second lesson, and one dollar for the third lesson. But I will charge you only twenty-five cents for the last lesson."

Uncle Jake answered: "All right, boss, I hires you on

dem terms. But boss! I wants yer to be sure an' give me dat las' lesson first."

Soon after my work in connection with the removal of the capital was finished, I received an invitation which gave me great joy and which at the same time was a very pleasant surprise. This was a letter from General Armstrong, inviting me to return to Hampton at the next Commencement to deliver what was called the "postgraduate address." This was an honour which I had not dreamed of receiving. With much care I prepared the best address that I was capable of. I chose for my subject "The Force That Wins."

As I returned to Hampton for the purpose of delivering this address, I went over much of the same ground —now, however, covered entirely by railroad—that I had traversed nearly six years before, when I first sought entrance into Hampton Institute as a student. Now I was able to ride the whole distance in the train. I was constantly contrasting this with my first journey to Hampton. I think I may say, without seeming egotism, that it is seldom that five years have wrought such a change in the life and aspirations of an individual.

At Hampton I received a warm welcome from teachers and students. I found that during my absence from Hampton the institute each year had been getting closer to the real needs and conditions of our people; that the industrial teaching, as well as that of the academic department, had greatly improved. The plan of the school was not modelled after that of any other institution then in existence, but every improvement was made under the magnificent leadership of General Armstrong solely with the view of meeting and helping the needs of our people as they presented themselves at the time. Too often, it seems to me, in missionary and educational work among undeveloped races, people yield to the temptation of doing that which was done a hundred years before, or is being done in other communities a thousand miles away. The temptation often is to run each individual through a certain educational mould, regardless of the condition of the subject or the end to be accomplished. This was not so at Hampton Institute.

The address which I delivered on Commencement Day seems to have pleased every one, and many kind and encouraging words were spoken to me regarding it. Soon

after my return to my home in West Virginia, where I
had planned to continue teaching, I was again surprised
to receive a letter from General Armstrong, asking me to
return to Hampton, partly as a teacher and partly to
pursue some supplementary studies. This was in the summer of 1879. Soon after I began my first teaching in West
Virginia I had picked out four of the brightest and most
promising of my pupils, in addition to my two brothers,
to whom I have already referred, and had given them special attention, with the view of having them go to Hampton. They had gone there, and in each case the teachers
had found them so well prepared that they entered advanced classes. This fact, it seems, led to my being called
back to Hampton as a teacher. One of the young men
that I sent to Hampton in this way is now Dr. Samuel E.
Courtney, a successful physician in Boston, and a member of the School Board of that city.

About this time the experiment was being tried for the
first time, by General Armstrong, of educating Indians at
Hampton. Few people then had any confidence in the
ability of the Indians to receive education and to profit
by it. General Armstrong was anxious to try the experiment systematically on a large scale. He secured from the
reservations in the Western states over one hundred wild
and for the most part perfectly ignorant Indians, the
greater proportion of whom were young men. The special
work which the General desired me to do was to be a
sort of "house father" to the Indian young men—that is,
I was to live in the building with them and have the
charge of their discipline, clothing, rooms, and so on. This
was a very tempting offer, but I had become so much
absorbed in my work in West Virginia that I dreaded to
give it up. However, I tore myself away from it. I did
not know how to refuse to perform any service that General Armstrong desired of me.

On going to Hampton, I took up my residence in a
building with about seventy-five Indian youths. I was the
only person in the building who was not a member of
their race. At first I had a good deal of doubt about my
ability to succeed. I knew that the average Indian felt
himself above the white man, and, of course, he felt himself far above the Negro, largely on account of the fact
of the Negro having submitted to slavery—a thing which
the Indian would never do. The Indians, in the Indian

territory, owned a large number of slaves during the days of slavery. Aside from this, there was a general feeling that the attempt to educate and civilize the red men at Hampton would be a failure. All this made me proceed very cautiously, for I felt keenly the great responsibility. But I was determined to succeed. It was not long before I had the complete confidence of the Indians, and not only this, but I think I am safe in saying that I had their love and respect. I found that they were about like any other human being; that they responded to kind treatment and resented ill-treatment. They were continually planning to do something that would add to my happiness and comfort. The things that they disliked most, I think, were to have their long hair cut, to give up wearing their blankets, and to cease smoking; but no white American ever thinks that any other race is wholly civilized until he wears the white man's clothes, eats the white man's food, speaks the white man's language, and professes the white man's religion.

When the difficulty of learning the English language was subtracted, I found that in the matter of learning trades and in mastering academic studies there was little difference between the coloured and Indian students. It was a constant delight to me to note the interest which the coloured students took in trying to help the Indians in every way possible. There were a few of the coloured students who felt that the Indians ought not to be admitted to Hampton, but these were in the minority. Whenever they were asked to do so, the Negro students gladly took the Indians as room-mates, in order that they might teach them to speak English and to acquire civilized habits.

I have often wondered if there was a white institution in this country whose students would have welcomed the incoming of more than a hundred companions of another race in the cordial way that these black students at Hampton welcomed the red ones. How often I have wanted to say to white students that they lift themselves up in proportion as they help to lift others, and the more unfortunate the race, and the lower in the scale of civilization, the more does one raise one's self by giving the assistance.

This reminds me of a conversation which I once had with the Hon. Frederick Douglass. At one time, Mr. Douglass was travelling in the state of Pennsylvania, and was forced, on account of his colour, to ride in the baggage-car, in spite of the fact that he had paid the same price for his passage that the other passengers had paid. When some of the white passengers went into the baggage-car to console Mr. Douglass, and one of them said to him: "I am sorry, Mr. Douglass, that you have been degraded in this manner," Mr. Douglass straightened himself up on the box upon which he was sitting, and replied: "They cannot degrade Frederick Douglass. The soul that is within me no man can degrade. I am not the one that is being degraded on account of this treatment, but those who are inflicting it upon me."

In one part of our country, where the law demands the separation of the races on the railroad trains, I saw at one time a rather amusing instance which showed how difficult it sometimes is to know where the black begins and the white ends.

There was a man who was well known in his community as a Negro, but who was so white that even an expert would have hard work to classify him as a black man. This man was riding in the part of the train set aside for the coloured passengers. When the train conductor reached him, he showed at once that he was perplexed. If the man was a Negro, the conductor did not want to send him into the white people's coach: at the same time, if he was a white man, the conductor did not want to insult him by asking him if he was a Negro. The official looked him over carefully, examining his hair, eyes, nose, and hands, but still seemed puzzled. Finally, to solve the difficulty, he stooped over and peeped at the man's feet. When I saw the conductor examining the feet of the man in question, I said to myself, "That will settle it"; and so it did, for the trainman promptly decided that the passenger was a Negro, and let him remain where he was. I congratulated myself that my race was fortunate in not losing one of its members.

My experience has been that the time to test a true gentleman is to observe him when he is in contact with individuals of a race that is less fortunate than his own. This is illustrated in no better way than by observing the conduct of the old-school type of Southern gentleman when he is in contact with his former slaves or their descendants.

An example of what I mean is shown in a story told of George Washington, who, meeting a coloured man in the road once, who politely lifted his hat, lifted his own in return. Some of his white friends who saw the incident criticised Washington for his action. In reply to their criticism George Washington said: "Do you suppose that I am going to permit a poor, ignorant, coloured man to be more polite than I am?"

While I was in charge of the Indian boys at Hampton, I had one or two experiences which illustrate the curious workings of caste in America. One of the Indian boys was taken ill, and it became my duty to take him to Washington, deliver him over to the Secretary of the Interior, and get a receipt for him, in order that he might be returned to his Western reservation. At that time I was rather ignorant of the ways of the world. During my journey to Washington, on a steamboat, when the bell rang for dinner, I was careful to wait and not enter the dining room until after the greater part of the passengers had finished their meal. Then, with my charge. I went to the dining saloon. The man in charge politely informed me that the Indian could be served, but that I could not. I never could understand how he knew just where to draw the colour line, since the Indian and I were about the same complexion. The steward, however, seemed to be an expert in this matter. I had been directed by the authorities at Hampton to stop at a certain hotel in Washington with my charge, but when I went to this hotel the clerk stated that he would be glad to receive the Indian into the house, but said that he could not accommodate me.

An illustration of something of this same feeling came under my observation afterward. I happened to find myself in a town in which so much excitement and indignation were being expressed that it seemed likely for a time that there would be a lynching. The occasion of the trouble was that a dark-skinned man had stopped at the local hotel. Investigation, however, developed the fact that this individual was a citizen of Morocco, and that while travelling in this country he spoke the English language. As soon as it was learned that he was not an American Negro, all the signs of indignation disappeared. The man who was the innocent cause of the excitement, though, found it prudent after that not to speak English.

At the end of my first year with the Indians there came

another opening for me at Hampton, which, as I look back over my life now, seems to have come providentially, to help to prepare me for my work at Tuskegee later. General Armstrong had found out that there was quite a number of young coloured men and women who were intensely in earnest in wishing to get an education, but who were prevented from entering Hampton Institute because they were too poor to be able to pay any portion of the cost of their board, or even to supply themselves with books. He conceived the idea of starting a night-school in connection with the Institute, into which a limited number of the most promising of these young men and women would be received, on condition that they were to work for ten hours during the day, and attend school for two hours at night. They were to be paid something above the cost of their board for their work. The greater part of their earnings was to be reserved in the school's treasury as a fund to be drawn on to pay their board when they had become students in the day-school, after they had spent one or two years in the night-school. In this way they would obtain a start in their books and a knowledge of some trade or industry, in addition to the other far-reaching benefits of the institution.

General Armstrong asked me to take charge of the night-school, and I did so. At the beginning of this school there were about twelve strong, earnest men and women who entered the class. During the day the greater part of the young men worked in the school's sawmill, and the young women worked in the laundry. The work was not easy in either place, but in all my teaching I never taught pupils who gave me such genuine satisfaction as these did. They were good students, and mastered their work thoroughly. They were so much in earnest that only the ringing of the retiring-bell would make them stop studying, and often they would urge me to continue the lessons after the usual hour for going to bed had come.

These students showed so much earnestness both in their hard work during the day, as well as in their application to their studies at night, that I gave them the name of "The Plucky Class"—a name which soon grew popular and spread throughout the institution. After a student had been in the night-school long enough to prove what was in him, I gave him a printed certificate which read something like this:—

"This is to certify that James Smith is a member of The Plucky Class of the Hampton Institute, and is in good and regular standing."

The students prized these certificates highly, and they added greatly to the popularity of the night-school. Within a few weeks this department had grown to such an extent that there were about twenty-five students in attendance. I have followed the course of many of these twenty-five men and women ever since then, and they are now holding important and useful positions in nearly every part of the South. The night-school at Hampton, which started with only twelve students, now numbers between three and four hundred, and is one of the permanent and most important features of the institution.

CHAPTER VII

Early Days at Tuskegee

During the time that I had charge of the Indians and the night-school at Hampton, I pursued some studies myself, under the direction of the instructors there. One of these instructors was the Rev. Dr. H. B. Frissell, the present Principal of the Hampton Institute, General Armstrong's successor.

In May, 1881, near the close of my first year in teaching the night-school, in a way that I had not dared expect, the opportunity opened for me to begin my life-work. One night in the chapel, after the usual chapel exercises were over, General Armstrong referred to the fact that he had received a letter from some gentlemen in Alabama asking him to recommend some one to take charge of what was to be a normal school for the coloured people in the little town of Tuskegee in that state. These gentlemen seemed to take it for granted that no coloured man suitable for the position could be secured, and they were expecting the General to recommend a white man for the place. The next day General Armstrong sent for me to come to his office, and, much to my surprise, asked me if I thought I could fill the position in Alabama. I told him that I would be willing to try. Accordingly, he wrote to the people who had applied to him for the information, that he did not know of any white man to suggest, but if they would be willing to take a coloured man, he had one whom he could recommend. In this letter he gave them my name.

Several days passed before anything more was heard about the matter Some time afterward, one Sunday evening during the chapel exercises, a messenger came in and handed the general a telegram. At the end of the exercises he read the telegram to the school. In substance, these were its words: "Booker T. Washington will suit us. Send him at once."

There was a great deal of joy expressed among the students and teachers, and I received very hearty congratulations. I began to get ready at once to go to Tuskegee. I went by way of my old home in West Virginia, where I remained for several days, after which I proceeded to Tuskegee I found Tuskegee to be a town of about two thousand inhabitants, nearly one-half of whom were coloured. It was in what was known as the Black Belt of the South. In the county in which Tuskegee is situated the coloured people outnumbered the whites by about three to one. In some of the adjoining and near-by counties the proportion was not far from six coloured persons to one white.

I have often been asked to define the term "Black Belt." So far as I can learn, the term was first used to designate a part of the country which was distinguished by the colour of the soil. The part of the country possessing this thick, dark, and naturally rich soil was, of course, the part of the South where the slaves were most profitable, and consequently they were taken there in the largest numbers Later and especially since the war, the term seems to be used wholly in a political sense—that is, to designate the counties where the black people outnumbered the white.

Before going to Tuskegee I had expected to find there a building and all the necessary apparatus ready for me to begin teaching. To my disappointment, I found nothing of the kind. I did find, though, that which no costly building and apparatus can supply,—hundreds of hungry, earnest souls who wanted to secure knowledge.

Tuskegee seemed an ideal place for the school. It was in the midst of the great bulk of the Negro population, and was rather secluded, being five miles from the main line of railroad, with which it was connected by a short line. During the days of slavery, and since. the town had been a centre for the education of the white people. This was an added advantage, for the reason that I found the white people possessing a degree of culture and education that is not surpassed by many localities. While the coloured people were ignorant, they had not, as a rule degraded and weakened their bodies by vices such as are common to the lower class of people in the large cities. In general, I found the relations between the two races pleasant. For example, the largest, and I think at that time the only hardware store in the town was owned and operated jointly by a coloured man and a white man. This copartnership continued until the death of the white partner.

I found that about a year previous to my going to Tuskegee some of the coloured people who had heard something of the work of education being done at Hampton had applied to the state Legislature, through their representatives, for a small appropriation to be used in starting a normal school in Tuskegee. This request the Legislature had complied with to the extent of granting an annual appropriation of two thousand dollars. I soon learned, however, that this money could be used only for the payment of the salaries of the instructors. and that there was no provision for securing land. buildings, or apparatus. The task before me did not seem a very encouraging one. It seemed much like making bricks without straw. The coloured people were overjoyed, and were constantly offering their services in any way in which they could be of assistance in getting the school started.

My first task was to find a place in which to open the school. After looking the town over with some care, the most suitable place that could be secured seemed to be a rather dilapidated shanty near the coloured Methodist church, together with the church itself as a sort of assembly-room. Both the church and the shanty were in about as bad condition as was possible. I recall that during the first months of school that I taught in this building it was in such poor repair that, whenever it rained, one of the older students would very kindly leave his lessons to hold an umbrella over me while I heard the recitations of the others. I remember, also, that on more than one occasion my landlady held an umbrella over me while I ate breakfast.

At the time I went to Alabama the coloured people were taking considerable interest in politics, and they were very anxious that I should become one of them politically, in every respect They seemed to have a little distrust of strangers in this regard. I recall that one man, who seemed to have been designated by the others to look after my political destiny came to me on several occasions and said, with a good deal of earnestness. "We wants you to be sure to vote jes like we votes. We can't read de newspapers very much but we knows how to vote, an' we wants you to vote jes' like we votes." He added: "We watches de white man, and we keeps watching de white man till we finds out which way de white man's gwine to vote; an' when we finds out which way de white man's gwine to vote den we votes 'xactly de other way. Den we know we's right."

I am glad to add, however, that at the present time the disposition to vote against the white man merely because he is white is largely disappearing, and the race is learning to vote from principle, for what the voter considers to be for the best interests of both races.

I reached Tuskegee, as I have said, early in June, 1881. The first month I spent in finding accommodations for the school, and in travelling through Alabama, examining into the actual life of the people especially in the country districts, and in getting the school advertised among the class of people that I wanted to have attend it. The most of my travelling was done over the country roads, with a mule and a cart or a mule and a buggy wagon for conveyance. I ate and slept with the people, in their little cabins I saw their farms, their schools, their churches Since, in the case of the most of these visits, there had been no notice given in advance that a stranger was expected, I had the advantage of seeing the real, everyday life of the people.

In the plantation districts I found that, as a rule the whole family slept in one room, and that in addition to the immediate family there sometimes were relatives, or others not related to the family, who slept in the same room. On more than one occasion I went outside the house to get ready for bed, or to wait until the family had gone to bed. They usually contrived some kind of a place for me to sleep, either on the floor or in a special part of another's bed. Rarely was there any place provided in the cabin where one could bathe even the face and hands, but usually some provision was made for this outside the house, in the yard.

The common diet of the people was fat pork and corn bread. At times I have eaten in cabins where they had only corn bread and "black-eye peas" cooked in plain water. The people seemed to have no other idea than to live on this fat meat and corn bread,—the meat, and the meal of which the bread was made, having been bought at a high price at a store in town, notwithstanding the fact that the land all about the cabin homes could easily have been made to produce nearly every kind of garden vegetable that is raised anywhere in the country. Their one object seemed to be to plant nothing but cotton; and in many cases cotton was planted up to the very door of the cabin.

In these cabin homes I often found sewing-machines which had been bought, or were being bought, on instalments, frequently at a cost of as much as sixty dollars, or showy clocks for which the occupants of the cabins had paid twelve or fourteen dollars. I remember that on one occasion when I went into one of these cabins for dinner, when I sat down to the table for a meal with the four members of the family, I noticed that, while there were five of us at the table, there was but one fork for the five of us to use. Naturally there was an awkward pause on my part. In the opposite corner of that same cabin was an organ for which the people told me they were paying sixty dollars in monthly instalments. One fork, and a sixty-dollar organ!

In most cases the sewing-machine was not used, the clocks were so worthless that they did not keep correct time—and if they had, in nine cases out of ten there would have been no one in the family who could have told the time of day—while the organ, of course, was rarely used for want of a person who could play upon it.

In the case to which I have referred, where the family sat down to the table for the meal at which I was their guest, I could plainly see that this was an awkward and unusual proceeding, and was done in my honour. In most cases, when the family got up in the morning, for example, the wife would put a piece of meat in a frying-pan and put a lump of dough in a "skillet," as they called it. These utensils would be placed on the fire, and in ten or fifteen minutes breakfast would be ready. Frequently the husband would take his bread and meat in his hand and start for the field, eating as he walked. The mother would sit down in a corner and eat her breakfast, perhaps from a plate and perhaps directly from the "skillet" or frying-pan, while the children would eat their portion of the bread and meat while running about the yard. At certain seasons of the year, when meat was scarce, it was rarely that the children who were not old enough or strong enough to work in the fields would have the luxury of meat.

The breakfast over, and with practically no attention given to the house, the whole family would, as a general thing, proceed to the cotton-field. Every child that was large enough to carry a hoe was put to work, and the baby—for usually there was at least one baby—would be laid down at the end of the cotton row, so that its mother could give it a certain amount of attention when she had finished chopping her row. The noon meal and the supper were taken in much the same way as the breakfast.

All the days of the family would be spent after much this same routine, except Saturday and Sunday. On Saturday the whole family would spend at least half a day, and often a whole day, in town. The idea in going to town was, I suppose, to do shopping, but all the shopping that the whole family had money for could have been attended to in ten minutes by one person. Still, the whole family remained in town for most of the day, spending the greater part of the time in standing on the streets, the women, too often, sitting about somewhere smoking or dipping snuff. Sunday was usually spent in going to some big meeting. With few exceptions, I found that the crops were mortgaged in the counties where I went, and that the most of the coloured farmers were in debt. The state had not been able to build schoolhouses in the country districts, and, as a rule, the schools were taught in churches or in log cabins. More than once, while on my journeys, I found that there was no provision made in the house used for school purposes for heating the building during the winter, and consequently a fire had to be built in the yard, and teacher and pupils passed in and out of the house as they got cold or warm. With few exceptions, I found the teachers in these country schools to be miserably poor in preparation for their work, and poor in moral character. The schools were in session from three to five months. There was practically no apparatus in the schoolhouses, except that occasionally there was a rough blackboard. I recall that one day I went into a schoolhouse—or rather into an abandoned log cabin that was being used as a schoolhouse—and found five pupils who were studying a lesson from one book. Two of these, on the front seat, were using the book between them; behind these were two others peeping over the shoulders of the first two, and behind the four was a fifth little fellow who was peeping over the shoulders of all four.

What I have said concerning the character of the schoolhouses and teachers will also apply quite accurately as a description of the church buildings and the ministers.

I met some very interesting characters during my travels. As illustrating the peculiar mental processes of the country people, I remember that I asked one coloured man, who was about sixty years old, to tell me something of his history. He said that he had been born in Virginia, and sold into Alabama in 1845. I asked him how many were sold at the same time. He said, "There were five of us; myself and brother and three mules."

In giving all these descriptions of what I saw during my month of travel in the country around Tuskegee, I wish my readers to keep in mind the fact that there were many encouraging exceptions to the conditions which I have described. I have stated in such plain words what I saw, mainly for the reason that later I want to emphasize the encouraging changes that have taken place in the communities, not wholly by the work of the Tuskegee school but by that of other institutions as well.

CHAPTER VIII

Teaching School in a Stable and a Hen-house

I confess that what I saw during my month of travel and investigation left me with a very heavy heart. The work to be done in order to lift these people up seemed almost beyond accomplishing. I was only one person, and it seemed to me that the little effort which I could put forth could go such a short distance toward bringing about results. I wondered if I could accomplish anything, and if it were worth while for me to try.

Of one thing I felt more strongly convinced than ever, after spending this month in seeing the actual life of the coloured people, and that was that, in order to lift them up, something must be done more than merely to imitate New England education as it then existed. I saw more clearly than ever the wisdom of the system which General Armstrong had inaugurated at Hampton. To take the children of such people as I had been among for a month, and each day give them a few hours of mere book education, I felt would be almost a waste of time.

After consultation with the citizens of Tuskegee, I set July 4, 1881, as the day for the opening of the school in the little shanty and church which had been secured for its accommodation. The white people, as well as the coloured, were greatly interested in the starting of the new school, and the opening day was looked forward to with much earnest discussion. There were not a few white people in the vicinity of Tuskegee who looked with some disfavour upon the project. They questioned its value to the coloured people, and had a fear that it might result in bringing about trouble between the races. Some had the feeling that in proportion as the Negro received education, in the same proportion would his value decrease as an economic factor in the state. These people feared the result of education would be that the Negroes would leave the farms, and that it would be difficult to secure them for domestic service.

The white people who questioned the wisdom of starting this new school had in their minds pictures of what was called an educated Negro, with a high hat, imitation gold eye-glasses, a showy walking-stick, kid gloves, fancy boots, and what not—in a word, a man who was determined to live by his wits. It was difficult for these people to see how education would produce any other kind of a coloured man.

In the midst of all the difficulties which I encountered in getting the little school started, and since then through a period of nineteen years, there are two men among all the many friends of the school in Tuskegee upon whom I have depended constantly for advice and guidance; and the success of the undertaking is largely due to these men, from whom I have never sought anything in vain. I mention them simply as types. One is a white man and an ex-slaveholder, Mr. George W. Campbell; the other is a black man and an ex-slave, Mr. Lewis Adams. These were the men who wrote to General Armstrong for a teacher.

Mr. Campbell is a merchant and banker, and had had little experience in dealing with matters pertaining to education. Mr. Adams was a mechanic, and had learned the trades of shoemaking, harness-making, and tinsmithing during the days of slavery. He had never been to school a day in his life, but in some way he had learned to read and write while a slave. From the first, these two men saw clearly what my plan of education was, sympathized with me, and supported me in every effort. In the days which were darkest financially for the school, Mr. Campbell was never appealed to when he was not willing to extend all the aid in his power. I do not know two men, one an ex-slaveholder, one an ex-slave, whose advice and judgment I would feel more like following in everything which concerns the life and development of the school at Tuskegee than those of these two men.

I have always felt that Mr. Adams, in a large degree, derived his unusual power of mind from the training given his hands in the process of mastering well three trades during the days of slavery. If one goes to-day into any Southern town, and asks for the leading and most reliable coloured man in the community, I believe that in five cases out of ten he will be directed to a Negro who learned a trade during the days of slavery.

On the morning that the school opened, thirty students reported for admission. I was the only teacher. The students were about equally divided between the sexes. Most of them lived in Macon County, the county in which Tuskegee is situated and of which it is the county-seat. A great many more students wanted to enter the school, but it had been decided to receive only those who were above fifteen years of age, and who had previously received some education. The greater part of the thirty were public-school teachers, and some of them were nearly forty years of age. With the teachers came some of their former pupils, and when they were examined it was amusing to note that in several cases the pupil entered a higher class than did his former teacher. It was also in-

teresting to note how many big books some of them had
studied, and how many high-sounding subjects some of
them claimed to have mastered. The bigger the book and
the longer the name of the subject, the prouder they felt
of their accomplishment. Some had studied Latin, and one
or two Greek. This they thought entitled them to special
distinction.

In fact, one of the saddest things I saw during the
month of travel which I have described was a young
man, who had attended some high school, sitting down in
a one-room cabin, with grease on his clothing, filth all
around him, and weeds in the yard and garden, engaged
in studying French grammar.

The students who came first seemed to be fond of
memorizing long and complicated "rules" in grammar and
mathematics, but had little thought or knowledge of applying these rules to the everyday affairs of their life.
One subject which they liked to talk about, and tell me
that they had mastered, in arithmetic, was "banking and
discount," but I soon found out that neither they nor almost any one in the neighbourhood in which they lived
had ever had a bank account. In registering the names
of the students, I found that almost every one of them
had one or more middle initials. When I asked what the
"J" stood for, in the name of John J. Jones, it was explained to me that this was a part of his "entitles." Most of
the students wanted to get an education because they
thought it would enable them to earn more money as
school-teachers.

Notwithstanding what I had said about them in these
respects, I have never seen a more earnest and willing
company of young men and women than these students
were. They were all willing to learn the right thing as
soon as it was shown them what was right. I was determined to start them off on a solid and thorough foundation, so far as their books were concerned. I soon learned
that most of them had the merest smattering of the high-
sounding things that they had studied. While they could
locate the Desert of Sahara or the capital of China on an
artificial globe, I found out that the girls could not locate
the proper places for the knives and forks on an actual
dinner-table, or the places on which the bread and meat
should be set.

I had to summon a good deal of courage to take a

student who had been studying cube root and "banking and discount," and explain to him that the wisest thing for him to do first was thoroughly to master the multiplication table.

The number of pupils increased each week, until by the end of the first month there were nearly fifty. Many of them, however, said that, as they could remain only for two or three months, they wanted to enter a high class and get a diploma the first year if possible.

At the end of the first six weeks a new and rare face entered the school as a co-teacher. This was Miss Olivia A. Davidson, who later became my wife. Miss Davidson was born in Ohio, and received her preparatory education in the public schools of that state. When little more than a girl, she heard of the need of teachers in the South. She went to the state of Mississippi and began teaching there. Later she taught in the city of Memphis. While teaching in Mississippi, one of her pupils became ill with smallpox. Every one in the community was so frightened that no one would nurse the boy. Miss Davidson closed her school and remained by the bedside of the boy night and day until he recovered. While she was at her Ohio home on her vacation, the worst epidemic of yellow fever broke out in Memphis, Tenn., that perhaps has ever occurred in the South. When she heard of this, she at once telegraphed the Mayor of Memphis, offering her services as a yellow-fever nurse, although she had never had the disease.

Miss Davidson's experience in the South showed her that the people needed something more than mere book-learning. She heard of the Hampton system of education, and decided that this was what she wanted in order to prepare herself for better work in the South. The attention of Mrs. Mary Hemenway, of Boston, was attracted to her rare ability. Through Mrs. Hemenway's kindness and generosity, Miss Davidson, after graduating at Hampton, received an opportunity to complete a two years' course of training at the Massachusetts State Normal School at Framingham.

Before she went to Framingham, some one suggested to Miss Davidson that, since she was so very light in colour, she might find it more comfortable not to be known as a coloured woman in this school in Massachusetts. She at once replied that under no circumstances and for no considerations would she consent to deceive any one in regard to her racial identity.

Soon after her graduation from the Framingham institution, Miss Davidson came to Tuskegee, bringing into the school many valuable and fresh ideas as to the best methods of teaching, as well as a rare moral character and a life of unselfishness that I think has seldom been equalled. No single individual did more toward laying the foundations of the Tuskegee Institute so as to insure the successful work that has been done there than Olivia A. Davidson.

Miss Davidson and I began consulting as to the future of the school from the first. The students were making progress in learning books and in developing their minds; but it became apparent at once that, if we were to make any permanent impression upon those who had come to us for training, we must do something besides teach them mere books. The students had come from homes where they had had no opportunities for lessons which would teach them how to care for their bodies. With few exceptions, the homes in Tuskegee in which the students boarded were but little improvement upon those from which they had come. We wanted to teach the students how to bathe; how to care for their teeth and clothing. We wanted to teach them what to eat, and how to eat it properly, and how to care for their rooms. Aside from this, we wanted to give them such a practical knowledge of some one industry, together with the spirit of industry, thrift, and economy, that they would be sure of knowing how to make a living after they had left us. We wanted to teach them to study actual things instead of mere books alone.

We found that most of our students came from the country districts, where agriculture in some form or other was the main dependence of the people. We learned that about eighty-five per cent of the coloured people in the Gulf states depended upon agriculture for their living. Since this was true, we wanted to be careful not to educate our students out of sympathy with agricultural life, so that they would be attracted from the country to the cities, and yield to the temptation of trying to live by their wits. We wanted to give them such an education as would fit a large proportion of them to be teachers, and at the same time cause them to return to the plantation districts and show the people there how to put new energy and new ideas into farming, as well as into the intellectual and moral and religious life of the people.

All these ideas and needs crowded themselves upon us with a seriousness that seemed well-nigh overwhelming. What were we to do? We had only the little old shanty and the abandoned church which the good coloured people of the town of Tuskegee had kindly loaned us for the accommodation of the classes. The number of students was increasing daily. The more we saw of them and the more we travelled through the country districts, the more we saw that our efforts were reaching, to only a partial degree, the actual needs of the people whom we wanted to lift up through the medium of the students whom we should educate and send out as leaders.

The more we talked with the students, who were then coming to us from several parts of the state, the more we found that the chief ambition among a large proportion of them was to get an education so that they would not have to work any longer with their hands.

This is illustrated by a story told of a coloured man in Alabama, who, one hot day in July, while he was at work in a cotton-field, suddenly stopped, and, looking toward the skies, said: "O Lawd, de cotton am so grassy, de work am so hard, and the sun am so hot dat I b'lieve dis darky am called to preach!"

About three months after the opening of the school, and at the time when we were in the greatest anxiety about our work, there came into the market for sale an old and abandoned plantation which was situated about a mile from the town of Tuskegee. The mansion house—or "big house," as it would have been called—which had been occupied by the owners during slavery, had been burned. After making a careful examination of this place, it seemed to be just the location that we wanted in order to make our work effective and permanent.

But how were we to get it? The price asked for it was very little—only five hundred dollars—but we had no money, and we were strangers in the town and had no credit. The owner of the land agreed to let us occupy the place if we could make a payment of two hundred and fifty dollars down, with the understanding that the re-

maining two hundred and fifty dollars must be paid within
a year. Although five hundred dollars was cheap for the
land, it was a large sum when one did not have any part
of it.

In the midst of the difficulty I summoned a great deal
of courage and wrote to my friend General J. F. B. Marshall, the Treasurer of the Hampton Institute, putting the
situation before him and beseeching him to lend me the two
hundred and fifty dollars on my own personal responsibility. Within a few days a reply came to the effect that
he had no authority to lend me money belonging to the
Hampton Institute, but that he would gladly lend me the
amount needed from his own personal funds.

I confess that the securing of this money in this way
was a great surprise to me, as well as a source of gratification. Up to that time I never had had in my possession so
much money as one hundred dollars at a time, and the
loan which I had asked General Marshall for seemed a
tremendously large sum to me. The fact of my being
responsible for the repaying of such a large amount of
money weighed very heavily upon me.

I lost no time in getting ready to move the school on to
the new farm. At the time we occupied the place there
were standing upon it a cabin, formerly used as the dining
room, an old kitchen, a stable, and an old hen-house.
Within a few weeks we had all of these structures in use.
The stable was repaired and used as a recitation-room,
and very presently the hen-house was utilized for the
same purpose.

I recall that one morning, when I told an old coloured
man who lived near, and who sometimes helped me, that
our school had grown so large that it would be necessary
for us to use the hen-house for school purposes, and that
I wanted him to help me give it a thorough cleaning out
the next day, he replied, in the most earnest manner:
"What you mean, boss? You sholy ain't gwine clean out
de hen-house in de *day*-time?"

Nearly all the work of getting the new location ready
for school purposes was done by the students after school
was over in the afternoon. As soon as we got the cabins
in condition to be used, I determined to clear up some
land so that we could plant a crop. When I explained my
plan to the young men, I noticed that they did not seem

to take to it very kindly. It was hard for them to see the connection between clearing land and an education. Besides, many of them had been school-teachers, and they questioned whether or not clearing land would be in keeping with their dignity. In order to relieve them from any embarrassment, each afternoon after school I took my axe and led the way to the woods. When they saw that I was not afraid or ashamed to work, they began to assist with more enthusiasm. We kept at the work each afternoon until we had cleared about twenty acres and had planted a crop.

In the meantime Miss Davidson was devising plans to repay the loan. Her first effort was made by holding festivals, or "suppers." She made a personal canvass among the white and coloured families in the town of Tuskegee, and got them to agree to give something, like a cake, a chicken, bread, or pies, that could be sold at the festival. Of course the coloured people were glad to give anything that they could spare, but I want to add that Miss Davidson did not apply to a single white family, so far as I now remember, that failed to donate something; and in many ways the white families showed their interest in the school.

Several of these festivals were held, and quite a little sum of money was raised. A canvass was also made among the people of both races for direct gifts of money, and most of those applied to gave small sums. It was often pathetic to note the gifts of the older coloured people, most of whom had spent their best days in slavery. Sometimes they would give five cents, sometimes twenty-five cents. Sometimes the contribution was a quilt, or a quantity of sugarcane. I recall one old coloured woman, who was about seventy years of age, who came to see me when we were raising money to pay for the farm. She hobbled into the room where I was, leaning on a cane. She was clad in rags; but they were clean. She said: "Mr. Washin'-ton, God knows I spent de bes' days of my life in slavery. God knows I's ignorant an' poor; but," she added, "I knows what you an' Miss Davidson is tryin' to do. I knows you is tryin' to make better men an' better women for de coloured race. I ain't got no money, but I wants you to take dese six eggs, what I's been savin' up, an' I wants you to put dese six eggs into de eddication of dese boys an' gals."

Since the work at Tuskegee started, it has been my privilege to receive many gifts for the benefit of the institution, but never any, I think, that touched me so deeply as this one.

CHAPTER IX

Anxious Days and Sleepless Nights

The coming of Christmas, that first year of our residence in Alabama, gave us an opportunity to get a farther insight into the real life of the people. The first thing that reminded us that Christmas had arrived was the "foreday" visits of scores of children rapping at our doors, asking for "Chris'mus gifts! Chris'mus gifts!" Between the hours of two o'clock and five o'clock in the morning I presume that we must have had a half-hundred such calls. This custom prevails throughout this portion of the South to-day.

During the days of slavery it was a custom quite generally observed throughout all the Southern states to give the coloured people a week of holiday at Christmas, or to allow the holiday to continue as long as the "yule log" lasted. The male members of the race, and often the female members, were expected to get drunk. We found that for a whole week the coloured people in and around Tuskegee dropped work the day before Christmas, and that it was difficult to get any one to perform any service from the time they stopped work until after the New Year. Persons who at other times did not use strong drink thought it quite the proper thing to indulge in it rather freely during the Christmas week. There was a widespread hilarity, and a free use of guns, pistols, and gunpowder generally. The sacredness of the season seemed to have been almost wholly lost sight of.

During this first Christmas vacation I went some distance from the town to visit the people on one of the large plantations. In their poverty and ignorance it was pathetic to see their attempts to get joy out of the season that in most parts of the country is so sacred and so dear to the heart. In one cabin I noticed that all that the five children had to remind them of the coming of Christ was a single bunch of firecrackers, which they had divided among them. In another cabin, where there were at least a half-dozen persons, they had only ten cents' worth of ginger-cakes, which had been bought in the store the day before. In another family they had only a few pieces of sugar-cane. In still another cabin I found nothing but a new jug of cheap, mean whiskey, which the husband and wife were making free use of, notwithstanding the fact that the husband was one of the local ministers. In a few instances I found that the people had gotten hold of some bright-coloured cards that had been designed for adver-tising purposes, and were making the most of those. In other homes some member of the family had bought a new pistol. In the majority of cases there was nothing to be seen in the cabin to remind one of the coming of the Saviour, except that the people had ceased work in the fields and were lounging about their homes. At night, during Christmas week, they usually had what they called a "frolic," in some cabin on the plantation. This meant a kind of rough dance, where there was likely to be a good deal of whiskey used, and where there might be some shooting or cutting with razors.

While I was making this Christmas visit I met an old coloured man who was one of the numerous local preach-ers, who tried to convince me, from the experience Adam had in the Garden of Eden, that God had cursed all labour, and that, therefore, it was a sin for any man to work. For that reason this man sought to do as little work as possible. He seemed at that time to be supremely happy, because he was living, as he expressed it, through one week that was free from sin.

In the school we made a special effort to teach our students the meaning of Christmas, and to give them lessons in its proper observance. In this we have been successful to a degree that makes me feel safe in saying that the season now has a new meaning, not only through all that immediate region, but, in a measure, wherever our graduates have gone.

At the present time one of the most satisfactory features of the Christmas and Thanksgiving seasons at Tuskegee is the unselfish and beautiful way in which our graduates and students spend their time in administering to the comfort and happiness of others, especially the unfortun-ate. Not long ago some of our young men spent a holiday

in rebuilding a cabin for a helpless coloured woman who
is about seventy-five years old. At another time I remember that I made it known in chapel, one night, that a
very poor student was suffering from cold, because he
needed a coat. The next morning two coats were sent to
my office for him.

I have referred to the disposition on the part of the
white people in the town of Tuskegee and vicinity to help
the school. From the first, I resolved to make the school
a real part of the community in which it was located. I
was determined that no one should have the feeling that
it was a foreign institution, dropped down in the midst
of the people, for which they had no responsibility and in
which they had no interest. I noticed that the very fact
that they had been asked to contribute toward the purchase of the land made them begin to feel as if it was
going to be their school, to a large degree. I noted that
just in proportion as we made the white people feel that
the institution was a part of the life of the community,
and that, while we wanted to make friends in Boston, for
example, we also wanted to make white friends in Tuskegee, and that we wanted to make the school of real service
to all the people, their attitude toward the school became
favourable.

Perhaps I might add right here, what I hope to demonstrate later, that, so far as I know, the Tuskegee school
at the present time has no warmer and more enthusiastic
friends anywhere than it has among the white citizens
of Tuskegee and throughout the state of Alabama and
the entire South. From the first, I have advised our people
in the South to make friends in every straightforward,
manly way with their next-door neighbour, whether he be
a black man or a white man. I have also advised them,
where no principle is at stake, to consult the interests
of their local communities, and to advise with their
friends in regard to their voting.

For several months the work of securing the money
with which to pay for the farm went on without ceasing.
At the end of three months enough was secured to repay
the loan of two hundred and fifty dollars to General Marshall, and within two months more we had secured the
entire five hundred dollars and had received a deed of the
one hundred acres of land. This gave us a great deal of
satisfaction. It was not only a source of satisfaction to

secure a permanent location for the school, but it was equally satisfactory to know that the greater part of the money with which it was paid for had been gotten from the white and coloured people in the town of Tuskegee. The most of this money was obtained by holding festivals and concerts, and from small individual donations.

Our next effort was in the direction of increasing the cultivation of the land, so as to secure some return from it, and at the same time give the students training in agriculture. All the industries at Tuskegee have been started in natural and logical order, growing out of the needs of a community settlement. We began with farming, because we wanted something to eat.

Many of the students, also, were able to remain in school but a few weeks at a time, because they had so little money with which to pay their board. Thus another object which made it desirable to get an industrial system started was in order to make it available as a means of helping the students to earn money enough so that they might be able to remain in school during the nine months' session of the school year.

The first animal that the school came into possession of was an old blind horse given us by one of the white citizens of Tuskegee. Perhaps I may add here that at the present time the school owns over two hundred horses, colts, mules, cows, calves, and oxen, and about seven hundred hogs and pigs, as well as a large number of sheep and goats.

The school was constantly growing in numbers, so much so that, after we had got the farm paid for, the cultivation of the land begun, and the old cabins which we had found on the place somewhat repaired, we turned our attention toward providing a large, substantial building. After having given a good deal of thought to the subject, we finally had the plans drawn for a building that was estimated to cost about six thousand dollars. This seemed to us a tremendous sum, but we knew that the school must go backward or forward, and that our work would mean little unless we could get hold of the students in their home life.

One incident which occurred about this time gave me a great deal of satisfaction as well as surprise. When it became known in the town that we were discussing the plans for a new, large building, a Southern white man who was operating a sawmill not far from Tuskegee came to me and said that he would gladly put all the lumber necessary to erect the building on the grounds, with no other guarantee for payment than my word that it would be paid for when we secured some money. I told the man frankly that at the time we did not have in our hands one dollar of the money needed. Notwithstanding this, he insisted on being allowed to put the lumber on the grounds. After we had secured some portion of the money we permitted him to do this.

Miss Davidson again began the work of securing in various ways small contributions for the new building from the white and coloured people in and near Tuskegee. I think I never saw a community of people so happy over anything as were the coloured people over the prospect of this new building. One day, when we were holding a meeting to secure funds for its erection, an old, ante-bellum coloured man came a distance of twelve miles and brought in his ox-cart a large hog. When the meeting was in progress, he rose in the midst of the company and said that he had no money which he could give, but said that he had raised two fine hogs, and that he had brought one of them as a contribution toward the expenses of the building. He closed his announcement by saying: "Any nigger that's got any love for his race, or any respect for himself, will bring a hog to the next meeting." Quite a number of men in the community also volunteered to give several days' work, each, toward the erection of the building.

After we had secured all the help that we could in Tuskegee, Miss Davidson decided to go North for the purpose of securing additional funds. For weeks she visited individuals and spoke in churches and before Sunday Schools and other organizations. She found this work quite trying, and often embarrassing. The school was not known, but she was not long in winning her way into the confidence of the best people in the North.

The first gift from any Northern person was received from a New York lady whom Miss Davidson met on the boat that was bringing her North. They fell into a conversation, and the Northern lady became so much interested in the effort being made at Tuskegee that before they parted Miss Davidson was handed a check for fifty dollars. For some time before our marriage, and also after it, Miss Davidson kept up the work of securing money in the North and in the South by interesting people by personal visits and through correspondence. At the same time she kept in close touch with the work at Tuskegee, as lady principal and classroom teacher. In addition to this, she worked among the older people in and near Tuskegee, and taught a Sunday school class in the town. She was never very strong, but never seemed happy unless she was giving all of her strength to the cause which she loved. Often, at night, after spending the day in going from door to door trying to interest persons in the work at Tuskegee, she would be so exhausted that she could not undress herself. A lady upon whom she called in Boston, afterward told me that at one time when Miss Davidson called to see her and sent up her card the lady was detained a little before she could see Miss Davidson, and when she entered the parlour she found Miss Davidson so exhausted that she had fallen asleep.

While putting up our first building, which was named Porter Hall, after Mr. A. H. Porter, of Brooklyn, N. Y., who gave a generous sum toward its erection, the need for money became acute. I had given one of our creditors a promise that upon a certain day he should be paid four hundred dollars. On the morning of that day we did not have a dollar. The mail arrived at the school at ten o'clock, and in this mail there was a check sent by Miss Davidson for exactly four hundred dollars. I could relate many instances of almost the same character. This four hundred dollars was given by two ladies in Boston. Two years later, when the work at Tuskegee had grown considerably, and when we were in the midst of a season when we were so much in need of money that the future looked doubtful and gloomy, the same two Boston ladies sent us six thousand dollars. Words cannot describe our surprise, or the encouragement that the gift brought to us. Perhaps I might add here that for fourteen years these same friends have sent us six thousand dollars each year.

As soon as the plans were drawn for the new building, the students began digging out the earth where the foundations were to be laid, working after the regular classes were over. They had not fully outgrown the idea that it was hardly the proper thing for them to use their hands, since they had come there, as one of them expressed it, "to be educated, and not to work." Gradually, though, I noted with satisfaction that a sentiment in favour of work was gaining ground. After a few weeks of hard work the foundations were ready, and a day was appointed for the laying of the corner-stone.

When it is considered that the laying of this corner-stone took place in the heart of the South, in the "Black Belt," in the centre of that part of our country that was most devoted to slavery; that at that time slavery had been abolished only about sixteen years; that only sixteen years before that no Negro could be taught from books without the teacher receiving the condemnation of the law or of public sentiment—when all this is considered, the scene that was witnessed on that spring day at Tuskegee was a remarkable one. I believe there are few places in the world where it could have taken place.

The principal address was delivered by the Hon. Waddy Thompson, the Superintendent of Education for the county. About the corner-stone were gathered the teachers, the students, their parents and friends, the county officials— who were white—and all the leading white men in that vicinity, together with many of the black men and women whom these same white people but a few years before had held a title to as property. The members of both races were anxious to exercise the privilege of placing under the corner-stone some memento.

Before the building was completed we passed through some very trying seasons. More than once our hearts were made to bleed, as it were, because bills were falling due that we did not have the money to meet. Perhaps no one who has not gone through the experience, month after month, of trying to erect buildings and provide equipment for a school when no one knew where the money was to come from, can properly appreciate the difficulties under which we laboured. During the first years at Tuskegee I recall that night after night I would roll and toss on my bed, without sleep, because of the anxiety and uncertainty which we were in regarding money. I knew that, in a large degree, we were trying an experiment—that of testing whether or not it was possible for Negroes to build up and control the affairs of a large educational institution. I knew that if we failed it would injure the whole race. I knew that the presumption was against us. I knew that in the case of white people beginning such an enterprise it would be taken for granted that they were going to succeed, but in our case I felt that people would be surprised if we succeeded. All this made a burden which pressed down on us, sometimes, it seemed, at the rate of a thousand pounds to the square inch.

In all our difficulties and anxieties, however, I never went to a white or a black person in the town of Tuskegee for any assistance that was in their power to render, without being helped according to their means. More than a dozen times, when bills figuring up into the hundreds of dollars were falling due, I applied to the white men of Tuskegee for small loans, often borrowing small amounts from as many as a half-dozen persons, to meet our obligations. One thing I was determined to do from the first, and that was to keep the credit of the school high, and this, I think I can say without boasting, we have done all through these years.

I shall always remember a bit of advice given me by Mr. George W. Campbell, the white man to whom I have referred as the one who induced General Armstrong to send me to Tuskegee. Soon after I entered upon the work Mr. Campbell said to me, in his fatherly way: "Washington, always remember that credit is capital."

At one time when we were in the greatest distress for money that we ever experienced, I placed the situation frankly before General Armstrong. Without hesitation he gave me his personal check for all the money which he had saved for his own use. This was not the only time that General Armstrong helped Tuskegee in this way. I do not think I have ever made this fact public before.

During the summer of 1882, at the end of the first year's work of the school, I was married to Miss Fannie N. Smith, of Malden, W. Va. We began keeping house in Tuskegee early in the fall. This made a home for our teachers, who now had been increased to four in number. My wife was also a graduate of the Hampton Institute. After earnest and constant work in the interests of the school, together with her housekeeping duties, my wife passed away in May, 1884. One child, Portia M. Washington, was born during our marriage.

From the first, my wife most earnestly devoted her

thoughts and time to the work of the school, and was completely one with me in every interest and ambition. She passed away, however, before she had an opportunity of seeing what the school was designed to be.

CHAPTER X

A Harder Task Than Making Bricks Without Straw

From the very beginning, at Tuskegee, I was determined to have the students do not only the agricultural and domestic work, but to have them erect their own buildings. My plan was to have them, while performing this service, taught the latest and best methods of labour, so that the school would not only get the benefit of their efforts, but the students themselves would be taught to see not only utility in labour, but beauty and dignity, would be taught, in fact, how to lift labour up from mere drudgery and toil, and would learn to love work for its own sake. My plan was not to teach them to work in the old way but to show them how to make the forces of nature—air, water, steam, electricity, horse-power—assist them in their labour.

At first many advised against the experiment of having the buildings erected by the labour of the students, but I was determined to stick to it. I told those who doubted the wisdom of the plan that I knew that our first buildings would not be so comfortable or so complete in their finish as buildings erected by the experienced hands of outside workmen, but that in the teaching of civilization, self-help, and self-reliance, the erection of the buildings by the students themselves would more than compensate for any lack of comfort or fine finish.

I further told those who doubted the wisdom of this plan, that the majority of our students came to us in poverty, from the cabins of the cotton, sugar, and rice plantations of the South, and that while I knew it would please the students very much to place them at once in finely constructed buildings I felt that it would be following out a more natural process of development to teach them how to construct their own buildings. Mistakes I knew would be made, but these mistakes would teach us valuable lessons for the future.

During the now nineteen years' existence of the Tuskegee school, the plan of having the buildings erected by student labour has been adhered to. In this time forty buildings, counting small and large, have been built, and all except four are almost wholly the product of student labour. As an additional result, hundreds of men are now scattered throughout the South who received their knowledge of mechanics while being taught how to erect these buildings. Skill and knowledge are now handed down from one set of students to another in this way, until at the present time a building of any description or size can be constructed wholly by our instructors and students, from the drawing of the plans to the putting in of the electric fixtures without going off the grounds for a single workman.

Not a few times, when a new student has been led into the temptation of marring the looks of some building by leadpencil marks or by the cuts of a jack-knife, I have heard an old student remind him: "Don't do that. That is our building. I helped put it up."

In the early days of the school I think my most trying experience was in the matter of brickmaking. As soon as we got the farm work reasonably well started, we directed our next efforts toward the industry of making bricks. We needed these for use in connection with the erection of our own buildings; but there was also another reason for establishing this industry. There was no brickyard in the town, and in addition to our own needs there was a demand for bricks in the general market.

I had always sympathized with the "Children of Israel," in their task of "making bricks without straw," but ours was the task of making bricks with no money and no experience.

In the first place, the work was hard and dirty, and it was difficult to get the students to help. When it came to brickmaking, their distaste for manual labour in connection with book education became especially manifest. It was not a pleasant task for one to stand in the mud-pit for hours, with the mud up to his knees. More than one man became disgusted and left the school.

We tried several locations before we opened up a pit that furnished brick clay. I had always supposed that brickmaking was very simple, but I soon found out by bitter experience that it required special skill and knowledge, particularly in the burning of the bricks. After a good deal of effort we moulded about twenty-five thousand bricks, and put them into a kiln to be burned. This kiln turned out to be a failure, because it was not properly constructed or properly burned. We began at once, however, on a second kiln. This, for some reason, also proved a failure. The failure of this kiln made it still more difficult to get the students to take any part in the work. Several of the teachers, however, who had been trained in the industries at Hampton, volunteered their services, and in some way we succeeded in getting a third kiln ready for burning. The burning of the kiln required about a week. Toward the latter part of the week, when it seemed as if we were going to have a good many thousand bricks in a few hours, in the middle of the night the kiln fell. For the third time we had failed.

The failure of this last kiln left me without a single dollar with which to make another experiment. Most of the teachers advised the abandoning of the effort to make bricks. In the midst of my troubles I thought of a watch which had come into my possession years before. I took this watch to the city of Montgomery, which was not far distant, and placed it in a pawn-shop. I secured cash upon it to the amount of fifteen dollars, with which to renew the brickmaking experiment. I returned to Tuskegee, and, with the help of the fifteen dollars, rallied our rather demoralized and discouraged forces and began a fourth attempt to make bricks. This time, I am glad to say, we were successful. Before I got hold of any money, the time-limit on my watch had expired, and I have never seen it since; but I have never regretted the loss of it.

Brickmaking has now become such an important industry at the school that last season our students manufactured twelve hundred thousand of first-class bricks, of a quality suitable to be sold to any market. Aside from this, scores of young men have mastered the brickmaking trade—both the making of bricks by hand and by machinery—and are now engaged in this industry in many parts of the South.

The making of these bricks taught me an important lesson in regard to the relations of the two races in the South. Many white people who had had no contact with the school, and perhaps no sympathy with it, came to us to buy bricks because they found out that ours were good bricks. They discovered that we were supplying a real want in the community. The making of these bricks caused many of the white residents of the neighbourhood to begin to feel that the education of the Negro was not making him worthless, but that in educating our students we were adding something to the wealth and comfort of the community. As the people of the neighbourhood came to us to buy bricks, we got acquainted with them; they traded with us and we with them. Our business interests became intermingled. We had something which they wanted; they had something which we wanted. This, in a large measure, helped to lay the foundation for the pleasant relations that have continued to exist between us and the white people in that section, and which now extend throughout the South.

Wherever one of our brickmakers has gone in the South, we find that he has something to contribute to the well-being of the community into which he has gone; something that has made the community feel that, in a degree, it is indebted to him, and perhaps, to a certain extent, dependent upon him. In this way pleasant relations between the races have been stimulated.

My experience is that there is something in human nature which always makes an individual recognize and reward merit, no matter under what colour of skin merit is found. I have found, too, that it is the visible, the tangible, that goes a long ways in softening prejudices. The actual sight of a first-class house that a Negro has built is ten times more potent than pages of discussion about a house that he ought to build, or perhaps could build.

The same principle of industrial education has been carried out in the building of our own wagons, carts, and buggies, from the first. We now own and use on our farm and about the school dozens of these vehicles, and every one of them has been built by the hands of the students. Aside from this, we help supply the local market with these vehicles. The supplying of them to the people in the community has had the same effect as the supplying of bricks, and the man who learns at Tuskegee to build and repair wagons and carts is regarded as a benefactor by both races in the community where he goes. The people with whom he lives and works are going to think twice before they part with such a man.

The individual who can do something that the world wants done will, in the end, make his way regardless of his race. One man may go into a community prepared to supply the people there with an analysis of Greek sentences. The community may not at that time be prepared for, or feel the need of, Greek analysis, but it may feel its need of bricks and houses and wagons. If the man can supply the need for those, then, it will lead eventually to a demand for the first product, and with the demand will come the ability to appreciate it and to profit by it.

About the time that we succeeded in burning our first kiln of bricks we began facing in an emphasized form the objection of the students to being taught to work. By this time it had gotten to be pretty well advertised throughout the state that every student who came to Tuskegee, no matter what his financial ability might be, must learn some industry. Quite a number of letters came from parents protesting against their children engaging in labour while they were in the school. Other parents came to the school to protest in person. Most of the new students brought a written or a verbal request from their parents to the effect that they wanted their children taught nothing but books. The more books, the larger they were, and the longer the titles printed upon them, the better pleased the students and their parents seemed to be.

I gave little heed to these protests, except that I lost no opportunity to go into as many parts of the state as I could, for the purpose of speaking to the parents, and showing them the value of industrial education. Besides, I talked to the students constantly on the subject. Notwithstanding the unpopularity of industrial work, the school continued to increase in numbers to such an extent that by the middle of the second year there was an attendance of about one hundred and fifty, representing almost all parts of the state of Alabama, and including a few from other states.

In the summer of 1882 Miss Davidson and I both went North and engaged in the work of raising funds for the completion of our new building. On my way North I stopped in New York to try to get a letter of recommendation from an officer of a missionary organization who had become somewhat acquainted with me a few years previous. This man not only refused to give me the letter, but advised me most earnestly to go back home at once, and not make an attempt to get money, for he was quite sure that I would never get more than enough to pay my travelling expenses. I thanked him for his advice, and proceeded on my journey.

The first place I went to in the North, was Northampton, Mass., where I spent nearly a half-day in looking for a coloured family with whom I could board, never dreaming that any hotel would admit me. I was greatly surprised when I found that I would have no trouble in being accommodated at a hotel.

We were successful in getting money enough so that on Thanksgiving Day of that year we held our first service in the chapel of Porter Hall, although the building was not completed.

In looking about for some one to preach the Thanksgiving sermon, I found one of the rarest men that it has ever been my privilege to know. This was the Rev. Robert C. Bedford, a white man from Wisconsin, who was then pastor of a little coloured Congregational church in Montgomery, Ala. Before going to Montgomery to look for some one to preach this sermon I had never heard of Mr. Bedford. He had never heard of me. He gladly consented to come to Tuskegee and hold the Thanksgiving service. It was the first service of the kind that the coloured people there had ever observed, and what a deep interest they manifested in it! The sight of the new building made it a day of Thanksgiving for them never to be forgotten.

Mr. Bedford consented to become one of the trustees of the school, and in that capacity, and as a worker for it, he has been connected with it for eighteen years. During this time he has borne the school upon his heart night and day, and is never so happy as when he is performing some service, no matter how humble, for it. He completely obliterates himself in everything, and looks only for permission to serve where service is most disagreeable, and where others would not be attracted. In all my relations with him he has seemed to me to approach as nearly to the spirit of the Master as almost any man I ever met.

A little later there came into the service of the school another man, quite young at the time, and fresh from Hampton, without whose service the school never could have become what it is. This was Mr. Warren Logan, who now for seventeen years has been the treasurer of the Institute, and the acting principal during my absence. He has always shown a degree of unselfishness and an amount of business tact, coupled with a clear judgment, that has kept the school in good condition no matter how long I have been absent from it. During all the financial stress through which the school had passed, his patience and faith in our ultimate success have not left him.

As soon as our first building was near enough to completion so that we could occupy a portion of it—which was near the middle of the second year of the school—we opened a boarding department. Students had begun coming from quite a distance, and in such increasing numbers that we felt more and more that we were merely skimming over the surface, in that we were not getting hold of the students in their home life.

We had nothing but the students and their appetites with which to begin a boarding department. No provision had been made in the new building for a kitchen and dining room; but we discovered that by digging out a large amount of earth from under the building we could make a partially lighted basement room that could be used for a kitchen and dining room. Again I called on the students to volunteer for work, this time to assist in digging out the basement. This they did, and in a few weeks we had a place to cook and eat in, although it was very rough and uncomfortable. Any one seeing the place now would never believe that it was once used for a dining room.

The most serious problem, though, was to get the boarding department started off in running order, with nothing to do with in the way of furniture, and with no money with which to buy anything. The merchants in the town would let us have what food we wanted on credit. In fact, in those earlier years I was constantly embarrassed because people seemed to have more faith in me than I had in myself. It was pretty hard to cook, however, without stoves, and awkward to eat without dishes. At first the cooking was done out-of-doors, in the old-fashioned, primitive style, in pots and skillets placed over a fire. Some of the carpenters' benches that had been used in the construction of the building were utilized for tables. As for dishes, there were too few to make it worth while to spend time in describing them.

No one connected with the boarding department seemed to have any idea that meals must be served at certain fixed and regular hours, and this was a source of great worry. Everything was so out of joint and so inconvenient that I feel safe in saying that for the first two weeks something was wrong at every meal. Either the meat was not done or had been burnt, or the salt had been left out of the bread, or the tea had been forgotten.

Early one morning I was standing near the dining-room door listening to the complaints of the students. The complaints that morning were especially emphatic and numerous, because the whole breakfast had been a failure. One of the girls who had failed to get any breakfast came out and went to the well to draw some water to drink to take the place of the breakfast which she had not been able to get. When she reached the well, she found that the rope was broken and that she could get no water. She turned from the well and said, in the most discouraged tone, not knowing that I was where I could hear her, "We can't even get water to drink at this school." I think no one remark ever came so near discouraging me as that one.

At another time, when Mr. Bedford—whom I have already spoken of as one of our trustees, and a devoted friend of the institution—was visiting the school, he was given a bedroom immediately over the dining room. Early in the morning he was awakened by a rather animated discussion between two boys in the dining room below. The discussion was over the question as to whose turn it was to use the coffee-cup that morning. One boy won the case by proving that for three mornings he had not had an opportunity to use the cup at all.

But gradually, by patience and hard work, we brought order out of chaos, just as will be true of any problem if we stick to it with patience and wisdom and earnest effort.

As I look back now over that part of our struggle, I am glad that we had it. I am glad that we endured all those discomforts and inconveniences. I am glad that our students had to dig out the place for their kitchen and dining room. I am glad that our first boarding-place was in that dismal, ill-lighted, and damp basement. Had we started in a fine, attractive, convenient room, I fear we would have "lost our heads" and become "stuck up." It means a great deal, I think, to start off on a foundation which one has made for one's self.

When our old students return to Tuskegee now, as they often do, and go into our large, beautiful, well-ventilated, and well-lighted dining room, and see tempting, well-cooked food—largely grown by the students themselves—and see tables, neat tablecloths and napkins, and vases of flowers upon the tables, and hear singing birds, and note that each meal is served exactly upon the minute, with no disorder, and with almost no complaint coming from the hundreds that now fill our dining room, they, too, often say to me that they are glad that we started as we did, and built ourselves up year by year, by a slow and natural process of growth.

CHAPTER XI

Making Their Beds Before They Could Lie on Them

A little later in the history of the school we had a visit from General J. F. B. Marshall, the Treasurer of the Hampton Institute, who had had faith enough to lend us the first two hundred and fifty dollars with which to make a payment down on the farm. He remained with us a week, and made a careful inspection of everything. He seemed well pleased with our progress, and wrote back interesting and encouraging reports to Hampton. A little later Miss Mary F. Mackie, the teacher who had given me the "sweeping" examination when I entered Hampton, came to see us, and still later General Armstrong himself came.

At the time of the visits of these Hampton friends the number of teachers at Tuskegee had increased considerably, and the most of the new teachers were graduates of the Hampton Institute. We gave our Hampton friends, especially General Armstrong, a cordial welcome. They were all surprised and pleased at the rapid progress that the school had made within so short a time. The coloured people from miles around came to the school to get a look at General Armstrong, about whom they had heard so much. The General was not only welcomed by the members of my own race, but by the Southern white people as well.

This first visit which General Armstrong made to Tuskegee gave me an opportunity to get an insight into his character such as I had not before had. I refer to his interest in the Southern white people. Before this, I had had the thought that General Armstrong, having fought the Southern white man, rather cherished a feeling of bitterness toward the white South, and was interested in helping only the coloured man there. But this visit convinced me that I did not know the greatness and the generosity of the man. I soon learned, by his visits to the Southern white people, and from his conversations with them, that he was as anxious about the prosperity and the happiness of the white race as the black. He cherished no bitterness against the South, and was happy when an opportunity offered for manifesting his sympathy. In all my acquaintance with General Armstrong I never heard him speak, in public or in private, a single bitter word against the white man in the South. From his example in this respect I learned the lesson that great men cultivate love, and that only little men cherish a spirit of hatred. I learned that assistance given to the weak makes the one who gives it strong; and that oppression of the unfortunate makes one weak.

It is now long ago that I learned this lesson from General Armstrong, and resolved that I would permit no man, no matter what his colour might be, to narrow and degrade my soul by making me hate him. With God's help, I believe that I have completely rid myself of any ill feeling toward the Southern white man for any wrong that he may have inflicted upon my race. I am made to feel just as happy now when I am rendering service to Southern white men as when the service is rendered to a member of my own race. I pity from the bottom of my heart any individual who is so unfortunate as to get into the habit of holding race prejudice.

The more I consider the subject, the more strongly I am convinced that the most harmful effect of the practice to which the people in certain sections of the South have felt themselves compelled to resort, in order to get rid of the force of the Negroes' ballot, is not wholly in the

wrong done to the Negro, but in the permanent injury to
the morals of the white man. The wrong to the Negro is
temporary, but to the morals of the white man the injury
is permanent. I have noted time and time again that when
an individual perjures himself in order to break the force
of the black man's ballot, he soon learns to practise dishonesty in other relations of life, not only where the
Negro is concerned, but equally so where a white man is
concerned. The white man who begins by cheating a Negro
usually ends by cheating a white man. The white man
who begins to break the law by lynching a Negro soon
yields to the temptation to lynch a white man. All this, it
seems to me, makes it important that the whole Nation
lend a hand in trying to lift the burden of ignorance
from the South.

Another thing that is becoming more apparent each
year in the development of education in the South is the
influence of General Armstrong's idea of education; and
this not upon the blacks alone, but upon the whites also.
At the present time there is almost no Southern state
that is not putting forth efforts in the direction of securing industrial education for its white boys and girls, and
in most cases it is easy to trace the history of these efforts back to General Armstrong.

Soon after the opening of our humble boarding department students began coming to us in still larger numbers. For weeks we not only had to contend with the
difficulty of providing board, with no money, but also with
that of providing sleeping accommodations. For this purpose we rented a number of cabins near the school. These
cabins were in a dilapidated condition, and during the
winter months the students who occupied them necessarily
suffered from the cold. We charged the students eight
dollars a month—all they were able to pay—for their
board. This included, besides board, room, fuel, and
washing. We also gave the students credit on their board
bills for all the work which they did for the school which
was of any value to the institution. The cost of tuition,
which was fifty dollars a year for each student, we had to
secure then, as now, wherever we could.

This small charge in cash gave us no capital with which
to start a boarding department. The weather during the
second winter of our work was very cold. We were not
able to provide enough bed-clothes to keep the students

warm. In fact, for some time we were not able to provide, except in a few cases, bedsteads and mattresses of any kind. During the coldest nights I was so troubled about the discomfort of the students that I could not sleep myself. I recall that on several occasions I went in the middle of the night to the shanties occupied by the young men, for the purpose of comforting them. Often I found some of them sitting huddled around a fire, with the one blanket which we had been able to provide wrapped around them, trying in this way to keep warm. During the whole night some of them did not attempt to lie down. One morning, when the night previous had been unusually cold, I asked those of the students in the chapel who thought that they had been frostbitten during the night to raise their hands. Three hands went up. Notwithstanding these experiences, there was almost no complaining on the part of the students. They knew that we were doing the best that we could for them. They were happy in the privilege of being permitted to enjoy any kind of opportunity that would enable them to improve their condition. They were constantly asking what they might do to lighten the burdens of the teachers.

I have heard it stated more than once, both in the North and in the South, that coloured people would not obey and respect each other when one member of the race is placed in a position of authority over others. In regard to this general belief and these statements, I can say that during the nineteen years of my experience at Tuskegee I never, either by word or act, have been treated with disrespect by any student or officer connected with the institution. On the other hand, I am constantly embarrassed by the many acts of thoughtful kindness. The students do not seem to want to see me carry a large book or a satchel or any kind of a burden through the grounds. In such cases more than one always offers to relieve me. I almost never go out of my office when the rain is falling that some student does not come to my side with an umbrella and ask to be allowed to hold it over me.

While writing upon this subject, it is a pleasure for me to add that in all my contact with the white people of the South I have never received a single personal insult. The white people in and near Tuskegee, to an especial degree, seem to count it a privilege to show me all the respect within their power, and often go out of their way to do this.

Not very long ago I was making a journey between Dallas (Texas) and Houston. In some way it became known in advance that I was on the train. At nearly every station at which the train stopped, numbers of white people, including in most cases the officials of the town, came aboard and introduced themselves and thanked me heartily for the work that I was trying to do for the South.

On another occasion, when I was making a trip from Augusta, Georgia, to Atlanta, being rather tired from much travel, I rode in a Pullman sleeper. When I went into the car, I found there two ladies from Boston whom I knew well. These good ladies were perfectly ignorant, it seems, of the customs of the South, and in the goodness of their hearts insisted that I take a seat with them in their section. After some hesitation I consented. I had been there but a few minutes when one of them, without my knowledge, ordered supper to be served to the three of us. This embarrassed me still further. The car was full of Southern white men, most of whom had their eyes on our party. When I found that supper had been ordered, I tried to contrive some excuse that would permit me to leave the section, but the ladies insisted that I must eat with them. I finally settled back in my seat with a sigh, and said to myself, "I am in for it now, sure."

To add further to the embarrassment of the situation, soon after the supper was placed on the table one of the ladies remembered that she had in her satchel a special kind of tea which she wished served, and as she said she felt quite sure the porter did not know how to brew it properly, she insisted upon getting up and preparing and serving it herself. At last the meal was over, and it seemed the longest one that I had ever eaten. When we were through, I decided to get myself out of the embarrassing situation and go into the smoking-room, where most of the men were by that time, to see how the land lay. In the meantime, however, it had become known in some way throughout the car who I was. When I went into the smoking-room I was never more surprised in my life than when each man, nearly every one of them a citizen of Georgia, came up and introduced himself to me and thanked me earnestly for the work that I was trying to do for the whole South. This was not flattery, because each one of these individuals knew that he had nothing to gain by trying to flatter me.

From the first I have thought to impress the students with the idea that Tuskegee is not my institution, or that of the officers, but that it is their institution, and that they have as much interest in it as any of the trustees or instructors. I have further sought to have them feel that I am at the institution as their friend and adviser, and not as their overseer. It has been my aim to have them speak with directness and frankness about anything that concerns the life of the school. Two or three times a year I ask the students to write me a letter criticising or making complaints or suggestions about anything connected with the institution. When this is not done, I have them meet me in the chapel for a heart-to-heart talk about the conduct of the school. There are no meetings with our students that I enjoy more than these, and none are more helpful to me in planning for the future. These meetings, it seems to me, enable me to get at the very heart of all that concerns the school. Few things help an individual more than to place responsibility upon him, and to let him know that you trust him. When I have read of labour troubles between employers and employees, I have often thought that many strikes and similar disturbances might be avoided if the employers would cultivate the habit of getting nearer to their employees, of consulting and advising with them, and letting them feel that the interests of the two are the same. Every individual responds to confidence, and this is not more true of any race than of the Negroes. Let them once understand that you are unselfishly interested in them, and you can lead them to any extent.

It was my aim from the first at Tuskegee to not only have the buildings erected by the students themselves, but to have them make their own furniture as far as was possible. I now marvel at the patience of the students while sleeping upon the floor while waiting for some kind of a bedstead to be constructed, or at their sleeping without any kind of a mattress while waiting for something that looked like a mattress to be made.

In the early days we had very few students who had been used to handling carpenter's tools, and the bedsteads made by the students then were very rough and very weak. Not unfrequently when I went into the students' rooms in the morning I would find at least two bedsteads lying about on the floor. The problem of providing mattresses was a difficult one to solve. We finally mastered this, however, by getting some cheap cloth and sewing pieces of this together so as to make large bags. These bags we filled with the pine straw—or, as it is sometimes called, pine needles—which we secured from the forests near by. I am glad to say that the industry of mattress-making has grown steadily since then, and has been improved to such an extent that at the present time it is an important branch of the work which is taught systematically to a number of our girls, and that the mattresses that now come out of the mattress shop at Tuskegee are about as good as those bought in the average store. For some time after the opening of the boarding department we had no chairs in the students' bedrooms or in the dining rooms. Instead of chairs we used stools which the students constructed by nailing together three pieces of rough board. As a rule, the furniture in the students' rooms during the early days of the school consisted of a bed, some stools, and sometimes a rough table made by the students. The plan of having the students make the furniture is still followed, but the number of pieces in a room has been increased, and the workmanship has so improved that little fault can be found with the articles now. One thing that I have always insisted upon at Tuskegee is that everywhere there should be absolute cleanliness. Over and over again the students were reminded in those first years—and are reminded now—that people would excuse us for our poverty, for our lack of comforts and conveniences, but that they would not excuse us for dirt.

Another thing that has been insisted upon at the school is the use of the tooth-brush. "The gospel of the tooth-brush," as General Armstrong used to call it, is a part of our creed at Tuskegee. No student is permitted to remain who does not keep and use a tooth-brush. Several times, in recent years, students have come to us who brought with them almost no other article except a tooth-brush. They had heard from the lips of older students about our insisting upon the use of this, and so, to make a good impression, they brought at least a tooth-brush with them. I remember that one morning, not long ago, I went with the lady principal on her usual morning tour of inspection of the girls' rooms. We found one room that contained three girls who had recently arrived at the school. When I asked them if they had tooth-brushes, one of the girls replied, pointing to a brush: "Yes, sir. That is our brush. We bought it together, yesterday." It did not take them long to learn a different lesson.

It has been interesting to note the effect that the use of the tooth-brush has had in bringing about a higher degree of civilization among the students. With few exceptions, I have noticed that, if we can get a student to the point where, when the first or second tooth-brush disappears, he of his own motion buys another, I have not been disappointed in the future of that individual. Absolute cleanliness of the body has been insisted upon from the first. The students have been taught to bathe as regularly as to take their meals. This lesson we began teaching before we had anything in the shape of a bathhouse. Most of the students came from plantation districts, and often we had to teach them how to sleep at night; that is, whether between the two sheets—after we got to the point where we could provide them two sheets—or under both of them. Naturally I found it difficult to teach them to sleep between two sheets when we were able to supply but one. The importance of the use of the nightgown received the same attention.

For a long time one of the most difficult tasks was to teach the students that all the buttons were to be kept on their clothes, and that there must be no torn places and no grease-spots. This lesson, I am pleased to be able to say, has been so thoroughly learned and so faithfully handed down from year to year by one set of students to another that often at the present time, when the students march out of the chapel in the evening and their dress is inspected, as it is every night, not one button is to be found missing.

CHAPTER XII

Raising Money

When we opened our boarding department, we provided rooms in the attic of Porter Hall, our first building, for

a number of girls. But the number of students, of both
sexes, continued to increase. We could find rooms outside the school grounds for many of the young men, but
the girls we did not care to expose in this way. Very soon
the problem of providing more rooms for the girls, as well
as a larger boarding department for all the students, grew
serious. As a result, we finally decided to undertake the
construction of a still larger building—a building that
would contain rooms for the girls and boarding accommodations for all.

After having had a preliminary sketch of the needed
building made, we found that it would cost about ten
thousand dollars. We had no money whatever with which
to begin; still we decided to give the needed building a
name. We knew we could name it, even though we were
in doubt about our ability to secure the means for its
construction. We decided to call the proposed building
Alabama Hall, in honour of the state in which we were
labouring. Again Miss Davidson began making efforts to
enlist the interest and help of the coloured and white people in and near Tuskegee. They responded willingly, in
proportion to their means. The students, as in the case
of our first building, Porter Hall, began digging out the
dirt in order to allow the laying of the foundations.

When we seemed at the end of our resources, so far as
securing money was concerned, something occurred which
showed the greatness of General Armstrong—something
which proved how far he was above the ordinary individual. When we were in the midst of great anxiety as to
where and how we were to get funds for the new building.
I received a telegram from General Armstrong asking me
if I could spend a month travelling with him through the
North, and asking me, if I could do so, to come to Hampton at once. Of course I accepted General Armstrong's
invitation, and went to Hampton immediately. On arriving
there I found that the General had decided to take a
quartette of singers through the North, and hold meetings
for a month in important cities, at which meetings he
and I were to speak. Imagine my surprise when the General told me, further, that these meetings were to be
held, not in the interests of Hampton but in the interests
of Tuskegee, and that the Hampton Institute was to be
responsible for all the expenses.

Although he never told me so in so many words, I

found out that General Armstrong took this method of introducing me to the people of the North, as well as for the sake of securing some immediate funds to be used in the erection of Alabama Hall. A weak and narrow man would have reasoned that all the money which came to Tuskegee in this way would be just so much taken from the Hampton Institute; but none of these selfish or short-sighted feelings ever entered the breast of General Armstrong. He was too big to be little, too good to be mean. He knew that the people in the North who gave money gave it for the purpose of helping the whole cause of Negro civilization, and not merely for the advancement of any one school. The General knew, too, that the way to strengthen Hampton was to make it a centre of unselfish power in the working out of the whole Southern problem.

In regard to the addresses which I was to make in the North, I recall just one piece of advice which the General gave me. He said: "Give them an idea for every word." I think it would be hard to improve upon this advice; and it might be made to apply to all public speaking. From that time to the present I have always tried to keep his advice in mind.

Meetings were held in New York, Brooklyn, Boston, Philadelphia, and other large cities, and at all of these meetings General Armstrong pleaded, together with myself, for help, not for Hampton, but for Tuskegee. At these meetings an especial effort was made to secure help for the building of Alabama Hall, as well as to introduce the school to the attention of the general public. In both these respects the meetings proved successful.

After that kindly introduction I began going North alone to secure funds. During the last fifteen years I have been compelled to spend a large proportion of my time away from the school, in an effort to secure money to provide for the growing needs of the institution. In my efforts to get funds I have had some experiences that may be of interest to my readers. Time and time again I have been asked, by people who are trying to secure money for philanthropic purposes, what rule or rules I followed to secure the interest and help of people who were able to contribute money to worthy objects. As far as the science of what is called begging can be reduced to rules, I would say that I have had but two rules. First, always to do my whole duty regarding making our work known to individuals and organizations; and, second, not to worry about the results. The second rule has been the hardest for me to live up to. When bills are on the eve of falling due, with not a dollar in hand with which to meet them, it is pretty difficult to learn not to worry, although I think I am learning more and more each year that all worry simply consumes, and to no purpose, just so much physical and mental strength that might otherwise be given to effective work. After considerable experience in coming into contact with wealthy and noted men, I have observed that those who have accomplished the greatest results are those who "keep under the body"; are those who never grow excited or lose self-control, but are always calm, self-possessed, patient, and polite. I think that President William McKinley is the best example of a man of this class that I have ever seen.

In order to be successful in any kind of undertaking, I think the main thing is for one to grow to the point where he completely forgets himself; that is, to lose himself in a great cause. In proportion as one loses himself in this way, in the same degree does he get the highest happiness out of his work.

My experience in getting money for Tuskegee has taught me to have no patience with those people who are always condemning the rich because they are rich, and because they do not give more to objects of charity. In the first place, those who are guilty of such sweeping criticisms do not know how many people would be made poor, and how much suffering would result, if wealthy people were to part all at once with any large proportion of their wealth in a way to disorganize and cripple great business enterprises. Then very few persons have any idea of the large number of applications for help that rich people are constantly being flooded with. I know wealthy people who receive as many as twenty calls a day for help. More than once, when I have gone into the offices of rich men, I have found half a dozen persons waiting to see them, and all come for the same purpose, that of securing money. And all these calls in person, to say nothing of the applications received through the mails. Very few people have any idea of the amount of money given away by persons who never permit their names to be known. I have often heard persons condemned for not giving away money, who, to my own knowledge, were giving away thousands of dollars every year so quietly that the world knew nothing about it.

As an example of this, there are two ladies in New York, whose names rarely appear in print, but who, in a quiet way, have given us the means with which to erect three large and important buildings during the last eight years. Besides the gift of these buildings, they have made other generous donations to the school. And they not only help Tuskegee, but they are constantly seeking opportunities to help other worthy causes.

Although it has been my privilege to be the medium through which a good many hundred thousand dollars have been received for the work at Tuskegee, I have always avoided what the world calls "begging." I often tell people that I have never "begged" any money, and that I am not a "beggar." My experience and observation have convinced me that persistent asking outright for money from the rich does not, as a rule, secure help. I have usually proceeded on the principle that persons who possess sense enough to earn money have sense enough to know how to give it away, and that the mere making known of the facts regarding Tuskegee, and especially the facts regarding the work of the graduates, has been more effective than outright begging. I think that the presentation of facts, on a high, dignified plane, is all the begging that most rich people care for.

While the work of going from door to door and from office to office is hard, disagreeable, and costly in bodily strength, yet it has some compensations. Such work gives one a rare opportunity to study human nature. It also has its compensations in giving one an opportunity to meet some of the best people in the world—to be more correct, I think I should say *the best* people in the world. When one takes a broad survey of the country, he will find that the most useful and influential people in it are those who take the deepest interest in institutions that exist for the purpose of making the world better.

At one time, when I was in Boston, I called at the door of a rather wealthy lady, and was admitted to the vestibule and sent up my card. While I was waiting for an answer, her husband came in, and asked me in the most abrupt manner what I wanted. When I tried to explain the object of my call, he became still more ungentlemanly in his words and manner; and finally grew so ex-

cited that I left the house without waiting for a reply
from the lady. A few blocks from that house I called to
see a gentleman who received me in the most cordial
manner. He wrote me his check for a generous sum, and
then, before I had had an opportunity to thank him, said:
"I am so grateful to you, Mr. Washington, for giving me
the opportunity to help a good cause. It is a privilege to
have a share in it. We in Boston are constantly indebted
to you for doing *our* work." My experience in securing
money convinces me that the first type of man is growing
more rare all the time, and that the latter type is increasing; that is, that, more and more, rich people are
coming to regard men and women who apply to them
for help for worthy objects, not as beggars, but as agents
for doing their work.

In the city of Boston I have rarely called upon an individual for funds that I have not been thanked for calling, usually before I could get an opportunity to thank
the donor for the money. In that city the donors seem
to feel, in a large degree, that an honour is being conferred upon them in their being permitted to give. Nowhere else have I met with, in so large a measure, this
fine and Christlike spirit as in the city of Boston, although there are many notable instances of it outside that
city. I repeat my belief that the world is growing in the
direction of giving. I repeat that the main rule by which
I have been guided in collecting money is to do my full
duty in regard to giving people who have money an opportunity to help.

In the early years of the Tuskegee school I walked the
streets or travelled country roads in the North for days
and days without receiving a dollar. Often it has happened, when during the week I had been disappointed
in not getting a cent from the very individuals from
whom I most expected help, and when I was almost
broken down and discouraged, that generous help has
come from someone who I had had little idea would
give at all.

I recall that on one occasion I obtained information
that led me to believe that a gentleman who lived about
two miles out in the country from Stamford, Conn.,
might become interested in our efforts at Tuskegee if our
conditions and needs were presented to him. On an unusually cold and stormy day I walked the two miles to see

him. After some difficulty I succeeded in securing an interview with him. He listened with some degree of interest to what I had to say, but did not give me anything. I could not help having the feeling that, in a measure, the three hours that I had spent in seeing him had been thrown away. Still, I had followed my usual rule of doing my duty. If I had not seen him, I should have felt unhappy over neglect of duty.

Two years after this visit a letter came to Tuskegee from this man, which read like this: "Enclosed I send you a New York draft for ten thousand dollars, to be used in furtherance of your work. I had placed this sum in my will for your school, but deem it wiser to give it to you while I live. I recall with pleasure your visit to me two years ago."

I can hardly imagine any occurrence which could have given me more genuine satisfaction than the receipt of this draft. It was by far the largest single donation which up to that time the school had ever received. It came at a time when an unusually long period had passed since we had received any money. We were in great distress because of lack of funds, and the nervous strain was tremendous. It is difficult for me to think of any situation that is more trying on the nerves than that of conducting a large institution, with heavy financial obligations to meet, without knowing where the money is to come from to meet these obligations from month to month.

In our case I felt a double responsibility, and this made the anxiety all the more intense. If the institution had been officered by white persons, and had failed, it would have injured the cause of Negro education; but I knew that the failure of our institution, officered by Negroes, would not only mean the loss of a school, but would cause people, in a large degree, to lose faith in the ability of the entire race. The receipt of this draft for ten thousand dollars, under all these circumstances, partially lifted a burden that had been pressing down upon me for days.

From the beginning of our work to the present I have always had the feeling, and lose no opportunity to impress our teachers with the same idea, that the school will always be supported in proportion as the inside of the institution is kept clean and pure and wholesome.

The first time I ever saw the late Collis P. Huntington, the great railroad man, he gave me two dollars for our

school. The last time I saw him, which was a few months
before he died, he gave me fifty thousand dollars toward
our endowment fund. Between these two gifts there were
others of generous proportions which came every year
from both Mr. and Mrs. Huntington.

Some people may say that it was Tuskegee's good luck
that brought to us this gift of fifty thousand dollars. No,
it was not luck. It was hard work. Nothing ever comes to
one, that is worth having, except as a result of hard work.
When Mr. Huntington gave me the first two dollars, I
did not blame him for not giving me more, but made up
my mind that I was going to convince him by tangible
results that we were worthy of larger gifts. For a dozen
years I made a strong effort to convince Mr. Huntington
of the value of our work. I noted that just in proportion
as the usefulness of the school grew, his donations increased. Never did I meet an individual who took a more
kindly and sympathetic interest in our school than did
Mr. Huntington. He not only gave money to us, but took
time in which to advise me, as a father would a son,
about the general conduct of the school.

More than once I have found myself in some pretty
tight places while collecting money in the North. The
following incident I have never related but once before,
for the reason that I feared that people would not believe
it. One morning I found myself in Providence, Rhode
Island, without a cent of money with which to buy breakfast. In crossing the street to see a lady from whom I
hoped to get some money, I found a bright new twenty-
five-cent piece in the middle of the streetcar track. I not
only had this twenty-five cents for my breakfast, but
within a few minutes I had a donation from the lady on
whom I had started to call.

At one of our Commencements I was bold enough to
invite the Rev. E. Winchester Donald, D.D., rector of
Trinity Church, Boston, to preach the Commencement
sermon. As we then had no room large enough to accommodate all who would be present, the place of meeting was under a large, improvised arbour, built partly of
brush and partly of rough boards. Soon after Dr. Donald
had begun speaking, the rain came down in torrents, and
he had to stop, while some one held an umbrella over
him.

The boldness of what I had done never dawned upon

me until I saw the picture made by the rector of Trinity Church standing before that large audience under an old umbrella, waiting for the rain to cease so that he could go on with his address.

It was not very long before the rain ceased and Dr. Donald finished his sermon; and an excellent sermon it was, too, in spite of the weather. After he had gone to his room, and had gotten the wet threads of his clothes dry, Dr. Donald ventured the remark that a large chapel at Tuskegee would not be out of place. The next day a letter came from two ladies who were then travelling in Italy, saying that they had decided to give us the money for such a chapel as we needed.

A short time ago we received twenty thousand dollars from Mr. Andrew Carnegie, to be used for the purpose of erecting a new library building. Our first library and reading-room were in a corner of a shanty, and the whole thing occupied a space about five by twelve feet. It required ten years of work before I was able to secure Mr. Carnegie's interest and help. The first time I saw him, ten years ago, he seemed to take but little interest in our school, but I was determined to show him that we were worthy of his help. After ten years of hard work I wrote him a letter reading as follows:

DECEMBER 15, 1900.

MR. ANDREW CARNEGIE, 5 W. FIFTY-FIRST ST., NEW YORK.

DEAR SIR: Complying with the request which you made of me when I saw you at your residence a few days ago, I now submit in writing an appeal for a library building for our institution.

We have 1100 students, 86 officers and instructors, together with their families, and about 200 coloured people living near the school, all of whom would make use of the library building.

We have over 12,000 books, periodicals, etc., gifts from our friends, but we have no suitable place for them, and we have no suitable reading-room.

Our graduates go to work in every section of the South, and whatever knowledge might be obtained in the library would serve to assist in the elevation of the whole Negro race.

Such a building as we need could be erected for about $20,000. All of the work for the building, such as brickmaking, brick-masonry, carpentry, blacksmithing, etc., would be done by the students. The money which you would give would not only supply the building, but the erection of the building would give a large number of students an opportunity to learn the building trades, and the students would use the money paid to them to keep themselves in school. I do not believe that a similar amount of money often could be made go so far in uplifting a whole race.

If you wish further information, I shall be glad to furnish it.

Yours truly,

BOOKER T. WASHINGTON, Principal.

The next mail brought back the following reply:

"I will be very glad to pay the bills for the library building as they are incurred, to the extent of twenty thousand dollars, and I am glad of this opportunity to show the interest I have in your noble work."

I have found that strict business methods go a long way in securing the interest of rich people. It has been my constant aim at Tuskegee to carry out, in our financial and other operations, such business methods as would be approved of by any New York banking house.

I have spoken of several large gifts to the school; but by far the greater proportion of the money that has built up the institution has come in the form of small donations from persons of moderate means. It is upon these small gifts, which carry with them the interest of hundreds of donors, that any philanthropic work must depend largely for its support. In my efforts to get money I have often been surprised at the patience and deep interest of the ministers, who are besieged on every hand and at all hours of the day for help. If no other consideration had convinced me of the value of the Christian life, the Christlike work which the Church of all denominations in America has done during the last thirty-five years for the elevation of the black man would have made me a Christian. In a large degree it has been the pennies, the nickels, and the dimes which have come from the Sunday-schools, the Christian Endeavour societies, and the missionary societies, as well as from the Church proper, that have helped to elevate the Negro at so rapid a rate.

This speaking of small gifts reminds me to say that very few Tuskegee graduates fail to send us an annual contribution. These contributions range from twenty-five cents up to ten dollars.

Soon after beginning our third year's work we were surprised to receive money from three special sources, and up to the present time we have continued to receive help from them. First, the State Legislature of Alabama increased its annual appropriation from two thousand dollars to three thousand dollars; I might add that still later it increased this sum to four thousand five hundred dollars a year. The effort to secure this increase was led by the Hon. M. F. Foster, the member of the Legislature from Tuskegee. Second, we received one thousand dollars from the John F. Slater Fund. Our work seemed to please the trustees of this fund, as they soon began increasing their annual grant. This has been added to from time to time until at present we receive eleven thousand dollars annually from this Fund. The other help to which I have referred came in the shape of an allowance from the Peabody Fund. This was at first five hundred dollars, but it has since been increased to fifteen hundred dollars.

The effort to secure help from the Slater and Peabody Funds brought me into contact with two rare men—men who have had much to do in shaping the policy for the education of the Negro. I refer to the Hon. J. L. M. Curry, of Washington, who is the general agent for these two funds, and Mr. Morris K. Jesup, of New York. Dr. Curry is a native of the South, an ex-Confederate soldier, yet I do not believe there is any man in the country who is more deeply interested in the highest welfare of the Negro than Dr. Curry, or one who is more free from race prejudice. He enjoys the unique distinction of possessing to an equal degree the confidence of the black man and the Southern white man. I shall never forget the first time I met him. It was in Richmond, Va., where he was then living. I had heard much about him. When I first went into his presence, trembling because of my youth and inexperience, he took me by the hand so cordially, and spoke such encouraging words, and gave me such helpful advice regarding the proper course to pursue, that I came to know him then, as I have known him ever since, as a high example of one who is constantly and unselfishly at work for the betterment of humanity.

Mr. Morris K. Jesup, the treasurer of the Slater Fund, I refer to because I know of no man of wealth and large and complicated business responsibilities who gives not only money but his time and thought to the subject of the proper method of elevating the Negro to the extent that is true of Mr. Jesup. It is very largely through his effort and influence that during the last few years the subject of industrial education has assumed the importance that it has, and been placed on its present footing.

CHAPTER XIII

Two Thousand Miles for a Five-minute Speech

Soon after the opening of our boarding department, quite a number of students who evidently were worthy, but who were so poor that they did not have any money to pay even the small charges at the school, began applying for admission. This class was composed of both men and women. It was a great trial to refuse admission to these applicants, and in 1884 we established a night-school to accommodate a few of them.

The night-school was organized on a plan similar to the one which I had helped to establish at Hampton. At first it was composed of about a dozen students. They were admitted to the night-school only when they had no money with which to pay any part of their board in the regular day-school. It was further required that they must work for ten hours during the day at some trade or industry, and study academic branches for two hours during the evening. This was the requirement for the first one or two years of their stay. They were to be paid something above the cost of their board, with the understanding that all of their earnings, except a very small part, were to be reserved in the school's treasury, to be used for paying their board in the regular day-school after they had entered that department. The night-school, started in this manner, has grown until there are at present four hundred and fifty-seven students enrolled in it alone.

There could hardly be a more severe test of a student's worth than this branch of the Institute's work. It is largely because it furnishes such a good opportunity to test the backbone of a student that I place such high value upon our night-school. Any one who is willing to work ten hours a day at the brick-yard, or in the laundry, through one or two years, in order that he or she may have the privilege of studying academic branches for two hours in the evening, has enough bottom to warrant being further educated.

After the student has left the night-school he enters the day-school, where he takes academic branches four days in a week, and works at his trade two days. Besides this he usually works at his trade during the three summer months. As a rule, after a student has succeeded in going through the night-school test, he finds a way to finish the regular course in industrial and academic training. No student, no matter how much money he may be able to command, is permitted to go through school without doing manual labour. In fact, the industrial work is now as popular as the academic branches. Some of the most successful men and women who have graduated from the institution obtained their start in the night-school.

While a great deal of stress is laid upon the industrial side of the work at Tuskegee, we do not neglect or overlook in any degree the religious and spiritual side. The school is strictly undenominational, but it is thoroughly Christian, and the spiritual training of the students is not neglected. Our preaching service, prayer-meetings, Sunday-school, Christian Endeavour Society, Young Men's Christian Association, and various missionary organizations, testify to this.

In 1885, Miss Olivia Davidson, to whom I have already referred as being largely responsible for the success of the school during its early history, and I were married. During our married life she continued to divide her time and strength between our home and the work for the school. She not only continued to work in the school at Tuskegee, but also kept up her habit of going North to secure funds. In 1889 she died, after four

years of happy married life and eight years of hard
and happy work for the school. She literally wore herself
out in her never ceasing efforts in behalf of the work
that she so dearly loved. During our married life there
were born to us two bright, beautiful boys, Booker Taliaferro and Ernest Davidson. The older of these, Booker, has
already mastered the brick-maker's trade at Tuskegee.

I have often been asked how I began the practice of
public speaking. In answer I would say that I never
planned to give any large part of my life to speaking in
public. I have always had more of an ambition to *do*
things than merely to talk *about* doing them. It seems
that when I went North with General Armstrong to speak
at the series of public meetings to which I have referred,
the President of the National Educational Association,
the Hon. Thomas W. Bicknell, was present at one of those
meetings and heard me speak. A few days afterward
he sent me an invitation to deliver an address at the next
meeting of the Educational Association. This meeting
was to be held in Madison, Wis. I accepted the invitation. This was, in a sense, the beginning of my public-
speaking career.

On the evening that I spoke before the Association
there must have been not far from four thousand persons
present. Without my knowing it, there were a large number of people present from Alabama, and some from the
town of Tuskegee. These white people afterward frankly
told me that they went to this meeting expecting to hear
the South roundly abused, but were pleasantly surprised
to find that there was no word of abuse in my address.
On the contrary, the South was given credit for all the
praiseworthy things that it had done. A white lady who
was teacher in a college in Tuskegee wrote back to the
local paper that she was gratified, as well as surprised, to
note the credit which I gave the white people of Tuskegee
for their help in getting the school started. This address
at Madison was the first that I had delivered that in any
large measure dealt with the general problem of the races.
Those who heard it seemed to be pleased with what I
said and with the general position that I took.

When I first came to Tuskegee, I determined that I
would make it my home, that I would take as much pride
in the right actions of the people of the town as any white
man could do, and that I would, at the same time, de-

plore the wrong-doing of the people as much as any white man. I determined never to say anything in a public address in the North that I would not be willing to say in the South. I early learned that it is a hard matter to convert an individual by abusing him, and that this is more often accomplished by giving credit for all the praiseworthy actions performed than by calling attention alone to all the evil done.

While pursuing this policy I have not failed, at the proper time and in the proper manner, to call attention, in no uncertain terms, to the wrongs which any part of the South has been guilty of. I have found that there is a large element in the South that is quick to respond to straightforward, honest criticism of any wrong policy. As a rule, the place to criticise the South, when criticism is necessary, is in the South—not in Boston. A Boston man who came to Alabama to criticise Boston would not effect so much good, I think, as one who had his word of criticism to say in Boston.

In this address at Madison I took the ground that the policy to be pursued with reference to the races was, by every honourable means, to bring them together and to encourage the cultivation of friendly relations, instead of doing that which would embitter. I further contended that, in relation to his vote, the Negro should more and more consider the interests of the community in which he lived, rather than seek alone to please some one who lived a thousand miles away from him and from his interests.

In this address I said that the whole future of the Negro rested largely upon the question as to whether or not he should make himself, through his skill, intelligence, and character, of such undeniable value to the community in which he lived that the community could not dispense with his presence. I said that any individual who learned to do something better than anybody else— learned to do a common thing in an uncommon manner— had solved his problem, regardless of the colour of his skin, and that in proportion as the Negro learned to produce what other people wanted and must have, in the same proportion would he be respected.

I spoke of an instance where one of our graduates had produced two hundred and sixty-six bushels of sweet potatoes from an acre of ground, in a community where the average production had been only forty-nine bushels to the acre. He had been able to do this by reason of his knowledge of the chemistry of the soil and by his knowledge of improved methods of agriculture. The white farmers in the neighbourhood respected him, and came to him for ideas regarding the raising of sweet potatoes. These white farmers honoured and respected him because he, by his skill and knowledge, had added something to the wealth and comfort of the community in which he lived. I explained that my theory of education for the Negro would not, for example, confine him for all time to farm life—to the production of the best and the most sweet potatoes—but that, if he succeeded in this line of industry, he could lay the foundations upon which his children and grandchildren could grow to higher and more important things in life.

Such, in brief, were some of the views I advocated in this first address dealing with the broad question of the relations of the two races, and since that time I have not found any reason for changing my views on any important point.

In my early life I used to cherish a feeling of ill will toward any one who spoke in bitter terms against the Negro, or who advocated measures that tended to oppress the black man or take from him opportunities for growth in the most complete manner. Now, whenever I hear any one advocating measures that are meant to curtail the development of another, I pity the individual who would do this. I know that the one who makes this mistake does so because of his lack of opportunity for the highest kind of growth. I pity him because I know that he is trying to stop the progress of the world, and because I know that in time the development and the ceaseless advance of humanity will make him ashamed of his weak and narrow position. One might as well try to stop the progress of a mighty railroad train by throwing his body across the track, as to try to stop the growth of the world in the direction of giving mankind more intelligence, more culture, more skill, more liberty, and in the direction of extending more sympathy and more brotherly kindness.

The address which I delivered at Madison, before the National Educational Association, gave me a rather wide introduction in the North, and soon after that opportunities began offering themselves for me to address audiences there.

I was anxious, however, that the way might also be opened for me to speak directly to a representative Southern white audience. A partial opportunity of this kind, one that seemed to me might serve as an entering wedge, presented itself in 1893, when the international meeting of Christian Workers was held at Atlanta, Ga. When this invitation came to me, I had engagements in Boston that seemed to make it impossible for me to speak in Atlanta. Still, after looking over my list of dates and places carefully, I found that I could take a train from Boston that would get me into Atlanta about thirty minutes before my address was to be delivered, and that I could remain in that city about sixty minutes before taking another train for Boston. My invitation to speak in Atlanta stipulated that I was to confine my address to five minutes. The question, then, was whether or not I could put enough into a five-minute address to make it worth while for me to make such a trip.

I knew that the audience would be largely composed of the most influential class of white men and women, and that it would be a rare opportunity for me to let them know what we were trying to do at Tuskegee, as well as to speak to them about the relations of the races. So I decided to make the trip. I spoke for five minutes to an audience of two thousand people, composed mostly of Southern and Northern whites. What I said seemed to be received with favour and enthusiasm. The Atlanta papers of the next day commented in friendly terms on my address, and a good deal was said about it in different parts of the country. I felt that I had in some degree accomplished my object—that of getting a hearing from the dominant class of the South.

The demands made upon me for public addresses continued to increase, coming in about equal numbers from my own people and from Northern whites. I gave as much time to these addresses as I could spare from the immediate work at Tuskegee. Most of the addresses in the North were made for the direct purpose of getting funds with which to support the school. Those delivered before the coloured people had for their main object the impressing upon them of the importance of industrial and technical education in addition to academic and religious training.

I now come to that one of the incidents in my life which seems to have excited the greatest amount of interest, and which perhaps went farther than anything else in giving me a reputation that in a sense might be called National. I refer to the address which I delivered at the opening of the Atlanta Cotton states and International Exposition, at Atlanta, Ga., September 18, 1895.

So much has been said and written about this incident, and so many questions have been asked me concerning the address, that perhaps I may be excused for taking up the matter with some detail. The five-minute address in Atlanta, which I came from Boston to deliver, was possibly the prime cause for an opportunity being given me to make the second address there. In the spring of 1895 I received a telegram from prominent citizens in Atlanta asking me to accompany a committee from that city to Washington for the purpose of appearing before a committee of Congress in the interest of securing Government help for the Exposition. The committee was composed of about twenty-five of the most prominent and most influential white men of Georgia. All the members of this committee were white men except Bishop Grant, Bishop Gaines, and myself. The mayor and several other city and state officials spoke before the committee. They were followed by the two coloured bishops. My name was the last on the list of speakers. I had never before appeared before such a committee, nor had I ever delivered any address in the capital of the Nation. I had many misgivings as to what I ought to say, and as to the impression that my address would make. While I cannot recall in detail what I said, I remember that I tried to impress upon the committee, with all the earnestness and plainness of any language that I could command, that if Congress wanted to do something which would assist in ridding the South of the race question and making friends between the two races, it should, in every proper way, encourage the material and intellectual growth of both races. I said that the Atlanta Exposition would present an opportunity for both races to show what advance they had made since freedom, and would at the same time afford encouragement to them to make still greater progress.

I tried to emphasize the fact that while the Negro should not be deprived by unfair means of the franchise, political agitation alone would not save him, and that back of the ballot he must have property, industry, skill, economy, intelligence, and character, and that no race without these elements could permanently succeed. I said that in granting the appropriation Congress could do something that would prove to be of real and lasting value to both races, and that it was the first great opportunity of the kind that had been presented since the close of the Civil War.

I spoke for fifteen or twenty minutes, and was surprised at the close of my address to receive the hearty congratulations of the Georgia committee and of the members of Congress who were present. The Committee was unanimous in making a favourable report, and in a few days the bill passed Congress. With the passing of this bill the success of the Atlanta Exposition was assured.

Soon after this trip to Washington the directors of the Exposition decided that it would be a fitting recognition of the coloured race to erect a large and attractive building which should be devoted wholly to showing the progress of the Negro since freedom. It was further decided to have the building designed and erected wholly by Negro mechanics. This plan was carried out. In design, beauty, and general finish the Negro Building was equal to the others on the grounds.

After it was decided to have a separate Negro exhibit, the question arose as to who should take charge of it. The officials of the Exposition were anxious that I should assume this responsibility, but I declined to do so, on the plea that the work at Tuskegee at that time demanded my time and strength. Largely at my suggestion, Mr. I. Garland Penn, of Lynchburg, Va., was selected to be at the head of the Negro department. I gave him all the aid that I could. The Negro exhibit, as a whole, was large and creditable. The two exhibits in this department which attracted the greatest amount of attention were those from the Hampton Institute and the Tuskegee Institute. The people who seemed to be the most surprised, as well as pleased, at what they saw in the Negro Building were the Southern white people.

As the day for the opening of the Exposition drew

near, the Board of Directors began preparing the programme for the opening exercises. In the discussion from
day to day of the various features of this programme, the
question came up as to the advisability of putting a member of the Negro race on for one of the opening addresses,
since the Negroes had been asked to take such a prominent part in the Exposition. It was argued, further, that
such recognition would mark the good feeling prevailing
between the two races. Of course there were those who
were opposed to any such recognition of the rights of
the Negro, but the Board of Directors, composed of men
who represented the best and most progressive element
in the South, had their way, and voted to invite a black
man to speak on the opening day. The next thing was
to decide upon the person who was thus to represent the
Negro race. After the question had been canvassed for
several days, the directors voted unanimously to ask
me to deliver one of the opening-day addresses, and in a
few days after that I received the official invitation.

The receiving of this invitation brought to me a
sense of responsibility that it would be hard for any one
not placed in my position to appreciate. What were my
feelings when this invitation came to me? I remembered
that I had been a slave; that my early years had been
spent in the lowest depths of poverty and ignorance, and
that I had had little opportunity to prepare me for such a
responsibility as this. It was only a few years before that
time that any white man in the audience might have
claimed me as his slave; and it was easily possible that
some of my former owners might be present to hear me
speak.

I knew, too, that this was the first time in the entire
history of the Negro that a member of my race had been
asked to speak from the same platform with white Southern men and women on any important National occasion.
I was asked now to speak to an audience composed of
the wealth and culture of the white South, the representatives of my former masters. I knew, too, that while
the greater part of my audience would be composed of
Southern people, yet there would be present a large number of Northern whites, as well as a great many men and
women of my own race.

I was determined to say nothing that I did not feel
from the bottom of my heart to be true and right. When

the invitation came to me, there was not one word of intimation as to what I should say or as to what I should omit. In this I felt that the Board of Directors had paid a tribute to me. They knew that by one sentence I could have blasted, in a large degree, the success of the Exposition. I was also painfully conscious of the fact that, while I must be true to my own race in my utterances, I had it in my power to make such an ill-timed address as would result in preventing any similar invitation being extended to a black man again for years to come. I was equally determined to be true to the North, as well as to the best element of the white South, in what I had to say.

The papers, North and South, had taken up the discussion of my coming speech, and as the time for it drew near this discussion became more and more widespread. Not a few of the Southern white papers were unfriendly to the idea of my speaking. From my own race I received many suggestions as to what I ought to say. I prepared myself as best I could for the address, but as the eighteenth of September drew nearer, the heavier my heart became, and the more I feared that my effort would prove a failure and a disappointment.

The invitation had come at a time when I was very busy with my school work, as it was the beginning of our school year. After preparing my address, I went through it, as I usually do with all those utterances which I consider particularly important, with Mrs. Washington, and she approved of what I intended to say. On the sixteenth of September, the day before I was to start for Atlanta, so many of the Tuskegee teachers expressed a desire to hear my address that I consented to read it to them in a body. When I had done so, and had heard their criticism and comments, I felt somewhat relieved, since they seemed to think well of what I had to say.

On the morning of September 17, together with Mrs. Washington and my three children, I started for Atlanta. I felt a good deal as I suppose a man feels when he is on his way to the gallows. In passing through the town of Tuskegee I met a white farmer who lived some distance out in the country. In a jesting manner this man said: "Washington, you have spoken before the Northern white people, the Negroes in the South, and to us country white people in the South; but in Atlanta, to-morrow, you will have before you the Northern whites, the Southern whites, and the Negroes all together. I am afraid that you have got yourself into a tight place." This farmer diagnosed the situation correctly, but his frank words did not add anything to my comfort.

In the course of the journey from Tuskegee to Atlanta both coloured and white people came to the train to point me out, and discussed with perfect freedom, in my hearing, what was going to take place the next day. We were met by a committee in Atlanta. Almost the first thing that I heard when I got off the train in that city was an expression something like this from an old coloured man near by: "Dat's de man of my race what's gwine to make a speech at de Exposition to-morrow. I'se sho' gwine to hear him."

Atlanta was literally packed, at the time, with people from all parts of this country, and with representatives of foreign governments, as well as with military and civic organizations. The afternoon papers had forecasts of the next day's proceedings in flaring headlines. All this tended to add to my burden. I did not sleep much that night. The next morning, before day, I went carefully over what I intended to say. I also kneeled down and asked God's blessing upon my effort. Right here, perhaps, I ought to add that I make it a rule never to go before an audience, on any occasion, without asking the blessing of God upon what I want to say.

I always make it a rule to make special preparation for each separate address. No two audiences are exactly alike. It is my aim to reach and talk to the heart of each individual audience, taking it into my confidence very much as I would a person. When I am speaking to an audience, I care little for how what I am saying is going to sound in the newspapers, or to another audience, or to an individual. At the time, the audience before me absorbs all my sympathy, thought, and energy.

Early in the morning a committee called to escort me to my place in the procession which was to march to the Exposition grounds. In this procession were prominent coloured citizens in carriages, as well as several Negro military organizations. I noted that the Exposition officials seemed to go out of their way to see that all of the coloured people in the procession were properly placed and properly treated. The procession was about three hours in reaching the Exposition grounds, and during all of this time the sun was shining down upon us disagreeably hot. When we reached the grounds, the heat, together with my nervous anxiety, made me feel as if I were about ready to collapse, and to feel that my address was not going to be a success. When I entered the audience-room, I found it packed with humanity from bottom to top, and there were thousands outside who could not get in.

The room was very large, and well suited to public speaking. When I entered the room, there were vigorous cheers from the coloured portion of the audience, and faint cheers from some of the white people. I had been told, while I had been in Atlanta, that while many white people were going to be present to hear me speak, simply out of curiosity, and that others who would be present would be in full sympathy with me, there was a still larger element of the audience which would consist of those who were going to be present for the purpose of hearing me make a fool of myself, or, at least, of hearing me say some foolish thing, so that they could say to the officials who had invited me to speak, "I told you so!"

One of the trustees of the Tuskegee Institute, as well as my personal friend, Mr. William H. Baldwin, Jr., was at the time General Manager of the Southern Railroad, and happened to be in Atlanta on that day. He was so nervous about the kind of reception that I would have, and the effect that my speech would produce, that he could not persuade himself to go into the building, but walked back and forth in the grounds outside until the opening exercises were over.

CHAPTER XIV

The Atlanta Exposition Address

The Atlanta Exposition, at which I had been asked to make an address as a representative of the Negro race, as stated in the last chapter, was opened with a short address from Governor Bullock. After other interesting exercises, including an invocation from Bishop Nelson, of Georgia, a dedicatory ode by Albert Howell, Jr., and

addresses by the President of the Exposition and Mrs.
Joseph Thompson, the President of the Woman's Board,
Governor Bullock introduced me with the words, "We
have with us to-day a representative of Negro enterprise
and Negro civilization."

When I arose to speak, there was considerable cheering,
especially from the coloured people. As I remember it
now, the thing that was uppermost in my mind was the
desire to say something that would cement the friendship
of the races and bring about hearty coöperation between
them. So far as my outward surroundings were concerned,
the only thing that I recall distinctly now is that when I
got up, I saw thousands of eyes looking intently into my
face. The following is the address which I delivered:—

MR. PRESIDENT AND GENTLEMEN OF THE BOARD OF DI-
RECTORS AND CITIZENS.

One-third of the population of the South is of the
Negro race. No enterprise seeking the material, civil, or
moral welfare of this section can disregard this element
of our population and reach the highest success. I but
convey to you, Mr. President and Directors, the sentiment of the masses of my race when I say that in no way
have the value and manhood of the American Negro been
more fittingly and generously recognized than by the
managers of this magnificent Exposition at every stage of
its progress. It is a recognition that will do more to cement the friendship of the two races than any occurrence
since the dawn of our freedom.

Not only this, but the opportunity here afforded will
awaken among us a new era of industrial progress. Ignorant and inexperienced, it is not strange that in the first
years of our new life we began at the top instead of at
the bottom; that a seat in Congress or the state legislature
was more sought than real estate or industrial skill; that
the political convention of stump speaking had more
attractions than starting a dairy farm or truck garden.

A ship lost at sea for many days suddenly sighted a
friendly vessel. From the mast of the unfortunate vessel
was seen a signal, "Water, water; we die of thirst!" The
answer from the friendly vessel at once came back, "Cast
down your bucket where you are." A second time the
signal, "Water, water; send us water!" ran up from the distressed vessel, and was answered, "Cast down your

bucket where you are." And a third and fourth signal for water was answered, "Cast down your bucket where you are." The captain of the distressed vessel, at last heeding the injunction, cast down his bucket, and it came up full of fresh, sparkling water from the mouth of the Amazon River. To those of my race who depend on bettering their condition in a foreign land or who underestimate the importance of cultivating friendly relations with the Southern white man, who is their next-door neighbour, I would say: "Cast down your bucket where you are"—cast it down in making friends in every manly way of the people of all races by whom we are surrounded.

Cast it down in agriculture, mechanics, in commerce, in domestic service, and in the professions. And in this connection it is well to bear in mind that whatever other sins the South may be called to bear, when it comes to business, pure and simple, it is in the South that the Negro is given a man's chance in the commercial world, and in nothing is this Exposition more eloquent than in emphasizing this chance. Our greatest danger is that in the great leap from slavery to freedom we may overlook the fact that the masses of us are to live by the productions of our hands, and fail to keep in mind that we shall prosper in proportion as we learn to dignify and glorify common labour and put brains and skill into the common occupations of life; shall prosper in proportion as we learn to draw the line between the superficial and the substantial, the ornamental gewgaws of life and the useful. No race can prosper till it learns that there is as much dignity in tilling a field as in writing a poem. It is at the bottom of life we must begin, and not at the top. Nor should we permit our grievances to overshadow our opportunities.

To those of the white race who look to the incoming of those of foreign birth and strange tongue and habits for the prosperity of the South, were I permitted I would repeat what I say to my own race, "Cast down your bucket where you are." Cast it down among the eight millions of Negroes whose habits you know, whose fidelity and love you have tested in days when to have proved treacherous meant the ruin of your firesides. Cast down your bucket among these people who have, without strikes and labour wars, tilled your fields, cleared your forests, builded your railroads and cities, and brought forth treasures from the bowels of the earth, and helped make possible this magnificent representation of the progress of the South. Casting down your bucket among my people, helping and encouraging them as you are doing on these grounds, and to education of head, hand, and heart, you will find that they will buy your surplus land, make blossom the waste places in your fields, and run your factories. While doing this, you can be sure in the future, as in the past, that you and your families will be surrounded by the most patient, faithful, law-abiding, and unresentful people that the world has seen. As we have proved our loyalty to you in the past, in nursing your children, watching by the sickbed of your mothers and fathers, and often following them with tear-dimmed eyes to their graves, so in the future, in our humble way, we shall stand by you with a devotion that no foreigner can approach, ready to lay down our lives, if need be, in defence of yours, interlacing our industrial, commercial, civil, and religious life with yours in a way that shall make the interests of both races one. In all things that are purely social we can be as separate as the fingers, yet one as the hand in all things essential to mutual progress.

There is no defence or security for any of us except in the highest intelligence and development of all. If anywhere there are efforts tending to curtail the fullest growth of the Negro, let these efforts be turned into stimulating, encouraging, and making him the most useful and intelligent citizen. Effort or means so invested will pay a thousand per cent interest. These efforts will be twice blessed—"blessing him that gives and him that takes."

There is no escape through law of man or God from the inevitable:—

> The laws of changeless justice bind
> Oppressor with oppressed;
> And close as sin and suffering joined
> We march to fate abreast.

Nearly sixteen millions of hands will aid you in pulling the load upward, or they will pull against you the load downward. We shall constitute one-third and more of the ignorance and crime of the South, or one-third its intelligence and progress; we shall contribute one-third to the business and industrial prosperity of the South, or we shall prove a veritable body of death, stagnating, depressing, retarding every effort to advance the body politic.

Gentlemen of the Exposition, as we present to you our humble effort at an exhibition of our progress, you must not expect overmuch. Starting thirty years ago with ownership here and there in a few quilts and pumpkins and chickens (gathered from miscellaneous sources), remember the path that has led from these to the inventions and production of agricultural implements, buggies, steamengines, newspapers, books, statuary, carving, paintings, the management of drug-stores and banks, has not been trodden without contact with thorns and thistles. While we take pride in what we exhibit as a result of our independent efforts, we do not for a moment forget that our part in this exhibition would fall far short of your expectations but for the constant help that has come to our educational life, not only from the Southern states, but especially from Northern philanthropists, who have made their gifts a constant stream of blessing and encouragement.

The wisest among my race understand that the agitation of questions of social equality is the extremest folly, and that progress in the enjoyment of all the privileges that will come to us must be the result of severe and constant struggle rather than of artificial forcing. No race that has anything to contribute to the markets of the world is long in any degree ostracized. It is important and right that all privileges of the law be ours, but it is vastly more important that we be prepared for the exercises of these privileges. The opportunity to earn a dollar in a factory just now is worth infinitely more than the opportunity to spend a dollar in an opera-house.

In conclusion, may I repeat that nothing in thirty years has given us more hope and encouragement, and drawn us so near to you of the white race, as this opportunity offered by the Exposition; and here bending, as it were, over the altar that represents the results of the struggles of your race and mine, both starting practically emptyhanded three decades ago. I pledge that in your effort to work out the great and intricate problem which God has laid at the doors of the South, you shall have at all times the patient, sympathetic help of my race; only let this be constantly in mind, that, while from representations in these buildings of the product of field, of forest, of mine, of factory, letters, and art, much good will come,

yet far above and beyond material benefits will be that
higher good, that, let us pray God, will come, in a blotting
out of sectional differences and racial animosities and
suspicions, in a determination to administer absolute justice, in a willing obedience among all classes to the man-
dates of law. This, then, coupled with our material prosperity, will bring into our beloved South a new heaven and
a new earth.

The first thing that I remember, after I had finished
speaking, was that Governor Bullock rushed across the
platform and took me by the hand, and that others did
the same. I received so many and such hearty congratulations that I found it difficult to get out of the building. I
did not appreciate to any degree, however, the impression which my address seemed to have made, until the
next morning, when I went into the business part of the
city. As soon as I was recognized, I was surprised to find
myself pointed out and surrounded by a crowd of men
who wished to shake hands with me. This was kept up on
every street on to which I went, to an extent which embarrassed me so much that I went back to my boarding-
place. The next morning I returned to Tuskegee. At the
station in Atlanta, and at almost all of the stations at which
the train stopped between that city and Tuskegee, I found
a crowd of people anxious to shake hands with me.

The papers in all parts of the United States published
the address in full, and for months afterward there were
complimentary editorial references to it. Mr. Clark Howell,
the editor of the Atlanta *Constitution*, telegraphed to a
New York paper, among other words, the following, "I do
not exaggerate when I say that Professor Booker T. Washington's address yesterday was one of the most notable
speeches, both as to character and as to the warmth of its
reception, ever delivered to a Southern audience. The
address was a revelation. The whole speech is a platform
upon which blacks and whites can stand with full justice
to each other."

The Boston *Transcript* said editorially: "The speech of
Booker T. Washington at the Atlanta Exposition, this week,
seems to have dwarfed all the other proceedings and the
Exposition itself. The sensation that it has caused in the
press has never been equalled."

I very soon began receiving all kinds of propositions

from lecture bureaus, and editors of magazines and papers, to take the lecture platform, and to write articles. One lecture bureau offered me fifty thousand dollars, or two hundred dollars a night and expenses, if I would place my services at its disposal for a given period. To all these communications I replied that my life-work was at Tuskegee; and that whenever I spoke it must be in the interests of the Tuskegee school and my race, and that I would enter into no arrangements that seemed to place a mere commercial value upon my services.

Some days after its delivery I sent a copy of my address to the President of the United States, the Hon. Grover Cleveland. I received from him the following autographed reply:—

> GRAY GABLES, BUZZARD'S BAY, MASS.,
> OCTOBER 6, 1895.
>
> BOOKER T. WASHINGTON, ESQ.:
>
> MY DEAR SIR: I thank you for sending me a copy of your address delivered at the Atlanta Exposition.
>
> I thank you with much enthusiasm for making the address. I have read it with intense interest, and I think the Exposition would be fully justified if it did not do more than furnish the opportunity for its delivery. Your words cannot fail to delight and encourage all who wish well for your race; and if our coloured fellow-citizens do not from your utterances gather new hope and form new determinations to gain every valuable advantage offered them by their citizenship, it will be strange indeed.
>
> Yours very truly,
> GROVER CLEVELAND.

Later I met Mr. Cleveland, for the first time, when, as President, he visited the Atlanta Exposition. At the request of myself and others he consented to spend an hour in the Negro Building, for the purpose of inspecting the Negro exhibit and of giving the coloured people in attendance an opportunity to shake hands with him. As soon as I met Mr. Cleveland I became impressed with his simplicity, greatness, and rugged honesty. I have met him many times since then, both at public functions and at his

private residence in Princeton, and the more I see of
him the more I admire him. When he visited the Negro
Building in Atlanta he seemed to give himself up wholly,
for that hour, to the coloured people. He seemed to be as
careful to shake hands with some old coloured "auntie"
clad partially in rags, and to take as much pleasure in
doing so, as if he were greeting some millionaire. Many of
the coloured people took advantage of the occasion to get
him to write his name in a book or on a slip of paper.
He was as careful and patient in doing this as if he were
putting his signature to some great state document.

Mr. Cleveland has not only shown his friendship for me
in many personal ways, but has always consented to do
anything I have asked of him for our school. This he has
done, whether it was to make a personal donation or to
use his influence in securing the donations of others. Judg-
ing from my personal acquaintance with Mr. Cleveland, I
do not believe that he is conscious of possessing any colour
prejudice. He is too great for that. In my contact with
people I find that, as a rule, it is only the little, narrow
people who live for themselves, who never read good
books, who do not travel, who never open up their souls
in a way to permit them to come into contact with other
souls—with the great outside world. No man whose vision is bounded by colour can come into contact with
what is highest and best in the world. In meeting men, in
many places, I have found that the happiest people are
those who do the most for others; the most miserable
are those who do the least. I have also found that few
things, if any, are capable of making one so blind and
narrow as race prejudice. I often say to our students, in
the course of my talks to them on Sunday evenings in the
chapel, that the longer I live and the more experience I
have of the world, the more I am convinced that, after all,
the one thing that is most worth living for—and dying for,
if need be—is the opportunity of making some one else
more happy and more useful.

The coloured people and the coloured newspapers at
first seemed to be greatly pleased with the character of
my Atlanta address, as well as with its reception. But
after the first burst of enthusiasm began to die away, and
the coloured people began reading the speech in cold
type, some of them seemed to feel that they had been
hypnotized. They seemed to feel that I had been too lib-

eral in my remarks toward the Southern whites, and that I had not spoken out strongly enough for what they termed the "rights" of the race. For a while there was a reaction, so far as a certain element of my own race was concerned, but later these reactionary ones seemed to have been won over to my way of believing and acting.

While speaking of changes in public sentiment, I recall that about ten years after the school at Tuskegee was established, I had an experience that I shall never forget. Dr. Lyman Abbott, then the pastor of Plymouth Church, and also editor of the *Outlook* (then the *Christian Union*), asked me to write a letter for his paper giving my opinion of the exact condition, mental and moral, of the coloured ministers in the South, as based upon my observations. I wrote the letter, giving the exact facts as I conceived them to be. The picture painted was a rather black one—or, since I am black, shall I say "white"? It could not be otherwise with a race but a few years out of slavery, a race which had not had time or opportunity to produce a competent ministry.

What I said soon reached every Negro minister in the country, I think, and the letters of condemnation which I received from them were not few. I think that for a year after the publication of this article every association and every conference or religious body of any kind, of my race, that met, did not fail before adjourning to pass a resolution condemning me, or calling upon me to retract or modify what I had said. Many of these organizations went so far in their resolutions as to advise parents to cease sending their children to Tuskegee. One association even appointed a "missionary" whose duty it was to warn the people against sending their children to Tuskegee. This missionary had a son in the school, and I noticed that, whatever the "missionary" might have said or done with regard to others, he was careful not to take his son away from the institution. Many of the coloured papers, especially those that were the organs of religious bodies, joined in the general chorus of condemnation or demands for retraction.

During the whole time of the excitement, and through all the criticism, I did not utter a word of explanation or retraction. I knew that I was right, and that time and the sober second thought of the people would vindicate me. It was not long before the bishops and other church leaders began to make a careful investigation of the conditions of the ministry, and they found out that I was right. In fact, the oldest and most influential bishop in one branch of the Methodist Church said that my words were far too mild. Very soon public sentiment began making itself felt, in demanding a purifying of the ministry. While this is not yet complete by any means, I think I may say, without egotism, and I have been told by many of our most influential ministers, that my words had much to do with starting a demand for the placing of a higher type of men in the pulpit. I have had the satisfaction of having many who once condemned me thank me heartily for my frank words.

The change of the attitude of the Negro ministry, so far as regards myself, is so complete that at the present time I have no warmer friends among any class than I have among the clergymen. The improvement in the character and life of the Negro ministers is one of the most gratifying evidences of the progress of the race. My experience with them, as well as other events in my life, convinced me that the thing to do, when one feels sure that he has said or done the right thing, and is condemned, is to stand still and keep quiet. If he is right, time will show it.

In the midst of the discussion which was going on concerning my Atlanta speech, I received the letter which I give below, from Dr. Gilman, the President of Johns Hopkins University, who had been made chairman of the judges of award in connection with the Atlanta Exposition:—

JOHNS HOPKINS UNIVERSITY, BALTIMORE,
President's Office, September 30, 1895.

DEAR MR. WASHINGTON: Would it be agreeable to you to be one of the Judges of Award in the Department of Education at Atlanta? If so, I shall be glad to place your name upon the list. A line by telegraph will be welcomed.

Yours very truly,
D. C. GILMAN.

I think I was even more surprised to receive this invitation than I had been to receive the invitation to speak at the opening of the Exposition. It was to be a part of my duty, as one of the jurors, to pass not only upon the exhibits of the coloured schools, but also upon those of the white schools. I accepted the position, and spent a month in Atlanta in performance of the duties which it entailed. The board of jurors was a large one, consisting in all of sixty members. It was about equally divided between Southern white people and Northern white people. Among them were college presidents, leading scientists and men of letters, and specialists in many subjects. When the group of jurors to which I was assigned met for organization, Mr. Thomas Nelson Page, who was one of the number, moved that I be made secretary of that division, and the motion was unanimously adopted. Nearly half of our division were Southern people. In performing my duties in the inspection of the exhibits of white schools I was in every case treated with respect, and at the close of our labours I parted from my associates with regret.

I am often asked to express myself more freely than I do upon the political condition and the political future of my race. These recollections of my experience in Atlanta give me the opportunity to do so briefly. My own belief is, although I have never before said so in so many words, that the time will come when the Negro in the South will be accorded all the political rights which his ability, character, and material possessions entitle him to. I think, though, that the opportunity to freely exercise such political rights will not come in any large degree through outside or artificial forcing, but will be accorded to the Negro by the Southern white people themselves, and that they will protect him in the exercise of those rights. Just as soon as the South gets over the old feeling that it is being forced by "foreigners," or "aliens," to do something which it does not want to do, I believe that the change in the direction that I have indicated is going to begin. In fact, there are indications that it is already beginning in a slight degree.

Let me illustrate my meaning. Suppose that some months before the opening of the Atlanta Exposition there had been a general demand from the press and public platform outside the South that a Negro be given a place on the opening programme, and that a Negro be placed upon the board of jurors of award. Would any such recognition of the race have taken place? I do not think so. The Atlanta officials went as far as they did because they felt it to be a pleasure, as well as a duty, to reward what they considered merit in the Negro race. Say what we will, there is something in human nature which we cannot blot out, which makes one man, in the end, recognize and reward merit in another, regardless of colour or race.

I believe it is the duty of the Negro—as the greater part of the race is already doing—to deport himself modestly in regard to political claims, depending upon the slow but sure influences that proceed from the possession of property, intelligence, and high character for the full recognition of his political rights. I think that the according of the full exercise of political rights is going to be a matter of natural, slow growth, not an over-night, gourd-vine affair. I do not believe that the Negro should cease voting, for a man cannot learn the exercise of self-government by ceasing to vote any more than a boy can learn to swim by keeping out of the water, but I do believe that in his voting he should more and more be influenced by those of intelligence and character who are his next-door neighbours.

I know coloured men who, through the encouragement, help, and advice of Southern white people, have accumulated thousands of dollars' worth of property, but who, at the same time, would never think of going to those same persons for advice concerning the casting of their ballots. This, it seems to me, is unwise and unreasonable, and should cease. In saying this I do not mean that the Negro should truckle, or not vote from principle, for the instant he ceases to vote from principle he loses the confidence and respect of the Southern white man even.

I do not believe that any state should make a law that permits an ignorant and poverty-stricken white man to vote, and prevents a black man in the same condition from voting. Such a law is not only unjust, but it will react, as all unjust laws do, in time; for the effect of such a law is to encourage the Negro to secure education and property, and at the same time it encourages the white man to remain in ignorance and poverty. I believe that in time, through the operation of intelligence and friendly race relations, all cheating at the ballot box in the South will cease. It will become apparent that the white man who begins by cheating a Negro out of his ballot soon learns to cheat a white man out of his, and that the man who does this ends his career of dishonesty by the theft of property or by some equally serious crime. In my opinion, the time will come when the South will encourage all of its citizens to vote. It will see that it pays better, from every standpoint, to have healthy, vigorous life than to have that political stagnation which always results when one-half of the population has no share and no interest in the Government.

As a rule, I believe in universal, free suffrage, but I believe that in the South we are confronted with peculiar conditions that justify the protection of the ballot in many of the states, for a while at least, either by an educational test, a property test, or by both combined; but whatever tests are required, they should be made to apply with equal and exact justice to both races.

CHAPTER XV

The Secret of Success in Public Speaking

As to how my address at Atlanta was received by the audience in the Exposition building, I think I prefer to let Mr. James Creelman, the noted war correspondent, tell. Mr. Creelman was present, and telegraphed the following account to the New York *World:*—

ATLANTA, SEPTEMBER 18.

While President Cleveland was waiting at Gray Gables to-day, to send the electric spark that started the machinery of the Atlanta Exposition, a Negro Moses stood before a great audience of white people and delivered an oration that marks a new epoch in the history of the South; and a body of Negro troops marched in a procession with the citizen soldiery of Georgia and Louisiana. The whole city is thrilling to-night with a realization of the extraordinary significance of these two unprecedented events. Nothing has happened since Henry Grady's immortal speech before the New England society in New York that indicates so profoundly the spirit of the New South, except, perhaps, the opening of the Exposition itself.

When Professor Booker T. Washington, Principal of an industrial school for coloured people in Tuskegee, Ala., stood on the platform of the Auditorium, with the sun shining over the heads of his auditors into his eyes, and with his whole face lit up with the fire of prophecy, Clark Howell, the successor to Henry Grady, said to me, "That man's speech is the beginning of a moral revolution in America."

It is the first time that a Negro has made a speech in the South on any important occasion before an audience composed of white men and women. It electrified the audience, and the response was as if it had come from the throat of a whirlwind.

Mrs. Thompson had hardly taken her seat when all eyes were turned on a tall tawny Negro sitting in the front row of the platform. It was Professor Booker T. Washington, President of the Tuskegee (Alabama) Normal and Industrial Institute, who must rank from this time forth as the foremost man of his race in America. Gilmore's Band played the "Star-Spangled Banner," and the audience cheered. The tune changed to "Dixie" and the audience roared with shrill "hi-yis." Again the music changed, this time to "Yankee Doodle," and the clamour lessened.

All this time the eyes of the thousands present looked straight at the Negro orator. A strange thing was to happen. A black man was to speak for his people, with none to interrupt him. As Professor Washington strode to the edge of the stage, the low, descending sun shot fiery rays through the windows into his face. A great shout greeted him. He turned his head to avoid the blinding light, and moved about the platform for relief. Then he turned his wonderful countenance to the sun without a blink of the eyelids, and began to talk.

There was a remarkable figure; tall, bony, straight as a Sioux chief, high forehead, straight nose, heavy jaws, and strong, determined mouth, with big white teeth, piercing eyes, and a commanding manner. The sinews stood out on his bronzed neck, and his muscular right arm swung high in the air, with a lead-pencil grasped in the clenched brown fist. His big feet were planted squarely, with the heels together and the toes turned out. His voice rang out clear and true, and he paused impressively as he made each point. Within ten minutes the multitude was in an uproar of enthusiasm—handkerchiefs were waved, canes were flourished, hats were tossed in the air. The fairest women of Georgia stood up and cheered. It was as if the orator had bewitched them.

And when he held his dusky hand high above his head, with the fingers stretched wide apart, and said to the white people of the South on behalf of his race, "In all things that are purely social we can be as separate as the fingers, yet one as the hand in all things essential to mutual progress," the great wave of sound dashed itself against the walls, and the whole audience was on its feet in a delirium of applause, and I thought at that moment of the night when Henry Grady stood among the curling wreaths of tobacco-smoke in Delmonico's banquet-hall and said, "I am a Cavalier among Roundheads."

I have heard the great orators of many countries, but not even Gladstone himself could have pleased a cause with more consummate power than did this angular Negro, standing in a nimbus of sunshine, surrounded by the men who once fought to keep his race in bondage. The roar might swell ever so high, but the expression of his earnest face never changed.

A ragged, ebony giant, squatted on the floor in one of the aisles, watched the orator with burning eyes and tremulous face until the supreme burst of applause came, and then the tears ran down his face. Most of the Negroes in the audience were crying, perhaps without knowing just why.

At the close of the speech Governor Bullock rushed across the stage and seized the orator's hand. Another shout greeted this demonstration, and for a few minutes the two men stood facing each other, hand in hand.

So far as I could spare the time from the immediate work at Tuskegee, after my Atlanta address, I accepted some of the invitations to speak in public which came to me, especially those that would take me into territory where I thought it would pay to plead the cause of my race, but I always did this with the understanding that I was to be free to talk about my life-work and the needs of my people. I also had it understood that I was not to speak in the capacity of a professional lecturer, or for mere commercial gain.

In my efforts on the public platform I never have been able to understand why people come to hear me speak. This question I never can rid myself of. Time and time again, as I have stood in the street in front of a building and have seen men and women passing in large numbers into the audience-room where I was to speak, I have felt ashamed that I should be the cause of people—as it seemed to me—wasting a valuable hour of time. Some years ago I was to deliver an address before a literary society in Madison, Wis. An hour before the time set for me to speak, a fierce snow-storm began, and continued for several hours. I made up my mind that there would be no audience, and that I should not have to speak, but, as a matter of duty, I went to the church, and found it packed with people. The surprise gave me a shock that I did not recover from during the whole evening.

People often ask me if I feel nervous before speaking, or else they suggest that, since I speak so often, they suppose that I get used to it. In answer to this question I have to say that I always suffer intensely from nervousness before speaking. More than once, just before I was to make an important address, this nervous strain has been so great that I have resolved never again to speak in public. I not only feel nervous before speaking, but after I have finished I usually feel a sense of regret, because it seems to me as if I had left out of my address the main thing and the best thing that I had meant to say.

There is a great compensation, though, for this preliminary nervous suffering, that comes to me after I have been speaking for about ten minutes, and have come to feel that I have really mastered my audience, and that we have gotten into full and complete sympathy with each other. It seems to me that there is rarely such a combination of mental and physical delight in any effort as that which comes to a public speaker when he feels that he has a great audience completely within his control. There is a thread of sympathy and oneness that connects a public speaker with his audience, that is just as strong as though it was something tangible and visible. If in an audience of a thousand people there is one person who is not in sympathy with my views, or is inclined to be doubtful, cold, or critical, I can pick him out. When I have found him I usually go straight at him, and it is a great satisfaction to watch the process of his thawing out. I find that the most effective medicine for such individuals is administered at first in the form of a story, although I never tell an anecdote simply for the sake of telling one. That kind of thing, I think, is empty and hollow, and an audience soon finds it out.

I believe that one always does himself and his audience an injustice when he speaks merely for the sake of speaking. I do not believe that one should speak unless, deep down in his heart, he feels convinced that he has a message to deliver. When one feels, from the bottom of his feet to the top of his head, that he has something to say that is going to help some individual or some cause, then let him say it; and in delivering his message I do not believe that many of the artificial rules of elocution can, under such circumstances, help him very much. Although there are certain things, such as pauses, breathing, and pitch of voice, that are very important, none of these can take the place of *soul* in an address. When I have an address to deliver, I like to forget all about the rules for the proper use of the English language, and all about rhetoric and that sort of thing, and I like to make the audience forget all about these things, too.

Nothing tends to throw me off my balance so quickly, when I am speaking, as to have some one leave the room. To prevent this, I make up my mind, as a rule, that I will try to make my address so interesting, will try to state so many interesting facts one after another, that no one can leave. The average audience, I have come to believe, wants facts rather than generalities or sermonizing. Most people, I think, are able to draw proper conclusions if they are given the facts in an interesting form on which to base them.

As to the kind of audience that I like best to talk to, I would put at the top of the list an organization of strong, wide-awake business men, such, for example, as is found in Boston, New York, Chicago, and Buffalo. I have found no other audience so quick to see a point, and so responsive. Within the last few years I have had the privilege of speaking before most of the leading organizations of this kind in the large cities of the United States. The best time to get hold of an organization of business men is after a good dinner, although I think that one of the worst instruments of torture that was ever invented is the custom which makes it necessary for a speaker to sit through a fourteen-course dinner, every minute of the time feeling sure that his speech is going to prove a dismal failure and disappointment.

I rarely take part in one of these long dinners that I do not wish that I could put myself back in the little cabin where I was a slave boy, and again go through the experience there—one that I shall never forget—of getting molasses to eat once a week from the "big house." Our usual diet on the plantation was corn bread and pork, but on Sunday morning my mother was permitted to bring down a little molasses from the "big house" for her three children, and when it was received how I did wish that every day was Sunday! I would get my tin plate and hold it up for the sweet morsel, but I would always shut my eyes while the molasses was being poured out into the plate, with the hope that when I opened them I would be surprised to see how much I had got. When I opened my eyes I would tip the plate in one direction and another, so as to make the molasses spread all over it, in the full belief that there would be more of it and that it would last longer if spread out in this way. So strong are my childish impressions of those Sunday morning feasts that it would be pretty hard for any one to convince me that there is not more molasses on a plate when it is spread all over the plate than when it occupies a little corner—if there is a corner in a plate. At any rate, I have never believed in "cornering" syrup. My share of the syrup was usually about two tablespoonfuls, and those two spoonfuls of molasses were much more enjoyable to me than is a fourteen-course dinner after which I am to speak.

Next to a company of business men, I prefer to speak to an audience of Southern people, of either race, together or taken separately. Their enthusiasm and responsiveness are a constant delight. The "amens" and "dat's de truf" that come spontaneously from the coloured individuals are calculated to spur any speaker on to his best efforts. I think that next in order of preference I would place a college audience. It has been my privilege to deliver addresses at many of our leading colleges, including Harvard, Yale, Williams, Amherst, Fisk University, the University of Pennsylvania, Wellesley, the University of Michigan, Trinity College in North Carolina, and many others.

It has been a matter of deep interest to me to note the number of people who have come to shake hands with me after an address, who say that this is the first time they have ever called a Negro "Mister."

When speaking directly in the interests of the Tuskegee Institute, I usually arrange, some time in advance, a series of meetings in important centres. This takes me before churches, Sunday-schools, Christian Endeavour Societies, and men's and women's clubs. When doing this I sometimes speak before as many as four organizations in a single day.

Three years ago, at the suggestion of Mr. Morris K. Jesup, of New York, and Dr. J. L. M. Curry, the general agent of the fund, the trustees of the John F. Slater Fund voted a sum of money to be used in paying the expenses of Mrs. Washington and myself while holding a series of meetings among the coloured people in the large centres of Negro population, especially in the large cities of the ex-slaveholding states. Each year during the last three years we have devoted some weeks to this work. The plan that we have followed has been for me to speak in the morning to the ministers, teachers, and professional men. In the afternoon Mrs. Washington would speak to the women alone, and in the evening I spoke to a large mass-meeting. In almost every case the meetings have been attended not only by the coloured people in large numbers, but by the white people. In Chattanooga, Tenn., for example, there was present at the mass-meeting an audience of not less than three thousand persons, and I was informed that eight hundred of these were white. I have done no work that I really enjoyed more than this, or that I think has accomplished more good.

These meetings have given Mrs. Washington and myself an opportunity to get first-hand, accurate information as to the real condition of the race by seeing the people in their homes, their churches, their Sunday-schools, and their places of work, as well as in the prisons and dens of crime. These meetings also gave us an opportunity to see the relations that exist between the races. I never feel so hopeful about the race as I do after being engaged in a series of these meetings. I know that on such occasions there is much that comes to the surface that is superficial and deceptive, but I have had experience enough not to be deceived by mere signs and fleeting enthusiasms. I have taken pains to go to the bottom of things and get facts, in a cold, business-like manner.

I have seen the statement made lately, by one who claims to know what he is talking about, that, taking the whole Negro race into account, ninety per cent of the Negro women are not virtuous. There never was a baser falsehood uttered concerning a race or a statement made that was less capable of being proved by actual facts.

No one can come into contact with the race for twenty years, as I have done in the heart of the South, without being convinced that the race is constantly making slow but sure progress materially, educationally, and morally. One might take up the life of the worst element in New York City, for example, and prove almost anything he wanted to prove concerning the white man, but all will agree that this is not a fair test.

Early in the year 1897 I received a letter inviting me to deliver an address at the dedication of the Robert Gould Shaw monument in Boston. I accepted the invitation. It is not necessary for me, I am sure, to explain who Robert Gould Shaw was and what he did. The monument to his memory stands near the head of Boston Common, facing the State House. It is counted to be the most perfect piece of art of the kind to be found in the country.

The exercises connected with the dedication were held in Music Hall, in Boston, and the great hall was packed from top to bottom with one of the most distinguished audiences that ever assembled in the city. Among those present there were more persons representing the famous old anti-slavery element than it is likely will ever be brought together in the country again. The late Hon. Roger Wolcott, then Governor of Massachusetts, was the presiding officer, and on the platform with him were many other officials and hundreds of distinguished men. A report of the meeting which appeared in the Boston *Transcript* will describe it better than any words of mine could do:—

The core and kernel of yesterday's great noon meeting in honour of the Brotherhood of Man, in Music Hall, was the superb address of the Negro President of Tuskegee. "Booker T. Washington received his Harvard A. M. last June, the first of his race," said Governor Wolcott, "to receive an honorary degree from the oldest university in the land, and this for the wise leadership of his people." When Mr. Washington rose in the flag-filled, enthusiasm-warmed, patriotic, and glowing atmosphere of Music Hall, people felt keenly that here was the civic justification of the old abolition spirit of Massachusetts; in his person the proof of her ancient and indomitable faith; in his strong thought and rich oratory, the crown and glory of the old war days of suffering and strife. The scene was full of historic beauty and deep significance. "Cold" Boston was alive with the fire that is always hot in her heart for righteousness and truth. Rows and rows of people who are seldom seen at any public function, whole families of those who are certain to be out of town on a holiday, crowded the place to overflowing. The city was at her birthright fête in the persons of hundreds of her best citizens, men and women whose names and lives stand for the virtues that make for honourable civic pride.

Battle-music had filled the air. Ovation after ovation, applause warm and prolonged, had greeted the officers and friends of Colonel Shaw, the sculptor, St. Gaudens, the Memorial Committee, the Governor and his staff, and the Negro soldiers of the Fifty-fourth Massachusetts as they came upon the platform or entered the hall. Colonel Henry Lee, of Governor Andrew's old staff, had made a noble, simple presentation speech for the committee, paying tribute to Mr. John M. Forbes, in whose stead he served. Governor Wolcott had made his short, memorable speech, saying, "Fort Wagner marked an epoch in the history of a race, and called it into manhood." Mayor Quincy had received the monument for the city of Boston. The story of Colonel Shaw and his black regiment had been told in gallant words, and then, after the singing of

> Mine eyes have seen the glory
> Of the coming of the Lord.

Booker Washington arose. It was, of course, just the moment for him. The multitude, shaken out of its usual symphony-concert calm, quivered with an excitement that was not suppressed. A dozen times it had sprung to its feet to cheer and wave and hurrah, as one person. When this man of culture and voice and power, as well as a dark skin, began, and uttered the names of Stearns and of Andrew, feeling began to mount. You could see tears glisten in the eyes of the soldiers and civilians. When the orator turned to the coloured soldiers on the platform, to the colour-bearer of Fort Wagner, who smilingly bore still the flag he had never lowered even when wounded, and said, "To you, to the scarred and scattered remnants of the Fifty-fourth, who, with empty sleeve and wanting leg, have honoured this occasion with your presence, to you, your commander is not dead. Though Boston erected no monument and history recorded no story, in you and in the loyal race which you represent, Robert Gould Shaw would have a monument which time could not wear away," then came the climax of the emotion of the day and the hour. It was Roger Wolcott, as well as the Governor of Massachusetts, the individual representative of the people's sympathy as well as the chief magistrate, who had sprung first to his feet and cried, "Three cheers to Booker T. Washington!"

Among those on the platform was Sergeant William H. Carney, of New Bedford, Mass., the brave coloured officer who was the colour-bearer at Fort Wagner and held the American flag. In spite of the fact that a large part of his regiment was killed, he escaped, and exclaimed, after the battle was over, "The old flag never touched the ground."

This flag Sergeant Carney held in his hands as he sat on the platform, and when I turned to address the survivors of the coloured regiment who were present, and referred to Sergeant Carney, he rose, as if by instinct, and raised the flag. It has been my privilege to witness a good many satisfactory and rather sensational demonstrations in connection with some of my public addresses, but in dramatic effect I have never seen or experienced anything which equalled this. For a number of minutes the audience seemed to entirely lose control of itself.

In the general rejoicing throughout the country which followed the close of the Spanish-American war, peace celebrations were arranged in several of the large cities. I was asked by President William R. Harper, of the University of Chicago, who was chairman of the committee of invitations for the celebration to be held in the city of Chicago, to deliver one of the addresses at the celebration there. I accepted the invitation, and delivered two addresses there during the Jubilee week. The first of these, and the principal one, was given in the Auditorium on the evening of Sunday, October 16. This was the largest audience that I have ever addressed, in any part of the country; and besides speaking in the main Auditorium, I also addressed, that same evening, two overflow audiences in other parts of the city.

It was said that there were sixteen thousand persons in the Auditorium, and it seemed to me as if there were as many more on the outside trying to get in. It was impossible for any one to get near the entrance without the aid of a policeman. President William McKinley attended this meeting, as did also the members of his Cabinet, many foreign ministers, and a large number of army and navy officers, many of whom had distinguished themselves in the war which had just closed. The speakers, besides myself, on Sunday evening, were Rabbi Emil G. Hirsch, Father Thomas P. Hodnett, and Dr. John H. Barrows.

The Chicago *Times-Herald*, in describing the meeting, said of my address:—

He pictured the Negro choosing slavery rather than extinction; recalled Crispus Attucks shedding his blood at the beginning of the American Revolution, that white Americans might be free, while black Americans remained in slavery; rehearsed the conduct of the Negroes with Jackson at New Orleans; drew a vivid and pathetic picture of the Southern slaves protecting and supporting the families of their masters while the latter were fighting to perpetuate black slavery; recounted the bravery of coloured troops at Port Hudson and Forts Wagner and Pillow, and praised the heroism of the black regiments that stormed El Caney and Santiago to give freedom to the enslaved people of Cuba, forgetting, for the time being, the unjust discrimination that law and custom make against them in their own country.

In all of these things, the speaker declared, his race had chosen the better part. And then he made his eloquent appeal to the consciences of the white Americans: "When you have gotten the full story of the heroic conduct of the Negro in the Spanish-American war, have heard it from the lips of Northern soldier and Southern soldier, from ex-abolitionist and ex-masters, then decide within yourselves whether a race that is thus willing to die for its country should not be given the highest opportunity to live for its country."

The part of the speech which seemed to arouse the wildest and most sensational enthusiasm was that in which I thanked the President for his recognition of the Negro in his appointments during the Spanish-American war. The President was sitting in a box at the right of the stage. When I addressed him I turned toward the box, and as I finished the sentence thanking him for his generosity, the whole audience rose and cheered again and again, waving handkerchiefs and hats and canes, until the President arose in the box and bowed his acknowledgments. At that the enthusiasm broke out again, and the demonstration was almost indescribable.

One portion of my address at Chicago seemed to have been misunderstood by the Southern press, and some of the Southern papers took occasion to criticise me rather strongly. These criticisms continued for several weeks, until I finally received a letter from the editor of the *Age-Herald*, published in Birmingham, Ala., asking me if I would say just what I meant by this part of my address. I replied to him in a letter which seemed to satisfy my critics. In this letter I said that I had made it a rule never to say before a Northern audience anything that I would not say before an audience in the South. I said that I did not think it was necessary for me to go into extended explanations; if my seventeen years of work in the heart of the South had not been explanation enough, I did not see how words could explain. I said that I made the same plea that I had made in my address at Atlanta, for the blotting out of race prejudice in "commercial and civil relations." I said that what is termed social recognition was a question which I never discussed and then I quoted from my Atlanta address what I had said there in regard to that subject.

In meeting crowds of people at public gatherings, there is one type of individual that I dread. I mean the crank. I have become so accustomed to these people now that I can pick them out at a distance when I see them elbowing their way up to me. The average crank has a long beard, poorly cared for, a lean, narrow face, and wears a black coat. The front of his vest and coat are slick with grease, and his trousers bag at the knees.

In Chicago, after I had spoken at a meeting, I met one of these fellows. They usually have some process for curing all of the ills of the world at once. This Chicago specimen had a patent process by which, he said, Indian corn could be kept through a period of three or four years, and he felt sure that if the Negro race in the South would, as a whole, adopt his process, it would settle the whole race question. It mattered nothing that I tried to convince him that our present problem was to teach the Negroes how to produce enough corn to last them through one year. Another Chicago crank had a scheme by which he wanted me to join him in an effort to close up all the National banks in the country. If that was done, he felt sure it would put the Negro on his feet.

The number of people who stand ready to consume one's time, to no purpose, is almost countless. At one time I spoke before a large audience in Boston in the evening. The next morning I was awakened by having a card brought to my room, and with it a message that some one was anxious to see me. Thinking that it must be something very important, I dressed hastily and went down. When I reached the hotel office I found a blank and innocent-looking individual waiting for me, who coolly remarked: "I heard you talk at a meeting last night. I rather liked your talk, and so I came in this morning to hear you talk some more."

I am often asked how it is possible for me to superintend the work at Tuskegee and at the same time be so much away from the school. In partial answer to this I would say that I think I have learned, in some degree at least, to disregard the old maxim which says, "Do not get others to do that which you can do yourself." My motto, on the other hand, is "Do not do that which others can do as well."

One of the most encouraging signs in connection with the Tuskegee school is found in the fact that the organization is so thorough that the daily work of the school is not dependent upon the presence of any one individual. The whole executive force, including instructors and clerks, now numbers eighty-six. This force is so organized and subdivided that the machinery of the school goes on day by day like clockwork. Most of our teachers have been connected with the institution for a number of years, and are as much interested in it as I am. In my absence, Mr. Warren Logan, the treasurer, who has been at the school seventeen years, is the executive. He is efficiently supported by Mrs. Washington, and by my faithful secretary, Mr. Emmett J. Scott, who handles the bulk of my correspondence and keeps me in daily touch with the life of the school, and who also keeps me informed of whatever takes place in the South that concerns the race. I owe more to his tact, wisdom, and hard work than I can describe.

The main executive work of the school, whether I am at Tuskegee or not, centres in what we call the executive council. This council meets twice a week, and is composed of the nine persons who are at the head of the nine departments of the school. For example: Mrs. B. K. Bruce, the Lady Principal, the widow of the late ex-senator Bruce, is a member of the council, and represents in it all that pertains to the life of the girls at the school. In addition to the executive council there is a financial committee of six, that meets every week and decides upon the expenditures for the week. Once a month, and sometimes oftener, there is a general meeting of all the instructors. Aside from these there are innumerable smaller meetings, such as that of the instructors in the Phelps Hall Bible Training School, or of the instructors in the agricultural department.

In order that I may keep in constant touch with the life of the institution, I have a system of reports so arranged that a record of the school's work reaches me every day in the year, no matter in what part of the country I am. I know by these reports even what students are excused from school, and why they are excused— whether for reasons of ill health or otherwise. Through the medium of these reports I know each day what the income of the school in money is; I know how many gallons of milk and how many pounds of butter come from the dairy; what the bill of fare for the teachers and students is; whether a certain kind of meat was boiled or baked, and whether certain vegetables served in the dining room were bought from a store or procured from our own farm. Human nature I find to be very much the same the world over, and it is sometimes not hard to yield to the temptation to go to a barrel of rice that has come from the store—with the grain all prepared to go into the pot—rather than to take the time and trouble to go to the field and dig and wash one's own sweet potatoes, which might be prepared in a manner to take the place of the rice.

I am often asked how, in the midst of so much work, a large part of which is before the public, I can find time for any rest or recreation, and what kind of recreation or sports I am fond of. This is rather a difficult question to answer. I have a strong feeling that every individual owes it to himself, and to the cause which he is serving, to keep a vigorous, healthy body, with the nerves steady and strong, prepared for great efforts and prepared for disappointments and trying positions. As far as I can, I make it a rule to plan for each day's work—not merely to go through with the same routine of daily duties, but to get rid of the routine work as early in the day as possible, and then to enter upon some new or advance work. I make it a rule to clear my desk every day, before leaving my office, of all correspondence and memoranda, so that on the morrow I can begin a *new* day of work. I make it a rule never to let my work drive me, but to so master it, and keep it in such complete control, and to keep so far ahead of it, that I will be the master instead of the servant. There is a physical and mental and spiritual enjoyment that comes from a consciousness of being the absolute master of one's work, in all its details, that is very satisfactory and inspiring. My experience teaches me that, if one learns to follow this plan, he gets a freshness of body and vigour of mind out of work that goes a long way toward keeping him strong and healthy. I believe that when one can grow to the point where he loves his work, this gives him a kind of strength that is most valuable.

When I begin my work in the morning, I expect to have a successful and pleasant day of it, but at the same time I prepare myself for unpleasant and unexpected hard places. I prepare myself to hear that one of our school buildings is on fire, or has burned, or that some disagreeable accident has occurred, or that some one has abused me in a public address or printed article, for something that I have done or omitted to do, or for something that he had heard that I had said—probably something that I had never thought of saying.

In nineteen years of continuous work I have taken but one vacation. That was two years ago, when some of my friends put the money into my hands and forced Mrs. Washington and myself to spend three months in Europe. I have said that I believe it is the duty of every one to keep his body in good condition. I try to look after the little ills, with the idea that if I take care of the little ills the big ones will not come. When I find myself unable to sleep well, I know that something is wrong. If I find any part of my system the least weak, and not performing its duty, I consult a good physician. The ability to sleep well, at any time and in any place, I find of great advantage. I have so trained myself that I can lie down for a nap of fifteen or twenty minutes, and get up refreshed in body and mind.

I have said that I make it a rule to finish up each day's work before leaving it. There is, perhaps, one exception to this. When I have an unusually difficult question to decide—one that appeals strongly to the emotions —I find it a safe rule to sleep over it for a night, or to wait until I have had an opportunity to talk it over with my wife and friends.

As to my reading; the most time I get for solid reading is when I am on the cars. Newspapers are to me a constant source of delight and recreation. The only trouble is that I read too many of them. Fiction I care little for. Frequently I have to almost force myself to read a novel that is on every one's lips. The kind of reading that I have the greatest fondness for is biography. I like to be sure that I am reading about a real man or a real thing. I think I do not go too far when I say that I have read nearly every book and magazine article that has been written about Abraham Lincoln. In literature he is my patron saint.

Out of the twelve months in a year I suppose that, on an average, I spend six months away from Tuskegee. While my being absent from the school so much unquestionably has its disadvantages, yet there are at the same time some compensations. The change of work brings a certain kind of rest. I enjoy a ride of a long distance on the cars, when I am permitted to ride where I can be comfortable. I get rest on the cars, except when the inevitable individual who seems to be on every train approaches me with the now familiar phrase: "Isn't this Booker Washington? I want to introduce myself to you." Absence from the school enables me to lose sight of the unimportant details of the work, and study it in a broader and more comprehensive manner than I could do on the grounds. This absence also brings me into contact with the best work being done in educational lines, and into contact with the best educators in the land.

But, after all this is said, the time when I get the most solid rest and recreation is when I can be at Tuskegee, and, after our evening meal is over, can sit down, as is our custom, with my wife and Portia and Booker and Davidson, my three children, and read a story, or each take turns in telling a story. To me there is nothing on earth equal to that, although what is nearly equal to it is to go with them for an hour or more, as we like to do on Sunday afternoons, into the woods, where we can live for a while near the heart of nature, where no one can disturb or vex us, surrounded by pure air, the trees, the shrubbery, the flowers, and the sweet fragrance that springs from a hundred plants, enjoying the chirp of the crickets and the songs of the birds. This is solid rest.

My garden, also, what little time I can be at Tuskegee, is another source of rest and enjoyment. Somehow I like, as often as possible, to touch nature, not something that is artificial or an imitation, but the real thing. When I can leave my office in time so that I can spend thirty or forty minutes in spading the ground, in planting seeds, in digging about the plants, I feel that I am coming into contact with something that is giving me strength for the many duties and hard places that await me out in the big world. I pity the man or woman who has never learned to enjoy nature and to get strength and inspiration out of it.

Aside from the large number of fowls and animals kept

by the school, I keep individually a number of pigs and fowls of the best grades, and in raising these I take a great deal of pleasure. I think the pig is my favourite animal. Few things are more satisfactory to me than a high-grade Berkshire or Poland China pig.

Games I care little for. I have never seen a game of football. In cards I do not know one card from another. A game of old-fashioned marbles with my two boys, once in a while, is all I care for in this direction. I suppose I would care for games now if I had had any time in my youth to give to them, but that was not possible.

CHAPTER XVI

Europe

In 1893 I was married to Miss Margaret James Murray, a native of Mississippi, and a graduate of Fisk University, in Nashville, Tenn., who had come to Tuskegee as a teacher several years before, and at the time we were married was filling the position of Lady Principal. Not only is Mrs. Washington completely one with me in the work directly connected with the school, relieving me of many burdens and perplexities, but aside from her work on the school grounds, she carries on a mothers' meeting in the town of Tuskegee, and a plantation work among the women, children, and men who live in a settlement connected with a large plantation about eight miles from Tuskegee. Both the mothers' meeting and the plantation work are carried on, not only with a view to helping those who are directly reached, but also for the purpose of furnishing object-lessons in these two kinds of work that may be followed by our students when they go out into the world for their own life-work.

Aside from these two enterprises, Mrs. Washington is also largely responsible for a woman's club at the school which brings together, twice a month, the women who live on the school grounds and those who live near, for the discussion of some important topic. She is also the President of what is known as the Federation of Southern Coloured Women's Clubs, and is Chairman of the

Executive Committee of the National Federation of Coloured Women's Clubs.

Portia, the oldest of my three children, has learned dressmaking. She has unusual ability in instrumental music. Aside from her studies at Tuskegee, she has already begun to teach there.

Booker Taliaferro is my next oldest child. Young as he is, he has already nearly mastered the brickmason's trade. He began working at this trade when he was quite small, dividing his time between this and class work; and he has developed great skill in the trade and a fondness for it. He says that he is going to be an architect and brickmason. One of the most satisfactory letters that I have ever received from any one came to me from Booker, last summer. When I left home for the summer, I told him that he must work at his trade half of each day, and that the other half of the day he could spend as he pleased. When I had been away from home two weeks, I received the following letter from him:

TUSKEGEE, ALABAMA.

MY DEAR PAPA: Before you left home you told me to work at my trade half of each day. I like my work so much that I want to work at my trade all day. Besides, I want to earn all the money I can, so that when I go to another school I shall have money to pay my expenses.

Your son,
BOOKER.

My youngest child, Ernest Davidson Washington, says that he is going to be a physician. In addition to going to school, where he studies books and has manual training, he regularly spends a portion of his time in the office of our resident physician, and has already learned to do many of the duties which pertain to a doctor's office.

The thing in my life which brings me the keenest regret is that my work in connection with public affairs keeps me for so much of the time away from my family, where, of all places in the world, I delight to be. I always envy the individual whose life-work is so laid that he can spend his evenings at home. I have sometimes thought that people who have this rare privilege do not appreciate it as they should. It is such a rest and relief to get away from crowds of people, and handshaking, and travelling, and get home, even if it be for but a very brief while.

Another thing at Tuskegee out of which I get a great deal of pleasure and satisfaction is in the meeting with our students, and teachers, and their families, in the chapel for devotional exercises every evening at half-past eight, the last thing before retiring for the night. It is an inspiring sight when one stands on the platform there and sees before him eleven or twelve hundred earnest young men and women; and one cannot but feel that it is a privilege to help to guide them to a higher and more useful life.

In the spring of 1899 there came to me what I might describe as almost the greatest surprise of my life. Some good ladies in Boston arranged a public meeting in the interests of Tuskegee, to be held in the Hollis Street Theatre. This meeting was attended by large numbers of the best people of Boston, of both races. Bishop Lawrence presided. In addition to an address made by myself, Mr. Paul Lawrence Dunbar read from his poems, and Dr. W. E. B. DuBois read an original sketch.

Some of those who attended this meeting noticed that I seemed unusually tired, and some little time after the close of the meeting, one of the ladies who had been interested in it asked me in a casual way if I had ever been to Europe. I replied that I never had. She asked me if I had ever thought of going, and I told her no; that it was something entirely beyond me. This conversation soon passed out of my mind, but a few days afterward I was informed that some friends in Boston, including Mr. Francis J. Garrison, had raised a sum of money sufficient to pay all the expenses of Mrs. Washington and myself during a three or four months' trip to Europe. It was added with emphasis that we *must* go. A year previous to this Mr. Garrison had attempted to get me to promise to go to Europe for a summer's rest, with the understanding that he would be responsible for raising the money among his friends for the expenses of the trip. At that time such a journey seemed so entirely foreign to anything that I should ever be able to undertake that I confess I did not give the matter very serious attention;

but later Mr. Garrison joined his efforts to those of the ladies whom I have mentioned, and when their plans were made known to me Mr. Garrison not only had the route mapped out, but had, I believe, selected the steamer upon which we were to sail.

The whole thing was so sudden and so unexpected that I was completely taken off my feet. I had been at work steadily for eighteen years in connection with Tuskegee, and I had never thought of anything else but ending my life in that way. Each day the school seemed to depend upon me more largely for its daily expenses, and I told these Boston friends that, while I thanked them sincerely for their thoughtfulness and generosity, I could not go to Europe, for the reason that the school could not live financially while I was absent. They then informed me that Mr. Henry L. Higginson, and some other good friends who I know do not want their names made public, were then raising a sum of money which would be sufficient to keep the school in operation while I was away. At this point I was compelled to surrender. Every avenue of escape had been closed.

Deep down in my heart the whole thing seemed more like a dream than like reality, and for a long time it was difficult for me to make myself believe that I was actually going to Europe. I had been born and largely reared in the lowest depths of slavery, ignorance, and poverty. In my childhood I had suffered for want of a place to sleep, for lack of food, clothing, and shelter. I had not had the privilege of sitting down to a dining-table until I was quite well grown. Luxuries had always seemed to me to be something meant for white people, not for my race. I had always regarded Europe, and London, and Paris, much as I regard heaven. And now could it be that I was actually going to Europe? Such thoughts as these were constantly with me.

Two other thoughts troubled me a good deal. I feared that people who heard that Mrs. Washington and I were going to Europe might not know all the circumstances, and might get the idea that we had become, as some might say, "stuck up," and were trying to "show off." I recalled that from my youth I had heard it said that too often, when people of my race reached any degree of success, they were inclined to unduly exalt themselves; to try and ape the wealthy, and in so doing to lose their heads. The fear that people might think this of us haunted me a good deal. Then, too, I could not see how my conscience would permit me to spare the time from my work and be happy. It seemed mean and selfish to me to be taking a vacation while others were at work, and while there was so much that needed to be done. From the time I could remember, I had always been at work, and I did not see how I could spend three or four months in doing nothing. The fact was that I did not know how to take a vacation.

Mrs. Washington had much the same difficulty in getting away, but she was anxious to go because she thought that I needed the rest. There were many important National questions bearing upon the life of the race which were being agitated at that time, and this made it all the harder for us to decide to go. We finally gave our Boston friends our promise that we would go, and then they insisted that the date of our departure be set as soon as possible. So we decided upon May 10. My good friend Mr. Garrison kindly took charge of all the details necessary for the success of the trip, and he, as well as other friends, gave us a great number of letters of introduction to people in France and England, and made other arrangements for our comfort and convenience abroad. Good-bys were said at Tuskegee, and we were in New York May 9, ready to sail the next day. Our daughter Portia, who was then studying in South Framingham, Mass., came to New York to see us off. Mr. Scott, my secretary, came with me to New York, in order that I might clear up the last bit of business before I left. Other friends also came to New York to see us off. Just before we went on board the steamer another pleasant surprise came to us in the form of a letter from two generous ladies, stating that they had decided to give us the money with which to erect a new building to be used in properly housing all our industries for girls at Tuskegee.

We were to sail on the *Friesland,* of the Red Star Line, and a beautiful vessel she was. We went on board just before noon, the hour of sailing. I had never before been on board a large ocean steamer, and the feeling which took possession of me when I found myself there is rather hard to describe. It was a feeling, I think, of awe mingled with delight. We were agreeably surprised to find that

the captain, as well as several of the other officers, not
only knew who we were, but was expecting us and gave
us a pleasant greeting. There were several passengers whom
we knew, including Senator Sewell, of New Jersey, and
Edward Marshall, the newspaper correspondent. I had
just a little fear that we would not be treated civilly by
some of the passengers. This fear was based upon what I
had heard other people of my race, who had crossed the
ocean, say about unpleasant experiences in crossing the
ocean in American vessels. But in our case, from the
captain down to the most humble servant, we were treated
with the greatest kindness. Nor was this kindness confined to those who were connected with the steamer; it
was shown by all the passengers also. There were not a
few Southern men and women on board, and they were
as cordial as those from other parts of the country.

As soon as the last good-bys were said, and the steamer
had cut loose from the wharf, the load of care, anxiety,
and responsibility which I had carried for eighteen years
began to lift itself from my shoulders at the rate, it
seemed to me, of a pound a minute. It was the first time
in all those years that I had felt, even in a measure, free
from care; and my feeling of relief it is hard to describe
on paper. Added to this was the delightful anticipation
of being in Europe soon. It all seemed more like a dream
than like a reality.

Mr. Garrison had thoughtfully arranged to have us have
one of the most comfortable rooms on the ship. The
second or third day out I began to sleep, and I think
that I slept at the rate of fifteen hours a day during the
remainder of the ten days' passage. Then it was that I
began to understand how tired I really was. These long
sleeps I kept up for a month after we landed on the other
side. It was such an unusual feeling to wake up in the
morning and realize that I had no engagements; did not
have to take a train at a certain hour; did not have an
appointment to meet some one, or to make an address,
at a certain hour. How different all this was from some
of the experiences that I have been through when travelling, when I have sometimes slept in three different beds
in a single night!

When Sunday came, the captain invited me to conduct
the religious services, but, not being a minister, I declined. The passengers, however, began making requests

that I deliver an address to them in the dining-saloon some time during the voyage, and this I consented to do. Senator Sewell presided at this meeting. After ten days of delightful weather, during which I was not seasick for a day, we landed at the interesting old city of Antwerp, in Belgium.

The next day after we landed happened to be one of those numberless holidays which the people of those countries are in the habit of observing. It was a bright, beautiful day. Our room in the hotel faced the main public square, and the sights there—the people coming in from the country with all kinds of beautiful flowers to sell, the women coming in with their dogs drawing large, brightly polished cans filled with milk, the people streaming into the cathedral—filled me with a sense of newness that I had never before experienced.

After spending some time in Antwerp, we were invited to go with a party of a half-dozen persons on a trip through Holland. This party included Edward Marshall and some American artists who had come over on the same steamer with us. We accepted the invitation, and enjoyed the trip greatly. I think it was all the more interesting and instructive because we went for most of the way on one of the slow, old-fashioned canal-boats. This gave us an opportunity of seeing and studying the real life of the people in the country districts. We went in this way as far as Rotterdam, and later went to The Hague, where the Peace Conference was then in session, and where we were kindly received by the American representatives.

The thing that impressed itself most on me in Holland was the thoroughness of the agriculture and the excellence of the Holstein cattle. I never knew, before visiting Holland, how much it was possible for people to get out of a small plot of ground. It seemed to me that absolutely no land was wasted. It was worth a trip to Holland, too, just to get a sight of three or four hundred fine Holstein cows grazing in one of those intensely green fields.

From Holland we went to Belgium, and made a hasty trip through that country, stopping at Brussels, where we visited the battlefield of Waterloo. From Belgium we went direct to Paris, where we found that Mr. Theodore Stanton, the son of Mrs. Elizabeth Cady Stanton, had kindly provided accommodations for us. We had barely got settled in Paris before an invitation came to me from the University Club of Paris to be its guest at a banquet which was soon to be given. The other guests were ex-President Benjamin Harrison and Archbishop Ireland, who were in Paris at the time. The American Ambassador, General Horace Porter, presided at the banquet. My address on this occasion seemed to give satisfaction to those who heard it. General Harrison kindly devoted a large portion of his remarks at dinner to myself and to the influence of the work at Tuskegee on the American race question. After my address at this banquet other invitations came to me, but I declined the most of them, knowing that if I accepted them all, the object of my visit would be defeated. I did, however, consent to deliver an address in the American chapel the following Sunday morning, and at this meeting General Harrison, General Porter, and other distinguished Americans were present.

Later we received a formal call from the American Ambassador, and were invited to attend a reception at his residence. At this reception we met many Americans, among them Justices Fuller and Harlan, of the United States Supreme Court. During our entire stay of a month in Paris, both the American Ambassador and his wife, as well as several other Americans, were very kind to us.

While in Paris we saw a good deal of the now famous American Negro painter, Mr. Henry C. Tanner, whom we had formerly known in America. It was very satisfactory to find how well known Mr. Tanner was in the field of art, and to note the high standing which all classes accorded to him. When we told some Americans that we were going to the Luxembourg Palace to see a painting by an American Negro, it was hard to convince them that a Negro had been thus honoured. I do not believe that they were really convinced of the fact until they saw the picture for themselves. My acquaintance with Mr. Tanner reënforced in my mind the truth which I am constantly trying to impress upon our students at Tuskegee—and on our people throughout the country, as far as I can reach them with my voice—that any man, regardless of colour, will be recognized and rewarded just in proportion as he learns to do something well—learns to do it better than some one else—however humble the thing may be. As I have said, I believe that my race will succeed in proportion as it learns to do a common thing in an uncommon manner; learns to do a thing so thoroughly that no one can improve upon what it has done; learns to make its services of indispensable value. This was the spirit that inspired me in my first effort at Hampton, when I was given the opportunity to sweep and dust that schoolroom. In a degree I felt that my whole future life depended upon the thoroughness with which I cleaned that room, and I was determined to do it so well that no one could find any fault with the job. Few people ever stopped, I found, when looking at his pictures, to inquire whether Mr. Tanner was a Negro painter, a French painter, or a German painter. They simply knew that he was able to produce something which the world wanted—a great painting—and the matter of his colour did not enter into their minds. When a Negro girl learns to cook, to wash dishes, to sew, to write a book, or a Negro boy learns to groom horses, or to grow sweet potatoes, or to produce butter, or to build a house, or to be able to practise medicine, as well or better than some one else, they will be rewarded regardless of race or colour. In the long run, the world is going to have the best, and any difference in race, religion, or previous history will not long keep the world from what it wants.

I think that the whole future of my race hinges on the question as to whether or not it can make itself of such indispensable value that the people in the town and the state where we reside will feel that our presence is necessary to the happiness and well-being of the community. No man who continues to add something to the material, intellectual, and moral well-being of the place in which he lives is long left without proper reward. This is a great human law which cannot be permanently nullified.

The love of pleasure and excitement which seems in a large measure to possess the French people impressed itself upon me. I think they are more noted in this respect than is true of the people of my own race. In point of morality and moral earnestness I do not believe that the French are ahead of my own race in America. Severe competition and the great stress of life have led them to learn to do things more thoroughly, and to exercise greater economy; but time, I think, will bring my race to the same point. In the matter of truth and high honour I do not believe that the average Frenchman is ahead of the American Negro; while so far as mercy and kindness to dumb animals go, I believe that my race is far ahead. In fact, when I left France, I had more faith in the future of the black man in America than I had ever possessed.

From Paris we went to London, and reached there early in July, just about the height of the London social season. Parliament was in session and there was a great deal of gaiety. Mr. Garrison and other friends had provided us with a large number of letters of introduction, and they had also sent letters to other persons in different parts of the United Kingdom, apprising these people of our coming. Very soon after reaching London we were flooded with invitations to attend all manner of social functions, and a great many invitations came to me asking that I deliver public addresses. The most of these invitations I declined, for the reason that I wanted to rest. Neither were we able to accept more than a small proportion of the other invitations. The Rev. Dr. Brooke Herford and Mrs. Herford, whom I had known in Boston, consulted with the American Ambassador, the Hon. Joseph Choate, and arranged for me to speak at a public meeting to be held in Exeter Hall. Mr. Choate kindly consented to preside. The meeting was largely attended. There were many distinguished persons present, among them several members of Parliament, including Mr. James Bryce, who spoke at the meeting. What the American Ambassador said in introducing me, as well as a synopsis of what I said, was widely published in England and in the American papers at the time. Dr. and Mrs. Herford gave Mrs. Washington and myself a reception, at which we had the privilege of meeting some of the best people in England. Throughout our stay in London Ambassador Choate was most kind and attentive to us. At the Ambassador's reception I met, for the first, time, Mark Twain.

We were the guests several times of Mr. T. Fisher Unwin, the daughter of the English statesman, Richard Cobden. It seemed as if both Mr. and Mrs. Unwin could not do enough for our comfort and happiness. Later, for nearly a week, we were the guests of the daughter of John Bright, now Mrs. Clark, of Street, England. Both Mr. and Mrs. Clark, with their daughter, visited us at Tuskegee the next year. In Birmingham, England, we were the guests for several days of Mr. Joseph Sturge, whose father was a great abolitionist and friend of Whittier and Garrison. It was a great privilege to meet throughout England those who had known and honoured the late William Lloyd Garrison, the Hon. Frederick Douglass, and other abolitionists. The English abolitionists with whom we came in contact never seemed to tire of talking about these two Americans. Before going to England I had had no proper conception of the deep interest displayed by the abolitionists of England in the cause of freedom, nor did I realize the amount of substantial help given by them.

In Bristol, England, both Mrs. Washington and I spoke at the Women's Liberal Club. I was also the principal speaker at the Commencement exercises of the Royal College for the Blind. These exercises were held in the Crystal Palace, and the presiding officer was the late Duke of Westminster, who was said to be, I believe, the richest man in England, if not in the world. The Duke, as well as his wife and their daughter, seemed to be pleased with what I said, and thanked me heartily. Through the kindness of Lady Aberdeen, my wife and I were enabled to go with a party of those who were attending the International Congress of Women, then in session in London, to see Queen Victoria, at Windsor Castle, where, afterward, we were all the guests of her Majesty at tea. In our party was Miss Susan B. Anthony, and I was deeply impressed with the fact that one did not often get an opportunity to see, during the same hour, two women so remarkable in different ways as Susan B. Anthony and Queen Victoria.

In the House of Commons, which we visited several times, we met Sir Henry M. Stanley. I talked with him about Africa and its relation to the American Negro, and after my interview with him I became more convinced than ever that there was no hope of the American Negro's improving his condition by emigrating to Africa.

On various occasions Mrs. Washington and I were the guests of Englishmen in their country homes, where I think, one sees the Englishman at his best. In one thing, at least, I feel sure that the English are ahead of Americans, and that is, that they have learned how to get more out of life. The home life of the English seems to me to be about as perfect as anything can be. Everything moves like clockwork. I was impressed, too, with the deference that the servants show to their "masters" and "mistresses," —terms which I suppose would not be tolerated in America. The English servant expects, as a rule, to be nothing but a servant, and so he perfects himself in the art to a degree that no class of servants in America has yet reached. In our country, the servant expects to become, in a few years, a "master" himself. Which system is preferable? I will not venture an answer.

Another thing that impressed itself upon me throughout England was the high regard that all classes have for law and order, and the ease and thoroughness with which everything is done. The Englishmen, I found, took plenty of time for eating, as for everything else. I am not sure if, in the long run, they do not accomplish as much or more than rushing, nervous Americans do.

My visit to England gave me a higher regard for the nobility than I had had. I had no idea that they were so generally loved and respected by the masses, nor had I any correct conception of how much time and money they spent in works of philanthropy, and how much real heart they put into this work. My impression had been that they merely spent money freely and had a "good time."

It was hard for me to get accustomed to speaking to English audiences. The average Englishman is so serious, and is so tremendously in earnest about everything, that when I told a story that would have made an American audience roar with laughter, the Englishmen simply looked me straight in the face without even cracking a smile.

When the Englishman takes you into his heart and friendship, he binds you there as with cords of steel, and I do not believe that there are many other friendships that are so lasting or so satisfactory. Perhaps I can illustrate this point in no better way than by relating the following incident. Mrs. Washington and I were invited to attend a reception given by the Duke and Duchess of Sutherland, at Stafford House—said to be the finest house in London; I may add that I believe the Duchess of Sutherland is said to be the most beautiful woman in England. There must have been at least three hundred persons at this reception. Twice during the evening the Duchess sought us out for a conversation, and she asked me to write her when we got home, and tell her more about the work at Tuskegee. This I did. When Christmas came we were surprised and delighted to receive her photograph with her autograph on it. The correspondence has continued, and we now feel that in the Duchess of Sutherland we have one of our warmest friends.

After three months in Europe we sailed from Southampton in the steamship *St. Louis*. On this steamer there was a fine library that had been presented to the ship by the citizens of St. Louis, Mo. In this library I found a life of Frederick Douglass, which I began reading. I became especially interested in Mr. Douglass's description of the way he was treated on shipboard during his first or second visit to England. In this description, he told how he was not permitted to enter the cabin, but had to confine himself to the deck of the ship. A few minutes after I had finished reading this description I was waited on by a committee of ladies and gentlemen with the request that I deliver an address at a concert which was to be given the following evening. And yet there are people who are bold enough to say that race feeling in America is not growing less intense! At this concert the Hon. Benjamin B. Odell, Jr., the present governor of New York, presided. I was never given a more cordial hearing anywhere. A large proportion of the passengers were Southern people. After the concert some of the passengers proposed that a subscription be raised to help the work at Tuskegee, and the money to support several scholarships was the result.

While we were in Paris I was very pleasantly surprised to receive the following invitation from the citizens of West Virginia and of the city near which I had spent my boyhood days:—

CHARLESTON, W. VA., MAY 16, 1899.

PROFESSOR BOOKER T. WASHINGTON, PARIS, FRANCE.

DEAR SIR.: Many of the best citizens of West Virginia have united in liberal expressions of admiration and praise of your worth and work, and desire that on your return from Europe you should favour them with your presence and with the inspiration of your words. We most sincerely indorse this move, and on behalf of the citizens of Charleston extend to you our most cordial invitation to have you come to us, that we may honour you who have done so much by your life and work to honour us.

We are,   Very truly yours,
THE COMMON COUNCIL OF THE CITY OF CHARLESTON
By W. HERMAN SMITH, Mayor.

This invitation from the City Council of Charleston was accompanied by the following:—

PROFESSOR BOOKER T. WASHINGTON, PARIS, FRANCE.

DEAR SIR: We, the citizens of Charleston and West Virginia, desire to express our pride in you and the splendid career that you have thus far accomplished, and ask that we be permitted to show our pride and interest in a substantial way.

Your recent visit to your old home in our midst awoke within us the keenest regret that we were not permitted to hear you and render some substantial aid to your work before you left for Europe.

In view of the foregoing, we earnestly invite you to share the hospitality of our city upon your return from Europe, and give us the opportunity to hear you and put ourselves in touch with your work in a way that will be most gratifying to yourself, and that we may receive the inspiration of your words and presence.

An early reply to this invitation, with an indication of the time we may reach our city, will greatly oblige,

Yours very respectfully,

The Charleston *Daily Gazette;* The *Daily Mail Tribune;* G. W. Atkinson, Governor; E. L. Boggs, Secretary to Governor; Wm. M. O. Dawson, Secretary of State; L. M. La Follette, Auditor; J. R. Trotter, Superintendent of Schools; E. W. Wilson, ex-Governor; W. A. MacCorkle, ex-Governor; John Q. Dickinson, President Kanawha Valley Bank; L. Prichard, President Charleston National Bank; Geo. S. Couch, President Kanawha National Bank; Ed. Reid, Cashier Kanawha National Bank; Gen. S. Laidley, Superintendent City Schools; L. E. McWhorter, President Board of Education; Chas. K. Payne, wholesale merchant; and many others.

This invitation, coming as it did from the City Council, the state officers, and all the substantial citizens of both races of the community where I had spent my boyhood, and from which I had gone a few years before, unknown, in poverty and ignorance, in quest of an education, not only sur-

prised me, but almost unmanned me. I could not understand what I had done to deserve it all.

I accepted the invitation, and at the appointed day
was met at the railway station at Charleston by a committee headed by ex-Governor W. A. MacCorkle, and
composed of men of both races. The public reception was
held in the Opera-House at Charleston. The Governor of
the state, the Hon. George W. Atkinson, presided, and an
address of welcome was made by ex-Governor MacCorkle.
A prominent part in the reception was taken by the
coloured citizens. The Opera-House was filled with citizens
of both races, and among the white people were many for
whom I had worked when a boy. The next day Governor and Mrs. Atkinson gave me a public reception at the
State House, which was attended by all classes.

Not long after this the coloured people in Atlanta,
Georgia, gave me a reception at which the Governor of
the state presided, and a similar reception was given me
in New Orleans, which was presided over by the Mayor
of the city. Invitations came from many other places which
I was not able to accept.

CHAPTER XVII

Last Words

Before going to Europe some events came into my life
which were great surprises to me. In fact, my whole life
has largely been one of surprises. I believe that any man's
life will be filled with constant, unexpected encourage-
ments of this kind if he makes up his mind to do his
level best each day of his life—that is, tries to make each
day reach as nearly as possible the high-water mark of
pure, unselfish, useful living. I pity the man, black or
white, who has never experienced the joy and satisfaction
that come to one by reason of an effort to assist in making some one else more useful and more happy.

Six months before he died, and nearly a year after he
had been stricken with paralysis, General Armstrong expressed a wish to visit Tuskegee again before he passed
away. Notwithstanding the fact that he had lost the use
of his limbs to such an extent that he was practically

helpless, his wish was gratified, and he was brought to Tuskegee. The owners of the Tuskegee Railroad, white men living in the town, offered to run a special train without cost, out to the main station—Chehaw, five miles away—to meet him. He arrived on the school grounds about nine o'clock in the evening. Some one had suggested that we give the General "a pine-knot torchlight reception." This plan was carried out, and the moment that his carriage entered the school grounds he began passing between two lines of lighted and waving "fat pine" wood knots held by over a thousand students and teachers. The whole thing was so novel and surprising that the General was completely overcome with happiness. He remained a guest in my home for nearly two months, and, although almost wholly without the use of voice or limb, he spent nearly every hour in devising ways and means to help the South. Time and time again he said to me, during this visit, that it was not only the duty of the country to assist in elevating the Negro of the South, but the poor white man as well. At the end of his visit I resolved anew to devote myself more earnestly than ever to the cause which was so near his heart. I said that if a man in his condition was willing to think, work, and act, I should not be wanting in furthering in every possible way the wish of his heart.

The death of General Armstrong, a few weeks later, gave me the privilege of getting acquainted with one of the finest, most unselfish, and most attractive men that I have ever come in contact with. I refer to the Rev. Dr. Hollis B. Frissell, now the Principal of the Hampton Institute, and General Armstrong's successor. Under the clear, strong, and almost perfect leadership of Dr. Frissell, Hampton has had a career of prosperity and usefulness that is all that the General could have wished for. It seems to be the constant effort of Dr. Frissell to hide his own great personality behind that of General Armstrong—to make himself of "no reputation" for the sake of the cause.

More than once I have been asked what was the greatest surprise that ever came to me. I have little hesitation in answering that question. It was the following letter, which came to me one Sunday morning when I was sitting on the veranda of my home at Tuskegee, surrounded by my wife and three children:—

HARVARD UNIVERSITY, CAMBRIDGE, MAY 28, 1896.

PRESIDENT BOOKER T. WASHINGTON,

MY DEAR SIR: Harvard University desires to confer
on you at the approaching Commencement an honorary degree; but it is our custom to confer degrees
only on gentlemen who are present. Our Commencement occurs this year on June 24, and your presence
would be desirable from about noon till about five
o'clock in the afternoon. Would it be possible for
you to be in Cambridge on that day?

Believe me, with great regard,

Very truly yours,
CHARLES W. ELIOT.

This was a recognition that had never in the slightest
manner entered into my mind, and it was hard for me
to realize that I was to be honoured by a degree from the
oldest and most renowned university in America. As I
sat upon my veranda, with this letter in my hand, tears
came into my eyes. My whole former life—my life as a
slave on the plantation, my work in the coal-mine, the
times when I was without food and clothing, when I
made my bed under a sidewalk, my struggles for an education, the trying days I had had at Tuskegee, days when
I did not know where to turn for a dollar to continue
the work there, the ostracism and sometimes oppression
of my race,—all this passed before me and nearly overcame me.

I had never sought or cared for what the world calls
fame. I have always looked upon fame as something to be
used in accomplishing good. I have often said to my
friends that if I can use whatever prominence may have
come to me as an instrument with which to do good, I
am content to have it. I care for it only as a means to be
used for doing good, just as wealth may be used. The
more I come into contact with wealthy people, the more
I believe that they are growing in the direction of looking upon their money simply as an instrument which
God has placed in their hand for doing good with. I
never go to the office of Mr. John D. Rockefeller, who
more than once has been generous to Tuskegee, without

being reminded of this. The close, careful, and minute investigation that he always makes in order to be sure that every dollar that he gives will do the most good—an investigation that is just as searching as if he were investing money in a business enterprise—convinces me that the growth in this direction is most encouraging.

At nine o'clock, on the morning of June 24, I met President Eliot, the Board of Overseers of Harvard University, and the other guests, at the designated place on the university grounds, for the purpose of being escorted to Sanders Theatre, where the Commencement exercises were to be held and degrees conferred. Among others invited to be present for the purpose of receiving a degree at this time were General Nelson A. Miles, Dr. Bell, the inventor of the Bell telephone, Bishop Vincent, and the Rev. Minot J. Savage. We were placed in line immediately behind the President and the Board of Overseers, and directly afterward the Governor of Massachusetts, escorted by the Lancers, arrived and took his place in the line of march by the side of President Eliot. In the line there were also various other officers and professors, clad in cap and gown. In this order we marched to Sanders Theatre, where, after the usual Commencement exercises, came the conferring of the honorary degrees. This, it seems, is always considered the most interesting feature at Harvard. It is not known, until the individuals appear, upon whom the honorary degrees are to be conferred, and those receiving these honours are cheered by the students and others in proportion to their popularity. During the conferring of the degrees excitement and enthusiasm are at the highest pitch.

When my name was called, I rose, and President Eliot, in beautiful and strong English, conferred upon me the degree of Master of Arts. After these exercises were over, those who had received honorary degrees were invited to lunch with the President. After the lunch we were formed in line again, and were escorted by the Marshal of the day, who that year happened to be Bishop William Lawrence, through the grounds, where, at different points, those who had been honoured were called by name and received the Harvard yell. This march ended at Memorial Hall, where the alumni dinner was served. To see over a thousand strong men, representing all that is best in State, Church, business, and education, with the glow and enthusiasm of college loyalty and college pride,—which has, I think, a peculiar Harvard flavour,—is a sight that does not easily fade from memory.

Among the speakers after dinner were President Eliot, Governor Roger Wolcott, General Miles, Dr. Minot J. Savage, the Hon. Henry Cabot Lodge, and myself. When I was called upon, I said, among other things:—

It would in some measure relieve my embarrassment if I could, even in a slight degree, feel myself worthy of the great honour which you do me to-day. Why you have called me from the Black Belt of the South, from among my humble people, to share in the honors of this occasion, is not for me to explain; and yet it may not be inappropriate for me to suggest that it seems to me that one of the most vital questions that touch our American life is how to bring the strong, wealthy, and learned into helpful touch with the poorest, most ignorant, and humblest, and at the same time make one appreciate the vitalizing, strengthening influence of the other. How shall we make the mansions on yon Beacon Street feel and see the need of the spirits in the lowliest cabin in Alabama cotton-fields or Louisiana sugar-bottoms? This problem Harvard University is solving, not by bringing itself down, but by bringing the masses up.

If my life in the past has meant anything in the lifting up of my people and the bringing about of better relations between your race and mine, I assure you from this day it will mean doubly more. In the economy of God there is but one standard by which an individual can succeed—there is but one for a race. This country demands that every race shall measure itself by the American standard. By it a race must rise or fall, succeed or fail, and in the last analysis mere sentiment counts for little. During the next half-century and more, my race must continue passing through the severe American crucible. We are to be tested in our patience, our forbearance, our perseverance, our power to endure wrong, to withstand temptations, to economize, to acquire and use skill; in our ability to compete, to succeed in commerce, to disregard the superficial for the real, the appearance for the substance, to be great and yet small, learned and yet simple, high and yet the servant of all.

As this was the first time that a New England university had conferred an honorary degree upon a Negro, it was the occasion of much newspaper comment throughout the country. A correspondent of a New York paper said:—

When the name of Booker T. Washington was called, and he arose to acknowledge and accept, there was such an outburst of applause as greeted no other name except that of the popular soldier patriot, General Miles. The applause was not studied and stiff, sympathetic and condoling; it was enthusiasm and admiration. Every part of the audience from pit to gallery joined in, and a glow covered the cheeks of those around me, proving sincere appreciation of the rising struggle of an ex-slave and the work he has accomplished for his race.

A Boston paper said, editorially:—

In conferring the honorary degree of Master of Arts upon the Principal of Tuskegee Institute, Harvard University has honoured itself as well as the object of this distinction. The work which Professor Booker T. Washington has accomplished for the education, good citizenship and popular enlightenment in his chosen field of labour in the South entitles him to rank with our national benefactors. The university which can claim him on its list of sons, whether in regular course or *honoris causa,* may be proud.

It has been mentioned that Mr. Washington is the first of his race to receive an honorary degree from a New England university. This, in itself, is a distinction, but the degree was not conferred because Mr. Washington is a coloured man, or because he was born in slavery, but because he has shown, by his work for the elevation of the people of the Black Belt of the South, a genius and a broad humanity which count for greatness in any man, whether his skin be white or black.

Another Boston paper said:—

It is Harvard which, first among New England colleges, confers an honorary degree upon a black man.

No one who has followed the history of Tuskegee and its work can fail to admire the courage, persistence, and splendid common sense of Booker T. Washington. Well may Harvard honour the ex-slave, the value of whose services, alike to his race and country, only the future can estimate.

The correspondent of the New York *Times* wrote:—

All the speeches were enthusiastically received, but the coloured man carried off the oratorical honours, and the applause which broke out when he had finished was vociferous and long-continued.

Soon after I began work at Tuskegee I formed a resolution, in the secret of my heart, that I would try to build up a school that would be of so much service to the country that the President of the United States would one day come to see it. This was, I confess, rather a bold resolution, and for a number of years I kept it hidden in my own thoughts, not daring to share it with any one.

In November, 1897, I made the first move in this direction, and that was in securing a visit from a member of President McKinley's Cabinet, the Hon. James Wilson, Secretary of Agriculture. He came to deliver an address at the formal opening of the Slater-Armstrong Agricultural Building, our first large building to be used for the purpose of giving training to our students in agriculture and kindred branches.

In the fall of 1898 I heard that President McKinley was likely to visit Atlanta, Georgia, for the purpose of taking part in the Peace Jubilee exercises to be held there to commemorate the successful close of the Spanish-American war. At this time I had been hard at work, together with our teachers, for eighteen years, trying to build up a school that we thought would be of service to the Nation, and I determined to make a direct effort to secure a visit from the President and his Cabinet. I went to Washington, and I was not long in the city before I found my way to the White House. When I got there I found the waiting rooms full of people, and my heart began to sink, for I feared there would not be much chance of my seeing the President that day, if at all. But, at any rate, I got an opportunity to see Mr. J. Addison

Porter, the secretary to the President, and explained to him my mission. Mr. Porter kindly sent my card directly to the President, and in a few minutes word came from Mr. McKinley that he would see me.

How any man can see so many people of all kinds, with all kinds of errands, and do so much hard work, and still keep himself calm, patient, and fresh for each visitor in the way that President McKinley does, I cannot understand. When I saw the President he kindly thanked me for the work which we were doing at Tuskegee for the interests of the country. I then told him, briefly, the object of my visit. I impressed upon him the fact that a visit from the Chief Executive of the Nation would not only encourage our students and teachers, but would help the entire race. He seemed interested, but did not make a promise to go to Tuskegee, for the reason that his plans about going to Atlanta were not then fully made; but he asked me to call the matter to his attention a few weeks later.

By the middle of the following month the President had definitely decided to attend the Peace Jubilee at Atlanta. I went to Washington again and saw him, with a view of getting him to extend his trip to Tuskegee. On this second visit Mr. Charles W. Hare, a prominent white citizen of Tuskegee, kindly volunteered to accompany me, to reënforce my invitation with one from the white people of Tuskegee and the vicinity.

Just previous to my going to Washington the second time, the country had been excited, and the coloured people greatly depressed, because of several severe race riots which had occurred at different points in the South. As soon as I saw the President, I perceived that his heart was greatly burdened by reason of these race disturbances. Although there were many people waiting to see him, he detained me for some time, discussing the condition and prospects of the race. He remarked several times that he was determined to show his interest and faith in the race, not merely in the words, but by acts. When I told him that I thought that at that time scarcely anything would go farther in giving hope and encouragement to the race than the fact that the President of the Nation would be willing to travel one hundred and forty miles out of his way to spend a day at a Negro institution, he seemed deeply impressed.

While I was with the President, a white citizen of
Atlanta, a Democrat and an ex-slaveholder, came into
the room, and the President asked his opinion as to the
wisdom of his going to Tuskegee. Without hesitation the
Atlanta man replied that it was the proper thing for him
to do. This opinion was reënforced by that friend of
the race, Dr. J. L. M. Curry. The President promised that
he would visit our school on the 16th of December.

When it became known that the President was going
to visit our school, the white citizens of the town of
Tuskegee—a mile distant from the school—were as much
pleased as were our students and teachers. The white
people of the town, including both men and women,
began arranging to decorate the town, and to form themselves into committees for the purpose of coöperating
with the officers of our school in order that the distinguished visitor might have a fitting reception. I think I
never realized before this how much the white people of
Tuskegee and vicinity thought of our institution. During the days when we were preparing for the President's
reception, dozens of these people came to me and said
that, while they did not want to push themselves into
prominence, if there was anything they could do to
help, or to relieve me personally, I had but to intimate
it and they would be only too glad to assist. In fact, the
thing that touched me almost as deeply as the visit of
the President itself was the deep pride which all classes
of citizens in Alabama seemed to take in our work.

The morning of December 16th brought to the little
city of Tuskegee such a crowd as it had never seen before. With the President came Mrs. McKinley and all of
the Cabinet officers but one; and most of them brought
their wives or some members of their families. Several
prominent generals came, including General Shafter and
General Joseph Wheeler, who were recently returned from
the Spanish-American war. There was also a host of
newspaper correspondents. The Alabama Legislature was
in session at Montgomery at this time. This body passed
a resolution to adjourn for the purpose of visiting
Tuskegee. Just before the arrival of the President's party
the Legislature arrived, headed by the governor and other
state officials.

The citizens of Tuskegee had decorated the town from
the station to the school in a generous manner. In order

to economize in the matter of time, we arranged to
have the whole school pass in review before the President. Each student carried a stalk of sugar-cane with
some open bolls of cotton fastened to the end of it.
Following the students the work of all departments of
the school passed in review, displayed on "floats" drawn
by horses, mules, and oxen. On these floats we tried
to exhibit not only the present work of the school, but
to show the contrasts between the old methods of doing
things and the new. As an example, we showed the
old method of dairying in contrast with the improved
methods, the old methods of tilling the soil in contrast
with the new, the old methods of cooking and housekeeping in contrast with the new. These floats consumed
an hour and a half of time in passing.

In his address in our large, new chapel, which the
students had recently completed, the President said,
among other things:—

To meet you under such pleasant auspices and to
have the opportunity of a personal observation of your
work is indeed most gratifying. The Tuskegee Normal
and Industrial Institute is ideal in its conception, and
has already a large and growing reputation in the
country, and is not unknown abroad. I congratulate
all who are associated in this undertaking for the
good work which it is doing in the education of its
students to lead lives of honour and usefulness, thus
exalting the race for which it was established.

Nowhere, I think, could a more delightful location
have been chosen for this unique educational experiment, which has attracted the attention and won the
support even of conservative philanthropists in all
sections of the country.

To speak of Tuskegee without paying special tribute
to Booker T. Washington's genius and perseverance
would be impossible. The inception of this noble enterprise was his, and he deserves high credit for it.
His was the enthusiasm and enterprise which made its
steady progress possible and established in the institution its present high standard of accomplishment.
He has won a worthy reputation as one of the great
leaders of his race, widely known and much respected

at home and abroad as an accomplished educator, a great orator, and a true philanthropist.

The Hon. John D. Long, the Secretary of the Navy, said in part:—

I cannot make a speech to-day. My heart is too full—full of hope, admiration, and pride for my countrymen of both sections and both colours. I am filled with gratitude and admiration for your work, and from this time forward I shall have absolute confidence in your progress and in the solution of the problem in which you are engaged.

The problem, I say, has been solved. A picture has been presented to-day which should be put upon canvas with the pictures of Washington and Lincoln, and transmitted to future time and generations—a picture which the press of the country should spread broadcast over the land, a most dramatic picture, and that picture in this: The President of the United States standing on this platform, on one side the Governor of Alabama, on the other, completing the trinity, a representative of a race only a few years ago in bondage, the coloured President of the Tuskegee Normal and Industrial Institute.

God bless the President under whose majesty such a scene as that is presented to the American people. God bless the state of Alabama, which is showing that it can deal with this problem for itself. God bless the orator, philanthropist, and disciple of the Great Master—who, if he were on earth, would be doing the same work—Booker T. Washington.

Postmaster General Smith closed the address which he made with these words:—

We have witnessed many spectacles within the last few days. We have seen the magnificent grandeur and the magnificent achievements of one of the great metropolitan cities of the South. We have seen heroes of the war pass by in procession. We have seen floral parades. But I am sure my colleagues will agree with me in saying that we have witnessed no spectacle more impressive and more encouraging, more inspiring for our future, than that which we have witnessed here this morning.

Some days after the President returned to Washington I received the letter which follows:—

EXECUTIVE MANSION, WASHINGTON, DEC. 23, 1899.

DEAR SIR: By this mail I take pleasure in sending you engrossed copies of the souvenir of the visit of the President to your institution. These sheets bear the autographs of the President and the members of the Cabinet who accompanied him on the trip. Let me take this opportunity of congratulating you most heartily and sincerely upon the great success of the exercises provided for and entertainment furnished us under your auspices during our visit to Tuskegee. Every feature of the programme was perfectly executed and was viewed or participated in with the heartiest satisfaction by every visitor present. The unique exhibition which you gave of your pupils engaged in their industrial vocations was not only artistic but thoroughly impressive. The tribute paid by the President and his Cabinet to your work was none too high, and forms a most encouraging augury, I think, for the future prosperity of your institution. I cannot close without assuring you that the modesty shown by yourself in the exercises was most favourably commented upon by all the members of our party.

With best wishes for the continued advance of your most useful and patriotic undertaking, kind personal regards and the compliments of the season, believe me, always.

Very sincerely yours,
JOHN ADDISON PORTER.
Secretary to the President.

To PRESIDENT BOOKER T. WASHINGTON, Tuskegee Normal and Industrial Institute, Tuskegee, Ala.

Twenty years have now passed since I made the first humble effort at Tuskegee, in a broken-down shanty and an old hen-house, without owning a dollar's worth of property, and with but one teacher and thirty students. At the present time the institution owns twenty-three hundred acres of land, one thousand of which are under cultivation each year, entirely by student labour. There are now upon the grounds, counting large and small, sixty-six buildings; and all except four of these have been almost wholly erected by the labour of our students. While the students are at work upon the land and in erecting buildings, they are taught, by competent instructors, the latest methods of agriculture and the trades connected with building.

There are in constant operation at the school, in connection with thorough academic and religious training, thirty industrial departments. All of these teach industries at which our men and women can find immediate employment as soon as they leave the institution. The only difficulty now is that the demand for our graduates from both white and black people in the South is so great that we cannot supply more than one-half the persons for whom applications come to us. Neither have we the buildings nor the money for current expenses to enable us to admit to the school more than one-half the young men and women who apply to us for admission.

In our industrial teaching we keep three things in mind: first, that the student shall be so educated that he shall be enabled to meet conditions as they exist *now*, in the part of the South where he lives—in a word, to be able to do the thing which the world wants done; second, that every student who graduates from the school shall have enough skill, coupled with intelligence and moral character, to enable him to make a living for himself and others; third, to send every graduate out feeling and knowing that labour is dignified and beautiful—to make each one love labour instead of trying to escape it. In addition to the agricultural training which we give to young men, and the training given to our girls in all the usual domestic employments, we now train a number of girls in agriculture each year. These girls are taught gardening, fruit-growing, dairying, bee-culture, and poultry-raising.

While the institution is in no sense denominational,

we have a department known as the Phelps Hall Bible Training School, in which a number of Students are prepared for the ministry and other forms of Christian work, especially work in the country districts. What is equally important, each one of these students works half of each day at some industry, in order to get skill and the love of work, so that when he goes out from the institution he is prepared to set the people with whom he goes to labour a proper example in the matter of industry.

The value of our property is now over $700,000. If we add to this our endowment fund, which at present is $1,000,000, the value of the total property is now $1,700,000. Aside from the need for more buildings and for money for current expenses, the endowment fund should be increased to at least $3,000,000. The annual current expenses are now about $150,000. The greater part of this I collect each year by going from door to door and from house to house. All of our property is free from mortgage, and is deeded to an undenominational board of trustees who have the control of the institution.

From thirty students the number has grown to fourteen hundred, coming from twenty-seven states and territories, from Africa, Cuba, Puerto Rico, Jamaica, and other foreign countries. In our departments there are one hundred and ten officers and instructors; and if we add the families of our instructors, we have a constant population upon our grounds of not far from seventeen hundred people.

I have often been asked how we keep so large a body of people together, and at the same time keep them out of mischief. There are two answers: that the men and women who come to us for an education are in earnest; and that everybody is kept busy. The following outline of our daily work will testify to this:—

5 A.M., rising bell; 5.50 A.M., warning breakfast bell; 6 A.M., breakfast bell; 6.20 A.M., breakfast over; 6.20 to 6.50 A.M., rooms are cleaned; 6.50, work bell; 7.30, morning study hour; 8.20, morning school bell; 8.25, inspection of young men's toilet in ranks; 8.40, devotional exercises in chapel; 8.55, "five minutes with the daily news"; 9 A.M., class work begins; 12, class work

closes; 12.15 P.M., dinner; 1 P.M., work bell; 1.30 P.M., class work begins; 3.30 P.M., class work ends; 5.30 P.M., bell to "knock off" work; 6 P.M., supper; 7.10 P.M., evening prayers; 7.30 P.M., evening study hour; 8.45 P.M., evening study hour closes; 9.20 P.M., warning retiring bell; 9.30 P.M., retiring bell.

We try to keep constantly in mind the fact that the worth of the school is to be judged by its graduates. Counting those who have finished the full course, together with those who have taken enough training to enable them to do reasonably good work, we can safely say that at least six thousand men and women from Tuskegee are now at work in different parts of the South; men and women who, by their own example or by direct effort, are showing the masses of our race how to improve their material, educational, and moral and religious life. What is equally important, they are exhibiting a degree of common sense and self-control which is causing better relations to exist between the races and is causing the Southern white man to learn to believe in the value of educating the men and women of my race. Aside from this, there is the influence that is constantly being exerted through the mothers' meeting and the plantation work conducted by Mrs. Washington.

Wherever our graduates go, the changes which soon begin to appear in the buying of land, improving homes, saving money, in education, and in high moral character are remarkable. Whole communities are fast being revolutionized through the instrumentality of these men and women.

Ten years ago I organized at Tuskegee the first Negro Conference. This is an annual gathering which now brings to the school eight or nine hundred representative men and women of the race, who come to spend a day in finding out what the actual industrial, mental, and moral conditions of the people are, and in forming plans for improvement. Out from this central Negro Conference at Tuskegee have grown numerous state and local conferences which are doing the same kind of work. As a result of the influence of these gatherings, one delegate reported at the last annual meeting that ten families in his community had bought and paid for homes. On the day following the annual Negro Conference, there is held the "Workers' Conference." This is composed of officers and teachers who are engaged in educational work in the larger institutions in the South. The Negro Conference furnishes a rare opportunity for these workers to study the real condition of the rank and file of the people.

In the summer of 1900, with the assistance of such prominent coloured men as Mr. T. Thomas Fortune, who has always upheld my hands in every effort, I organized the National Negro Business League, which held its first meeting in Boston, and brought together for the first time a large number of the coloured men who are engaged in various lines of trade or business in different parts of the United States. Thirty states were represented at our first meeting. Out of this national meeting grew state and local business leagues.

In addition to looking after the executive side of the work at Tuskegee, and raising the greater part of the money for the support of the school, I cannot seem to escape the duty of answering at least a part of the calls which come to me unsought to address Southern white audiences and audiences of my own race, as well as frequent gatherings in the North. As to how much of my time is spent in this way, the following clipping from a Buffalo (N.Y.) paper will tell. This has reference to an occasion when I spoke before the National Educational Association in that city.

Booker T. Washington, the foremost educator among the coloured people of the world, was a very busy man from the time he arrived in the city the other night from the West and registered at the Iroquois. He had hardly removed the stains of travel when it was time to partake of supper. Then he held a public levee in the parlours of the Iroquois until eight o'clock. During that time he was greeted by over two hundred eminent teachers and educators from all parts of the United States. Shortly after eight o'clock he was driven in a carriage to Music Hall, and in one hour and a half he made two ringing addresses, to as many as five thousand people, on Negro education. Then Mr. Washington was taken in charge by a delegation of coloured citizens, headed by the Rev. Mr. Watkins, and hustled

off to a small informal reception, arranged in honour
of the visitor by the people of his race.

Nor can I, in addition to making these addresses, escape the duty of calling the attention of the South and
of the country in general, through the medium of the
press, to matters that pertain to the interests of both
races. This, for example, I have done in regard to the
evil habit of lynching. When the Louisiana State Constitutional Convention was in session, I wrote an open
letter to that body pleading for justice for the race. In
all such efforts I have received warm and hearty support
from the Southern newspapers, as well as from those in
all other parts of the country.

Despite superficial and temporary signs which might
lead one to entertain a contrary opinion, there was never
a time when I felt more hopeful for the race than I do
at the present. The great human love that in the end
recognizes and rewards merit is everlasting and universal. The outside world does not know, neither can it
appreciate, the struggle that is constantly going on in the
hearts of both the Southern white people and their former
slaves to free themselves from racial prejudice; and while
both races are thus struggling they should have the sympathy, the support, and the forbearance of the rest of the
world.

As I write the closing words of this autobiography I
find myself—not by design—in the city of Richmond,
Virginia: the city which only a few decades ago was the
capital of the Southern Confederacy, and where, about
twenty-five years ago, because of my poverty I slept night
after night under a sidewalk.

This time I am in Richmond as the guest of the coloured people of the city; and came at their request to
deliver an address last night to both races in the Academy
of Music, the largest and finest audience room in the
city. This was the first time that the coloured people had
ever been permitted to use this hall. The day before I
came, the City Council passed a vote to attend the meeting in a body to hear me speak. The state Legislature,
including the House of Delegates and the Senate, also
passed a unanimous vote to attend in a body. In the

presence of hundreds of coloured people, many distinguished white citizens, the City Council, the state Legislature, and state officials, I delivered my message, which was one of hope and cheer; and from the bottom of my heart I thanked both races for this welcome back to the state that gave me birth.

The Souls of
Black Folk

THE FORETHOUGHT

Herein lie buried many things which if read with patience may show the strange meaning of being black here at the dawning of the Twentieth Century. This meaning is not without interest to you, Gentle Reader; for the problem of the Twentieth Century is the problem of the color line. I pray you, then, receive my little book in all charity, studying my words with me, forgiving mistake and foible for sake of the faith and passion that is in me, and seeking the grain of truth hidden there.

I have sought here to sketch, in vague, uncertain outline, the spiritual world in which ten thousand Americans live and strive. First, in two chapters I have tried to show what emancipation meant to them, and what was its aftermath. In a third chapter I have pointed out the slow rise of personal leadership, and criticised candidly the leader who bears the chief burden of his race to-day. Then, in two other chapters I have sketched in swift outline the two worlds within and without the Veil, and thus have come to the central problem of training men for life. Venturing now into deeper detail, I have in two chapters studied the struggles of the massed millions of the black peasantry, and in another have sought to make clear the present relations of the sons of master and man. Leaving, then, the white world, I have stepped within the Veil, raising it that you may view faintly its deeper recesses,—the meaning of its religion, the passion of its human sorrow, and the struggle of its greater souls. All this I have ended with a tale twice told but seldom written, and a chapter of song.

Some of these thoughts of mine have seen the light before in other guise. For kindly consenting to their republication here, in altered and extended form, I must thank the publishers of the *Atlantic Monthly, The World's Work,* the *Dial, The New World,* and the *Annals of the American Academy of Political and Social Science.* Before each chapter, as now printed, stands a bar of the Sorrow Songs, —some echo of haunting melody from the only American music which welled up from black souls in the dark past. And, finally, need I add that I who speak here am bone of the bone and flesh of the flesh of them that live within the Veil?

ATLANTA, GA., FEB. 1, 1903.   W. E. B. Du B.

TO

BURGHARDT AND YOLANDE

THE LOST AND THE FOUND

I

Of Our Spiritual Strivings

O water, voice of my heart, crying in the sand,
  All night long crying with a mournful cry,
As I lie and listen, and cannot understand
    The voice of my heart in my side or the voice
        of the sea,
  O water, crying for rest, is it I, is it I?
  All night long the water is crying to me.

Unresting water, there shall never be rest
  Till the last moon droop and the last tide fail,
And the fire of the end begin to burn in the west;
    And the heart shall be weary and wonder and cry
        like the sea,
  All life long crying without avail,
    As the water all night long is crying to me.

ARTHUR SYMONS.

Between me and the other world there is ever an unasked question: unasked by some through feelings of
delicacy; by others through the difficulty of rightly framing it. All, nevertheless, flutter round it. They approach
me in a half-hesitant sort of way, eye me curiously or
compassionately, and then, instead of saying directly,
How does it feel to be a problem? they say, I know an
excellent colored man in my town; or, I fought at Me-
chanicsville; or, Do not these Southern outrages make
your blood boil? At these I smile, or am interested, or
reduce the boiling to a simmer, as the occasion may require. To the real question, How does it feel to be a
problem? I answer seldom a word.

And yet, being a problem is a strange experience,—
peculiar even for one who has never been anything else,
save perhaps in babyhood and in Europe. It is in the early
days of rollicking boyhood that the revelation first bursts
upon one, all in a day, as it were. I remember well
when the shadow swept across me. I was a little thing,
away up in the hills of New England, where the dark
Housatonic winds between Hoosac and Taghkanic to the
sea. In a wee wooden schoolhouse, something put it into
the boys' and girls' heads to buy gorgeous visiting-cards
—ten cents a package—and exchange. The exchange was
merry, till one girl, a tall newcomer, refused my card,—
refused it peremptorily, with a glance. Then it dawned
upon me with a certain suddenness that I was different
from the others; or like, mayhap, in heart and life and
longing, but shut out from their world by a vast veil. I
had thereafter no desire to tear down that veil, to creep
through; I held all beyond it in common contempt, and
lived above it in a region of blue sky and great wandering shadows. That sky was bluest when I could beat my
mates at examination-time, or beat them at a foot-race,
or even beat their stringy heads. Alas, with the years all
this fine contempt began to fade; for the worlds I longed
for, and all their dazzling opportunities, were theirs, not
mine. But they should not keep these prizes, I said; some,
all, I would wrest from them. Just how I would do it I
could never decide: by reading law, by healing the sick,
by telling the wonderful tales that swam in my head,—
some way. With other black boys the strife was not so
fiercely sunny: their youth shrunk into tasteless syco-
phancy, or into silent hatred of the pale world about
them and mocking distrust of everything white; or wasted
itself in a bitter cry, Why did God make me an outcast
and a stranger in mine own house? The shades of the
prison-house closed round about us all: walls strait and
stubborn to the whitest, but relentlessly narrow, tall, and
unscalable to sons of night who must plod darkly on in
resignation, or beat unavailing palms against the stone,
or steadily, half hopelessly, watch the streak of blue
above.

After the Egyptian and Indian, the Greek and Roman,
the Teuton and Mongolian, the Negro is a sort of seventh
son, born with a veil, and gifted with second-sight in
this American world,—a world which yields him no true

self-consciousness, but only lets him see himself through the revelation of the other world. It is a peculiar sensation, this double-consciousness, this sense of always looking at one's self through the eyes of others, of measuring one's soul by the tape of a world that looks on in amused contempt and pity. One ever feels his twoness,—an American, a Negro; two souls, two thoughts, two unreconciled strivings; two warring ideals in one dark body, whose dogged strength alone keeps it from being torn asunder.

The history of the American Negro is the history of this strife,—this longing to attain self-conscious manhood, to merge his double self into a better and truer self. In this merging he wishes neither of the older selves to be lost. He would not Africanize America, for America has too much to teach the world and Africa. He would not bleach his Negro soul in a flood of white Americanism, for he knows that Negro blood has a message for the world. He simply wishes to make it possible for a man to be both a Negro and an American, without being cursed and spit upon by his fellows, without having the doors of Opportunity closed roughly in his face.

This, then, is the end of his striving; to be a co-worker in the kingdom of culture, to escape both death and isolation, to husband and use his best powers and his latent genius. These powers of body and mind have in the past been strangely wasted, dispersed, or forgotten. The shadow of a mighty Negro past flits through the tale of Ethiopia the Shadowy and of Egypt the Sphinx. Throughout history, the powers of single black men flash here and there like falling stars, and die sometimes before the world has rightly gauged their brightness. Here in America, in the few days since Emancipation, the black man's turning hither and thither in hesitant and doubtful striving has often made his very strength to lose effectiveness, to seem like absence of power, like weakness. And yet it is not weakness,—it is the contradiction of double aims. The double-aimed struggle of the black artisan—on the one hand to escape white contempt for a nation of mere hewers of wood and drawers of water, and on the other hand to plough and nail and dig for a poverty-stricken horde—could only result in making him a poor craftsman, for he had but half a heart in either cause. By the poverty and ignorance of his people, the

Negro minister or doctor was tempted toward quackery
and demagogy; and by the criticism of the other world,
toward ideals that made him ashamed of his lowly tasks.
The would-be black *savant* was confronted by the paradox
that the knowledge his people needed was a twice-told
tale to his white neighbors, while the knowledge which
would teach the white world was Greek to his own flesh
and blood. The innate love of harmony and beauty that set the ruder souls of his people a-dancing and
a-singing raised but confusion and doubt in the soul of
the black artist; for the beauty revealed to him was the
soul-beauty of a race which his larger audience despised,
and he could not articulate the message of another people. This waste of double aims, this seeking to satisfy
two unreconciled ideals, has wrought sad havoc with the
courage and faith and deeds of ten thousand thousand
people,—has sent them often wooing false gods and in-
voking false means of salvation, and at times has even
seemed about to make them ashamed of themselves.

Away back in the days of bondage they thought to
see in one divine event the end of all doubt and disappointment; few men ever worshipped Freedom with half
such unquestioning faith as did the American Negro for
two centuries. To him, so far as he thought and dreamed,
slavery was indeed the sum of all villainies, the cause
of all sorrow, the root of all prejudice; Emancipation was
the key to a promised land of sweeter beauty than ever
stretched before the eyes of wearied Israelites. In song and
exhortation swelled one refrain—Liberty; in his tears and
curses the God he implored had Freedom in his right
hand. At last it came,—suddenly, fearfully, like a dream.
With one wild carnival of blood and passion came the
message in his own plaintive cadences:—

> "Shout, O children!
>   Shout, you're free!
> For God has bought your liberty!"

Years have passed away since then,—ten, twenty, forty;
forty years of national life, forty years of renewal and
development, and yet the swarthy spectre sits in its accustomed seat at the Nation's feast. In vain do we cry to
this our vastest social problem:—

"Take any shape but that, and my firm nerves
   Shall never tremble!"

The Nation has not yet found peace from its sins; the
freedman has not yet found in freedom his promised
land. Whatever of good may have come in these years of
change, the shadow of a deep disappointment rests upon
the Negro people,—a disappointment all the more bitter
because the unattained ideal was unbounded save by the
simple ignorance of a lowly people.

The first decade was merely a prolongation of the vain
search for freedom, the boon that seemed ever barely to
elude their grasp,—like a tantalizing will-o'-the-wisp,
maddening and misleading the headless host. The holo-
caust of war, the terrors of the Ku-Klux Klan, the lies of
carpet-baggers, the disorganization of industry, and the
contradictory advice of friends and foes, left the bewildered serf with no new watchword beyond the old cry
for freedom. As the time flew, however, he began to grasp
a new idea. The ideal of liberty demanded for its at-
tainment powerful means, and these the Fifteenth Amendment gave him. The ballot, which before he had looked
upon as a visible sign of freedom, he now regarded as
the chief means of gaining and perfecting the liberty
with which war had partially endowed him. And why
not? Had not votes made war and emancipated millions?
Had not votes enfranchised the freedmen? Was anything
impossible to a power that had done all this? A million
black men started with renewed zeal to vote themselves
into the kingdom. So the decade flew away, the revolution
of 1876 came, and left the half-free serf weary, wondering but still inspired. Slowly but steadily, in the
following years, a new vision began gradually to replace
the dream of political power,—a powerful movement,
the rise of another ideal to guide the unguided, another
pillar of fire by night after a clouded day. It was the
ideal of "book-learning"; the curiosity, born of compul-
sory ignorance, to know and test the power of the
cabalistic letters of the white man, the longing to know.
Here at last seemed to have been discovered the mountain path to Canaan; longer than the highway of Emancipation and law, steep and rugged, but straight, leading to
heights high enough to overlook life.

Up the new path the advance guard toiled, slowly,

heavily, doggedly; only those who have watched and guided the faltering feet, the misty minds, the dull understandings, of the dark pupils of these schools know how faithfully, how piteously, this people strove to learn. It was weary work. The cold statistician wrote down the inches of progress here and there, noted also where here and there a foot had slipped or some one had fallen. To the tired climbers, the horizon was ever dark, the mists were often cold, the Canaan was always dim and far away. If, however, the vistas disclosed as yet no goal, no resting-place, little but flattery and criticism, the journey at least gave leisure for reflection and self-examination; it changed the child of Emancipation to the youth with dawning self-consciousness, self-realization, self-respect. In those sombre forests of his striving his own soul rose before him, and he saw himself,—darkly as through a veil; and yet he saw in himself some faint revelation of his power, of his mission. He began to have a dim feeling that, to attain his place in the world, he must be himself, and not another. For the first time he sought to analyze the burden he bore upon his back, that deadweight of social degradation partially masked behind a half-named Negro problem. He felt his poverty; without a cent, without a home, without land, tools, or savings, he had entered into competition with rich, landed, skilled neighbors. To be a poor man is hard, but to be a poor race in a land of dollars is the very bottom of hardships. He felt the weight of his ignorance,—not simply of letters, but of life, of business, of the humanities; the accumulated sloth and shirking and awkwardness of decades and centuries shackled his hands and feet. Nor was his burden all poverty and ignorance. The red stain of bastardy, which two centuries of systemic legal defilement of Negro women had stamped upon his race, meant not only the loss of ancient African chastity, but also the hereditary weight of a mass of corruption from white adulterers, threatening almost the obliteration of the Negro home.

A people thus handicapped ought not to be asked to race with the world, but rather allowed to give all its time and thought to its own social problems. But alas! while sociologists gleefully count his bastards and his prostitutes, the very soul of the toiling, sweating black man is darkened by the shadow of a vast despair. Men call the shadow prejudice, and learnedly explain it as the natural defence of culture against barbarism, learning against ignorance, purity against crime, the "higher" against the "lower" races. To which the Negro cries Amen! and swears that to so much of this strange prejudice as is founded on just homage to civilization, culture, righteousness, and progress, he humbly bows and meekly does obeisance. But before that nameless prejudice that leaps beyond all this he stands helpless, dismayed, and well-nigh speechless; before that personal disrespect and mockery, the ridicule and systematic humiliation, the distortion of fact and wanton license of fancy, the cynical ignoring of the better and the boisterous welcoming of the worse, the all-pervading desire to inculcate disdain for everything black, from Toussaint to the devil,—before this there rises a sickening despair that would disarm and discourage any nation save that black host to whom "discouragement" is an unwritten word.

Lut the facing of so vast a prejudice could not but bring the inevitable self-questioning, self-disparagement, and lowering of ideals which ever accompany repression and breed in an atmosphere of contempt and hate. Whispering and portents came borne upon the four winds: Lo! we are diseased and dying, cried the dark hosts; we cannot write, our voting is vain; what need of education, since we must always cook and serve? And the Nation echoed and enforced this self-criticism, saying: Be content to be servants, and nothing more; what need of higher culture for half-men? Away with the black man's ballot, by force or fraud,—and behold the suicide of a race! Nevertheless, out of the evil came something of good,—the more careful adjustment of education to real life, the clearer perception of the Negroes' social responsibilities, and the sobering realization of the meaning of progress.

So dawned the time of *Sturm und Drang*: storm and stress to-day rocks our little boat on the mad waters of the world-sea; there is within and without the sound of conflict, the burning of body and rending of soul; inspiration strives with doubt, and faith with vain questionings. The bright ideals of the past,—physical freedom, political power, the training of brains and the training of hands,—all these in turn have waxed and waned, until even the last grows dim and overcast. Are they all wrong,

—all false? No, not that, but each alone was over-simple and incomplete,—the dreams of a credulous race-childhood, or the fond imaginings of the other world which does not know and does not want to know our power. To be really true, all these ideals must be melted and welded into one. The training of the schools we need to-day more than ever,—the training of deft hands, quick eyes and ears, and above all the broader, deeper, higher culture of gifted minds and pure hearts. The power of the ballot we need in sheer self-defence,—else what shall save us from a second slavery? Freedom, too, the long-sought, we still seek,—the freedom of life and limb, the freedom to work and think, the freedom to love and aspire. Work, culture, liberty,—all these we need, not singly but together, not successively but together, each growing and aiding each, and all striving toward that vaster ideal that swims before the Negro people, the ideal of human brotherhood, gained through the unify-ing ideal of Race; the ideal of fostering and developing the traits and talents of the Negro, not in opposition to or contempt for other races, but rather in large conformity to the greater ideals of the American Republic, in order that some day on American soil two world-races may give each to each those characteristics both so sadly lack. We the darker ones come even now not altogether empty-handed: there are to-day no truer exponents of the pure human spirit of the Declaration of Independence than the American Negroes; there is no true American music but the wild sweet melodies of the Negro slave; the American fairy tales and folk-lore are Indian and Afri-can; and, all in all, we black men seem the sole oasis of simple faith and reverence in a dusty desert of dollars and smartness. Will America be poorer if she replace her brutal dyspeptic blundering with light-hearted but de-termined Negro humility? or her coarse and cruel wit with loving jovial good-humor? or her vulgar music with the soul of the Sorrow Songs?

Merely a concrete test of the underlying principles of the great republic is the Negro Problem, and the spiritual striving of the freedmen's sons is the travail of souls whose burden is almost beyond the measure of their strength, but who bear it in the name of an historic race, in the name of this the land of their fathers' fa-thers, and in the name of human opportunity.

And now what I have briefly sketched in large outline let me on coming pages tell again in many ways, with loving emphasis and deeper detail, that men may listen to the striving in the souls of black folk.

I I

Of the Dawn of Freedom

Careless seems the great Avenger;
    History's lessons but record
One death-grapple in the darkness
    'Twixt old systems and the Word;
Truth forever on the scaffold,
    Wrong forever on the throne;
Yet that scaffold sways the future,
    And behind the dim unknown
Standeth God within the shadow
    Keeping watch above His own.

                           LOWELL.

The problem of the twentieth century is the problem of the color-line,—the relation of the darker to the lighter races of men in Asia and Africa, in America and the islands of the sea. It was a phase of this problem that caused the Civil War; and however much they who marched South and North in 1861 may have fixed on the technical points of union and local autonomy as a shibboleth, all nevertheless knew, as we know, that the question of Negro slavery was the real cause of the conflict. Curious it was, too, how this deeper question ever forced itself to the surface despite effort and disclaimer. No

sooner had Northern armies touched Southern soil than
this old question, newly guised, sprang from the earth,
—What shall be done with Negroes? Peremptory military commands, this way and that, could not answer the
query; the Emancipation Proclamation seemed but to
broaden and intensify the difficulties; and the War Amendments made the Negro problems of to-day.

It is the aim of this essay to study the period of
history from 1861 to 1872 so far as it relates to the
American Negro. In effect, this tale of the dawn of Freedom is an account of that government of men called
the Freedmen's Bureau,—one of the most singular and
interesting of the attempts made by a great nation to
grapple with vast problems of race and social condition.

The war has naught to do with slaves, cried Congress,
the President, and the Nation; and yet no sooner had
the armies, East and West, penetrated Virginia and Tennessee than fugitive slaves appeared within their lines.
They came at night, when the flickering camp-fires shone
like vast unsteady stars along the black horizon: old men
and thin, with gray and tufted hair; women, with frightened eyes, dragging whimpering hungry children; men
and girls, stalwart and gaunt,—a horde of starving vagabonds, homeless, helpless, and pitiable, in their dark
distress. Two methods of treating these newcomers seemed
equally logical to opposite sorts of minds. Ben Butler,
in Virginia, quickly declared slave property contraband
of war, and put the fugitives to work; while Fremont, in
Missouri, declared the slaves free under martial law.
Butler's action was approved, but Fremont's was hastily
countermanded, and his successor, Halleck, saw things differently. "Hereafter," he commanded, "no slaves should
be allowed to come into your lines at all; if any
come without your knowledge, when owners call for
them deliver them." Such a policy was difficult to enforce; some of the black refugees declared themselves
freemen, others showed that their masters had deserted
them, and still others were captured with forts and plantations. Evidently, too, slaves were a source of strength
to the Confederacy, and were being used as laborers and
producers. "They constitute a military resource," wrote
Secretary Cameron, late in 1861; "and being such, that
they should not be turned over to the enemy is too plain
to discuss." So gradually the tone of the army chiefs

changed; Congress forbade the rendition of fugitives, and Butler's "contrabands" were welcomed as military laborers. This complicated rather than solved the problem, for now the scattering fugitives became a steady stream, which flowed faster as the armies marched.

Then the long-headed man with care-chiselled face who sat in the White House saw the inevitable, and emancipated the slaves of rebels on New Year's, 1863. A month later Congress called earnestly for the Negro soldiers whom the act of July, 1862, had half grudgingly allowed to enlist. Thus the barriers were levelled and the deed was done. The stream of fugitives swelled to a flood, and anxious army officers kept inquiring: "What must be done with slaves, arriving almost daily? Are we to find food and shelter for women and children?"

It was a Pierce of Boston who pointed out the way, and thus became in a sense the founder of the Freedmen's Bureau. He was a firm friend of Secretary Chase; and when, in 1861, the care of slaves and abandoned lands developed upon the Treasury officials, Pierce was specially detailed from the ranks to study the conditions. First, he cared for the refugees at Fortress Monroe; and then, after Sherman had captured Hilton Head, Pierce was sent there to found his Port Royal experiment of making free workingmen out of slaves. Before his experiment was barely started, however, the problem of the fugitives had assumed such proportions that it was taken from the hands of the over-burdened Treasury Department and given to the army officials. Already centres of massed freedmen were forming at Fortress Monroe, Washington, New Orleans, Vicksburg and Corinth, Columbus, Ky., and Cairo, Ill., as well as at Port Royal. Army chaplains found here new and fruitful fields; "superintendents of contrabands" multiplied, and some attempt at systematic work was made by enlisting the able-bodied men and giving work to the others.

Then came the Freedmen's Aid societies, born of the touching appeals from Pierce and from these other centres of distress. There was the American Missionary Association, sprung from the *Amistad,* and now full-grown for work; the various church organizations, the National Freedmen's Relief Association, the American Freedmen's Union, the Western Freedmen's Aid Commission,—in all fifty or more active organizations, which sent clothes, money, school-books, and teachers southward. All they did was needed, for the destitution of the freedmen was often reported as "too appalling for belief," and the situation was daily growing worse rather than better.

And daily, too, it seemed more plain that this was no ordinary matter of temporary relief, but a national crisis; for here loomed a labor problem of vast dimensions. Masses of Negroes stood idle, or, if they worked spasmodically, were never sure of pay; and if perchance they received pay, squandered the new thing thoughtlessly. In these and other ways were camp-life and the new liberty demoralizing the freedmen. The broader economic organization thus clearly demanded sprang up here and there as accident and local conditions determined. Here it was that Pierce's Port Royal plan of leased plantations and guided workmen pointed out the rough way. In Washington the military governor, at the urgent appeal of the superintendent, opened confiscated estates to the cultivation of the fugitives, and there in the shadow of the dome gathered black farm villages. General Dix gave over estates to the freedmen of Fortress Monroe, and so on, South and West. The government and benevolent societies furnished the means of cultivation, and the Negro turned again slowly to work. The systems of control, thus started, rapidly grew, here and there, into strange little governments, like that of General Banks in Louisiana, with its ninety thousand black subjects, its fifty thousand guided laborers, and its annual budget of one hundred thousand dollars and more. It made out four thousand pay-rolls a year, registered all freedmen, inquired into grievances and redressed them, laid and collected taxes, and established a system of public schools. So, too, Colonel Eaton, the superintendent of Tennessee and Arkansas, ruled over one hundred thousand freedmen, leased and cultivated seven thousand acres of cotton land, and fed ten thousand paupers a year. In South Carolina was General Saxton, with his deep interest in black folk. He succeeded Pierce and the Treasury officials, and sold forfeited estates, leased abandoned plantations, encouraged schools, and received from Sherman, after that terribly picturesque march to the sea, thousands of the wretched camp followers.

Three characteristic things one might have seen in Sherman's raid through Georgia, which threw the new situation in shadowy relief: the Conqueror, the Conquered, and the Negro. Some see all significance in the grim front of the destroyer, and some in the bitter sufferers of the Lost Cause. But to me neither soldier nor fugitive speaks with so deep a meaning as that dark human cloud that clung like remorse on the rear of those swift columns, swelling at times to half their size, almost engulfing and choking them. In vain were they ordered back, in vain were bridges hewn from beneath their feet; on they trudged and writhed and surged, until they rolled into Savannah, a starved and naked horde of tens of thousands. There too came the characteristic military remedy: "The islands from Charleston south, the abandoned rice-fields along the rivers for thirty miles back from the sea, and the country bordering the St. John's River, Florida, are reserved and set apart for the settlement of Negroes now made free by act of war." So read the celebrated "Field-order Number Fifteen."

All these experiments, orders, and systems were bound to attract and perplex the government and the nation. Directly after the Emancipation Proclamation, Representative Eliot had introduced a bill creating a Bureau of Emancipation; but it was never reported. The following June a committee of inquiry, appointed by the Secretary of War, reported in favor of a temporary bureau for the "improvement, protection, and employment of refugee freedmen," on much the same lines as were afterwards followed. Petitions came in to President Lincoln from distinguished citizens and organizations, strongly urging a comprehensive and unified plan of dealing with the freedmen, under a bureau which should be "charged with the study of plans and execution of measures for easily guiding, and in every way judiciously and humanely aiding, the passage of our emancipated and yet to be emancipated blacks from the old condition of forced labor to their new state of voluntary industry."

Some half-hearted steps were taken to accomplish this, in part, by putting the whole matter again in charge of the special Treasury agents. Laws of 1863 and 1864 directed them to take charge of and lease abandoned lands for periods not exceeding twelve months, and to "provide in such leases, or otherwise, for the employment and general welfare" of the freedmen. Most of the army officers greeted this as a welcome relief from perplexing

"Negro affairs," and Secretary Fessenden, July 29, 1864, issued an excellent system of regulations, which were afterward closely followed by General Howard. Under Treasury agents, large quantities of land were leased in the Mississippi Valley, and many Negroes were employed; but in August, 1864, the new regulations were suspended for reasons of "public policy," and the army was again in control.

Meanwhile Congress had turned its attention to the subject; and in March the House passed a bill by a majority of two establishing a Bureau for Freedmen in the War Department. Charles Sumner, who had charge of the bill in the Senate, argued that freedmen and abandoned lands ought to be under the same department, and reported a substitute for the House bill attaching the Bureau to the Treasury Department. This bill passed, but too late for action by the House. The debates wandered over the whole policy of the administration and the general question of slavery, without touching very closely the specific merits of the measure in hand. Then the national election took place; and the administration, with a vote of renewed confidence from the country, addressed itself to the matter more seriously. A conference between the two branches of Congress agreed upon a carefully drawn measure which contained the chief provisions of Sumner's bill, but made the proposed organization a department independent of both the War and the Treasury officials. The bill was conservative, giving the new department "general superintendence of all freedmen." Its purpose was to "establish regulations" for them, protect them, lease them lands, adjust their wages, and appear in civil and military courts as their "next friend." There were many limitations attached to the powers thus granted, and the organization was made permanent. Nevertheless, the Senate defeated the bill, and a new conference committee was appointed. This committee reported a new bill, February 28, which was whirled through just as the session closed, and became the act of 1865 establishing in the War Department a "Bureau of Refugees, Freedmen, and Abandoned Lands."

This last compromise was a hasty bit of legislation, vague and uncertain in outline. A Bureau was created, "to continue during the present War of Rebellion, and for one year thereafter," to which was given "the supervision and management of all abandoned lands and the control of all subjects relating to refugees and freedmen," under "such rules and regulations as may be presented by the head of the Bureau and approved by the President." A Commissioner, appointed by the President and Senate, was to control the Bureau, with an office force not exceeding ten clerks. The President might also appoint assistant commissioners in the seceded States, and to all these offices military officials might be detailed at regular pay. The Secretary of War could issue rations, clothing, and fuel to the destitute, and all abandoned property was placed in the hands of the Bureau for eventual lease and sale to ex-slaves in forty-acre parcels.

Thus did the United States government definitely assume charge of the emancipated Negro as the ward of the nation. It was a tremendous undertaking. Here at a stroke of the pen was erected a government of millions of men, —and not ordinary men either, but black men emasculated by a peculiarly complete system of slavery, centuries old; and now, suddenly, violently, they come into a new birthright, at a time of war and passion, in the midst of the stricken and embittered population of their former masters. Any man might well have hesitated to assume charge of such a work, with vast responsibilities, indefinite powers, and limited resources. Probably no one but a soldier would have answered such a call promptly; and, indeed, no one but a soldier could be called, for Congress had appropriated no money for salaries and expenses.

Less than a month after the weary Emancipator passed to his rest, his successor assigned Major-Gen. Oliver O. Howard to duty as Commissioner of the new Bureau. He was a Maine man, then only thirty-five years of age. He had marched with Sherman to the sea, had fought well at Gettysburg, and but the year before had been assigned to the command of the Department of Tennessee. An honest man, with too much faith in human nature, little aptitude for business and intricate detail, he had had large opportunity of becoming acquainted at first hand with much of the work before him. And of that work it has been truly said that "no approximately correct history of civilization can ever be written which does not throw out in bold relief, as one of the great landmarks of political and social progress, the organization and administration of the Freedmen's Bureau."

On May 12, 1865, Howard was appointed; and he assumed the duties of his office promptly on the 15th, and began examining the field of work. A curious mess he looked upon: little despotisms, communistic experiments, slavery, peonage, business speculations, organized charity, unorganized almsgiving,—all reeling on under the guise of helping the freedmen, and all enshrined in the smoke and blood of war and the cursing and silence of angry men. On May 19 the new government—for a government it really was—issued its constitution; commissioners were to be appointed in each of the seceded states, who were to take charge of "all subjects relating to refugees and freedmen," and all relief and rations were to be given by their consent alone. The Bureau invited continued coöperation with benevolent societies, and declared: "It will be the object of all commissioners to introduce practicable systems of compensated labor," and to establish schools. Forthwith nine assistant commissioners were appointed. They were to hasten to their fields of work; seek gradually to close relief establishments, and make the destitute self-supporting; act as courts of law where there were no courts, or where Negroes were not recognized in them as free; establish the institution of marriage among ex-slaves, and keep records; see that freedmen were free to choose their employers, and help in making fair contracts for them; and finally, the circular said: "Simple good faith, for which we hope on all hands for those concerned in the passing away of slavery, will especially relieve the assistant commissioners in the discharge of their duties toward the freedmen, as well as promote the general welfare."

No sooner was the work thus started, and the general system and local organization in some measure begun, than two grave difficulties appeared which changed largely the theory and outcome of Bureau work. First, there were the abandoned lands of the South. It had long been the more or less definitely expressed theory of the North that all the chief problems of Emancipation might be settled by establishing the slaves on the forfeited lands of their masters,—a sort of poetic justice, said some. But this poetry done into solemn prose meant either wholesale confiscation of private property in the South, or vast appropriations. Now Congress had not approriated a cent, and no sooner did the proclamations of general amnesty appear than the eight hundred thousand acres of abandoned lands in the hands of the Freedmen's Bureau melted quickly away. The second difficulty lay in perfecting the local organization of the Bureau throughout the wide field of work. Making a new machine and sending out officials of duly ascertained fitness for a great work of social reform is no child's task; but this task was even harder, for a new central organization had to be fitted on a heterogeneous and confused but already existing system of relief and control of ex-slaves; and the agents available for this work must be sought for in an army still busy with war operations,—men in the very nature of the case ill fitted for delicate social work,—or among the questionable camp followers of an invading host. Thus, after a year's work, vigorously as it was pushed, the problem looked even more difficult to grasp and solve than at the beginning. Nevertheless, three things that year's work did, well worth the doing: it relieved a vast amount of physical suffering; it transported seven thousand fugitives from congested centres back to the farm; and, best of all, it inaugurated the crusade of the New England schoolma'am.

The annals of this Ninth Crusade are yet to be written,—the tale of a mission that seemed to our age far more quixotic than the quest of St. Louis seemed to his. Behind the mists of ruin and rapine waved the calico dresses of women who dared, and after the hoarse mouthings of the field guns rang the rhythm of the alphabet. Rich and poor they were, serious and curious. Bereaved now of a father, now of a brother, now of more than these, they came seeking a life work in planting New England schoolhouses among the white and black of the South. They did their work well. In that first year they taught one hundred thousand souls, and more.

Evidently, Congress must soon legislate again on the hastily organized Bureau, which had so quickly grown into wide significance and vast possibilities. An institution such as that was well-nigh as difficult to end as to begin. Early in 1866 Congress took up the matter, when Senator Trumbull, of Illinois, introduced a bill to extend the Bureau and enlarge its powers. This measure received, at the hands of Congress, far more thorough discussion and attention than its predecessor. The war cloud had thinned enough to allow a clearer conception of the work of Emancipation. The champions of the bill argued that the strengthening of the Freedmen's Bureau was still a military necessity; that it was needed for the proper carrying out of the Thirteenth Amendment, and was a work of sheer justice to the ex-slave, at a trifling cost to the government. The opponents of the measure declared that the war was over, and the necessity for war measures past; that the Bureau, by reason of its extraordinary powers, was clearly unconstitutional in time of peace, and was destined to irritate the South and pauperize the freedmen, at a final cost of possibly hundreds of millions. These two arguments were unanswered, and indeed unanswerable: the one that the extraordinary powers of the Bureau threatened the civil rights of all citizens; and the other that the government must have power to do what manifestly must be done, and that present abandonment of the freedmen meant their practical re-enslavement. The bill which finally passed enlarged and made permanent the Freedmen's Bureau. It was promptly vetoed by President Johnson as "unconstitutional," "unnecessary," and "extrajudicial," and failed of passage over the veto. Meantime, however, the breach between Congress and the President began to broaden, and a modified form of the lost bill was finally passed over the President's second veto, July 16.

The act of 1866 gave the Freedmen's Bureau its final form,—the form by which it will be known to posterity and judged of men. It extended the existence of the Bureau to July, 1868; it authorized additional assistant commissioners, the retention of army officers mustered out of regular service, the sale of certain forfeited lands to freedmen on nominal terms, the sale of Confederate public property for Negro schools, and a wider field of judicial interpretation and cognizance. The government of the unreconstructed South was thus put very largely in the hands of the Freedmen's Bureau, especially as in many cases the departmental military commander was now made also assistant commissioner. It was thus that the Freedmen's Bureau became a full-fledged government of men. It made laws, executed them and interpreted them; it laid and collected taxes, defined and punished crime, maintained and used military force, and dictated such measures as it thought necessary and proper for the accomplishment of its varied ends. Naturally, all these powers were not exercised continuously nor to their fullest extent; and yet, as General Howard has said, "scarcely any subject that has to be legislated upon in civil society failed, at one time or another, to demand the action of this singular Bureau."

To understand and criticise intelligently so vast a work, one must not forget an instant the drift of things in the later sixties. Lee had surrendered, Lincoln was dead, and Johnson and Congress were at loggerheads; the Thirteenth Amendment was adopted, the Fourteenth pending, and the Fifteenth declared in force in 1870. Guerrilla raiding, the ever-present flickering after-flame of war, was spending its forces against the Negroes, and all the Southern land was awakening as from some wild dream to poverty and social revolution. In a time of perfect calm, amid willing neighbors and streaming wealth, the social uplifting of four million slaves to an assured and self-sustaining place in the body politic and economic would have been a herculean task; but when to the inherent difficulties of so delicate and nice a social operation were added the spite and hate of conflict, the hell of war; when suspicion and cruelty were rife, and gaunt Hunger wept beside Bereavement,—in such a case, the work of any instrument of social regeneration was in large part foredoomed to failure. The very name of the Bureau stood for a thing in the South which for two centuries and better men had refused even to argue,—that life amid free Negroes was simply unthinkable, the maddest of experiments.

The agents that the Bureau could command varied all the way from unselfish philanthropists to narrow-minded busybodies and thieves; and even though it be true that the average was far better than the worst, it was the occasional fly that helped spoil the ointment.

Then amid all crouched the freed slave, bewildered between friend and foe. He had emerged from slavery,—not the worst slavery in the world, not a slavery that made all life unbearable, rather a slavery that had here and there something of kindliness, fidelity, and happiness,—but withal slavery, which, so far as human aspiration and desert were concerned, classed the black man and the ox together. And the Negro knew full well that, whatever their deeper convictions may have been, Southern men had fought with desperate energy to perpetuate this slavery under which the black masses, with half-articulate thought, had writhed and shivered. They welcomed freedom with a cry. They shrank from the master who still strove for their chains; they fled to the friends that had freed them, even though those friends stood ready to use them as a club for driving the recalcitrant South back into loyalty. So the cleft between the white and black South grew. Idle to say it never should have been; it was as inevitable as its results were pitiable. Curiously incongruous elements were left arrayed against each other,—the North, the government, the carpet-bagger, and the slave, here; and there, all the South that was white, whether gentleman or vagabond, honest man or rascal, lawless murderer or martyr to duty.

Thus it is doubly difficult to write of this period calmly, so intense was the feeling, so mighty the human passions that swayed and blinded men. Amid it all, two figures ever stand to typify that day to coming ages,—the one, a gray-haired gentleman, whose fathers had quit themselves like men, whose sons lay in nameless graves; who bowed to the evil of slavery because its abolition threatened untold ill to all; who stood at last, in the evening of life, a blighted, ruined form, with hate in his eyes;—and the other, a form hovering dark and mother-like; her awful face black with the mists of centuries, had aforetime quailed at that white master's command, had bent in love over the cradles of his sons and daughters, and closed in death the sunken eyes of his wife,—aye, too, at his behest had laid herself low to his lust, and borne a tawny man-child to the world, only to see her dark boy's limbs scattered to the winds by midnight marauders riding after "damned Niggers." These were the saddest sights of that woeful day; and no man clasped the hands of these two passing figures of the present-past; but, hating, they went to their long home, and, hating, their children's children live to-day.

Here, then, was the field of work for the Freedmen's Bureau; and since, with some hesitation, it was continued by the act of 1868 until 1869, let us look upon four years of its work as a whole. There were, in 1868, nine hundred Bureau officials scattered from Washington to Texas, ruling, directly and indirectly, many millions of men. The deeds of these rulers fall mainly under seven heads: the relief of physical suffering, the overseeing of the beginnings of free labor, the buying and selling of land, the establishment of schools, the paying of bounties, the administration of justice, and the financiering of all these activities.

Up to June, 1869, over half a million patients had been treated by Bureau physicians and surgeons, and sixty hospitals and asylums had been in operation. In fifty months twenty-one million free rations were distributed at a cost of over four million dollars. Next came the difficult question of labor. First, thirty thousand black men were transported from the refuges and relief stations back to the farms, back to the critical trial of a new way of working. Plain instructions went out from Washington: the laborers must be free to choose their employers, no fixed rate of wages was prescribed, and there was to be no peonage or forced labor. So far, so good; but where local agents differed *toto cœlo* in capacity and character, where the *personnel* was continually changing, the outcome was necessarily varied. The largest element of success lay in the fact that the majority of the freedmen were willing, even eager, to work. So labor contracts were written,—fifty thousand in a single State,—laborers advised, wages guaranteed, and employers supplied. In truth, the organization became a vast labor bureau,—not perfect, indeed, notably defective here and there, but on the whole successful beyond the dreams of thoughtful men. The two great obstacles which confronted the officials were the tyrant and the idler,—the slaveholder who was determined to perpetuate slavery under another name; and the freedman who regarded freedom as perpetual rest,—the Devil and the Deep Sea.

In the work of establishing the Negroes as peasant proprietors, the Bureau was from the first handicapped and at last absolutely checked. Something was done, and larger things were planned; abandoned lands were leased so long as they remained in the hands of the Bureau, and a total revenue of nearly half a million dollars derived from black tenants. Some other lands to which the nation had gained title were sold on easy terms, and public lands were opened for settlement to the very few freedmen who had tools and capital. But the vision of "forty acres and a mule"—the righteous and reasonable ambition to become a landholder, which the nation had all but categorically promised the freedmen—was destined in most cases to

bitter disappointment. And those men of marvellous hind-
sight who are today seeking to preach the Negro back to
the present peonage of the soil know well, or ought to
know, that the opportunity of binding the Negro peasant
willingly to the soil was lost on that day when the Commissioner of the Freedmen's Bureau had to go to South
Carolina and tell the weeping freedmen, after their years
of toil, that their land was not theirs, that there was a
mistake—somewhere. If by 1874 the Georgia Negro alone
owned three hundred and fifty thousand acres of land,
it was by grace of his thrift rather than by bounty of the
government.

The greatest success of the Freedmen's Bureau lay in
the planting of the free school among Negroes, and the
idea of free elementary education among all classes in the
South. It not only called the school-mistresses through the
benevolent agencies and built them school-houses, but it
helped discover and support such apostles of human
culture as Edmund Ware, Samuel Armstrong and Erastus
Cravath. The opposition to Negro education in the South
was at first bitter, and showed itself in ashes, insult, and
blood; for the South believed an educated Negro to be a
dangerous Negro. And the South was not wholly wrong;
for education among all kinds of men always has had, and
always will have, an element of danger and revolution, of
dissatisfaction and discontent. Nevertheless, men strive to
know. Perhaps some inkling of this paradox, even in the
unquiet days of the Bureau, helped the bayonets allay an
opposition to human training which still to-day lies
smouldering in the South, but not flaming. Fisk, Atlanta,
Howard, and Hampton were founded in these days, and
six million dollars were expended for educational work,
seven hundred and fifty thousand dollars of which the
freedmen themselves gave of their poverty.

Such contributions, together with the buying of land
and various other enterprises, showed that the ex-slave
was handling some free capital already. The chief initial
source of this was labor in the army, and his pay and
bounty as a soldier. Payments to Negro soldiers were
at first complicated by the ignorance of the recipients,
and the fact that the quotas of colored regiments from
Northern States were largely filled by recruits from the
South, unknown to their fellow soldiers. Consequently,
payments were accompanied by such frauds that Congress,

by joint resolution in 1867, put the whole matter in the hands of the Freedmen's Bureau. In two years six million dollars was thus distributed to five thousand claimants, and in the end the sum exceeded eight million dollars. Even in this system fraud was frequent; but still the work put needed capital in the hands of practical paupers, and some, at least, was well spent.

The most perplexing and least successful part of the Bureau's work lay in the exercise of its judicial functions. The regular Bureau court consisted of one representative of the employer, one of the Negro, and one of the Bureau. If the Bureau could have maintained a perfectly judicial attitude, this arrangement would have been ideal, and must in time have gained confidence; but the nature of its other activities and the character of its *personnel* prejudiced the Bureau in favor of the black litigants, and led without doubt to much injustice and annoyance. On the other hand, to leave the Negro in the hands of Southern courts was impossible. In a distracted land where slavery had hardly fallen, to keep the strong from wanton abuse of the weak, and the weak from gloating insolently over the half-shorn strength of the strong, was a thankless, hopeless task. The former masters of the land were peremptorily ordered about, seized, and imprisoned, and punished over and again, with scant courtesy from army officers. The former slaves were intimidated, beaten, raped, and butchered by angry and revengeful men. Bureau courts tended to become centres simply for punishing whites, while the regular civil courts tended to become solely institutions for perpetuating the slavery of blacks. Almost every law and method ingenuity could devise was employed by the legislatures to reduce the Negroes to serfdom,—to make them the slaves of the State, if not of individual owners; while the Bureau officials too often were found striving to put the "bottom rail on top," and gave the freedmen a power and independence which they could not yet use. It is all well enough for us of another generation to wax wise with advice to those who bore the burden in the heat of the day. It is full easy now to see that the man who lost home, fortune, and family at a stroke, and saw his land ruled by "mules and niggers," was really benefited by the passing of slavery. It is not difficult now to say to the young freedman, cheated and cuffed about who has seen his father's head beaten to a jelly and his own mother namelessly assaulted, that the meek shall inherit the earth. Above all, nothing is more convenient than to heap on the Freedmen's Bureau all the evils of that evil day, and damn it utterly for every mistake and blunder that was made.

All this is easy, but it is neither sensible nor just. Some one had blundered, but that was long before Oliver Howard was born; there was criminal aggression and heedless neglect, but without some system of control there would have been far more than there was. Had that control been from within, the Negro would have been re-enslaved, to all intents and purposes. Coming as the control did from without, perfect men and methods would have bettered all things; and even with imperfect agents and questionable methods, the work accomplished was not undeserving of commendation.

Such was the dawn of Freedom; such was the work of the Freedmen's Bureau, which, summed up in brief, may be epitomized thus: for some fifteen million dollars, beside the sums spent before 1865, and the dole of benevolent societies, this Bureau set going a system of free labor, established a beginning of peasant proprietorship, secured the recognition of black freedmen before courts of law, and founded the free common school in the South. On the other hand, it failed to begin the establishment of good-will between ex-masters and freedmen, to guard its work wholly from paternalistic methods which discouraged self-reliance, and to carry out to any considerable extent its implied promises to furnish the freedmen with land. Its successes were the result of hard work, supplemented by the aid of philanthropists and the eager striving of black men. Its failures were the result of bad local agents, the inherent difficulties of the work, and national neglect.

Such an institution, from its wide powers, great responsibilities, large control of moneys, and generally conspicuous position, was naturally open to repeated and bitter attack. It sustained a searching Congressional investigation at the instance of Fernando Wood in 1870. Its archives and few remaining functions were with blunt discourtesy transferred from Howard's control, in his absence, to the supervision of Secretary of War Belknap in 1872, on the Secretary's recommendation. Finally, in consequence of grave intimations of wrong-doing made by the Secretary and his subordinates, General Howard was court-martialed in 1874. In both of these trials the Commissioner of the Freedmen's Bureau was officially exonerated from any willful misdoing, and his work commended. Nevertheless, many unpleasant things were brought to light,—the methods of transacting the business of the Bureau were faulty; several cases of defalcation were proved, and other frauds strongly suspected; there were some business transactions which savored of dangerous speculation, if not dishonesty; and around it all lay the smirch of the Freedmen's Bank.

Morally and practically, the Freedmen's Bank was part of the Freedmen's Bureau, although it had no legal connection with it. With the prestige of the government back of it, and a directing board of unusual respectability and national reputation, this banking institution had made a remarkable start in the development of that thrift among black folk which slavery had kept them from knowing. Then in one sad day came the crash,—all the hard-earned dollars of the freedmen disappeared; but that was the least of the loss,—all the faith in saving went too, and much of the faith in men; and that was a loss that a Nation which to-day sneers at Negro shiftlessness has never yet made good. Not even ten additional years of slavery could have done so much to throttle the thrift of the freedmen as the mismanagement and bankruptcy of the series of savings banks chartered by the Nation for their especial aid. Where all the blame should rest, it is hard to say; whether the Bureau and the Bank died chiefly by reason of the blows of its selfish friends or the dark machinations of its foes, perhaps even time will never reveal, for here lies unwritten history.

Of the foes without the Bureau, the bitterest were those who attacked not so much its conduct or policy under the law as the necessity for any such institution at all. Such attacks came primarily from the Border States and the South; and they were summed up by Senator Davis, of Kentucky, when he moved to entitle the act of 1866 a bill "to promote strife and conflict between the white and black races . . . by a grant of unconstitutional power." The argument gathered tremendous strength South and North; but its very strength was its weakness. For, argued the plain common-sense of the nation, if it is unconstitutional, unpractical, and futile for the nation to stand guardian over its helpless wards,

then there is left but one alternative,—to make those
wards their own guardians by arming them with the
ballot. Moreover, the path of the practical politician
pointed the same way; for, argued this opportunist, if we
cannot peacefully reconstruct the South with white votes
we certainly can with black votes. So justice and force
joined hands.

The alternative thus offered the nation was not between full and restricted Negro suffrage; else every sensible man, black and white, would easily have chosen the
latter. It was rather a choice between suffrage and slavery,
after endless blood and gold had flowed to sweep human
bondage away. Not a single Southern legislature stood
ready to admit a Negro, under any conditions, to the
polls; not a single Southern legislature believed free Negro
labor was possible without a system of restrictions that
took all its freedom away; there was scarcely a white
man in the South who did not honestly regard emancipation as a crime, and its practical nullification as a duty.
In such a situation, the granting of the ballot to the black
man was a necessity, the very least a guilty nation could
grant a wronged race, and the only method of compelling
the South to accept the results of the war. Thus Negro
suffrage ended a civil war by beginning a race feud. And
some felt gratitude toward the race thus sacrificed in its
swaddling clothes on the altar of national integrity; and
some felt and feel only indifference and contempt.

Had political exigencies been less pressing, the opposition to government guardianship of Negroes less bitter,
and the attachment to the slave system less strong, the
social seer can well imagine a far better policy,—a permanent Freedmen's Bureau, with a national system of
Negro schools; a carefully supervised employment and
labor office; a system of impartial protection before the
regular courts; and such institutions for social betterment as savings-banks, land and building associations,
and social settlements. All this vast expenditure of money
and brains might have formed a great school of prospective citizenship, and solved in a way we have not yet
solved the most perplexing and persistent of the Negro
problems.

That such an institution was unthinkable in 1870 was
due in part to certain acts of the Freedmen's Bureau itself. It came to regard its work as merely temporary, and

Negro suffrage as a final answer to all present perplexities. The political ambition of many of its agents and *protégés* led it far afield into questionable activities, until the South, nursing its own deep prejudices, came easily to ignore all the good deeds of the Bureau and hate its very name with perfect hatred. So the Freedmen's Bureau died, and its child was the Fifteenth Amendment.

The passing of a great human institution before its work is done, like the untimely passing of a single soul, but leaves a legacy of striving for other men. The legacy of the Freedmen's Bureau is the heavy heritage of this generation. To-day, when new and vaster problems are destined to strain every fibre of the national mind and soul, would it not be well to count this legacy honestly and carefully? For this much all men know: despite compromise, war, and struggle, the Negro is not free. In the backwoods of the Gulf States, for miles and miles, he may not leave the plantation of his birth; in well-nigh the whole rural South the black farmers are peons, bound by law and custom to an economic slavery, from which the only escape is death or the penitentiary. In the most cultured sections and cities of the South the Negroes are a segregated servile caste, with restricted rights and privileges. Before the courts, both in law and custom, they stand on a different and peculiar basis. Taxation without representation is the rule of their political life. And the result of all this is, and in nature must have been, lawlessness and crime. That is the large legacy of the Freedmen's Bureau, the work it did not do because it could not.

I have seen a land right merry with the sun, where children sing, and rolling hills lie like passioned women wanton with harvest. And there in the King's Highway sat and sits a figure veiled and bowed, by which the traveller's footsteps hasten as they go. On the tainted air broods fear. Three centuries' thought has been the raising and unveiling of that bowed human heart, and now behold a century new for the duty and the deed. The problem of the Twentieth Century is the problem of the color-line.

III

Of Mr. Booker T. Washington And Others

From birth till death enslaved; in word, in deed,
    unmanned!

.   .   .   .

Hereditary bondsmen! Know ye not
Who would be free themselves must strike the blow?

BYRON.

Easily the most striking thing in the history of the
American Negro since 1876 is the ascendancy of Mr.
Booker T. Washington. It began at the time when war
memories and ideals were rapidly passing; a day of
astonishing commercial development was dawning; a sense
of doubt and hesitation overtook the freedmen's sons,—
then it was that his leading began. Mr. Washington came,
with a single definite programme, at the psychological
moment when the nation was a little ashamed of having
bestowed so much sentiment on Negroes, and was con-
centrating its energies on Dollars. His programme of industrial education, conciliation of the South, and submission and silence as to civil and political rights, was not
wholly original; the Free Negroes from 1830 up to war-
time had striven to build industrial schools, and the
American Missionary Association had from the first taught
various trades; and Price and others had sought a way
of honorable alliance with the best of the Southerners.
But Mr. Washington first indissolubly linked these things;

he put enthusiasm, unlimited energy, and perfect faith into this programme, and changed it from a by-path into a veritable Way of Life. And the tale of the methods by which he did this is a fascinating study of human life.

It startled the nation to hear a Negro advocating such a programme after many decades of bitter complaint; it startled and won the applause of the South, it interested and won the admiration of the North; and after a confused murmur of protest, it silenced if it did not convert the Negroes themselves.

To gain the sympathy and coöperation of the various elements comprising the white South was Mr. Washington's first task; and this, at the time Tuskegee was founded, seemed, for a black man, well-nigh impossible. And yet ten years later it was done in the word spoken at Atlanta: "In all things purely social we can be as separate as the five fingers, and yet one as the hand in all things essential to mutual progress." This "Atlanta Compromise" is by all odds the most notable thing in Mr. Washington's career. The South interpreted it in different ways: the radicals received it as a complete surrender of the demand for civil and political equality; the conservatives, as a generously conceived working basis for mutual understanding. So both approved it, and to-day its author is certainly the most distinguished Southerner since Jefferson Davis, and the one with the largest personal following.

Next to this achievement comes Mr. Washington's work in gaining place and consideration in the North. Others less shrewd and tactful had formerly essayed to sit on these two stools and had fallen between them; but as Mr. Washington knew the heart of the South from birth and training, so by singular insight he intuitively grasped the spirit of the age which was dominating the North. And so thoroughly did he learn the speech and thought of triumphant commercialism, and the ideals of material prosperity, that the picture of a lone black boy poring over a French grammar amid the weeds and dirt of a neglected home soon seemed to him the acme of absurdities. One wonders what Socrates and St. Francis of Assisi would say to this.

And yet this very singleness of vision and thorough oneness with his age is a mark of the successful man. It is as though Nature must needs make men narrow in order to give them force. So Mr. Washington's cult has gained unquestioning followers, his work has wonderfully prospered, his friends are legion, and his enemies are confounded. To-day he stands as the one recognized spokesman of his ten million fellows, and one of the most notable figures in a nation of seventy millions. One hesitates, therefore, to criticise a life which, beginning with so little, has done so much. And yet the time is come when one may speak in all sincerity and utter courtesy of the mistakes and shortcomings of Mr. Washington's career, as well as of his triumphs, without being thought captious or envious, and without forgetting that it is easier to do ill than well in the world.

The criticism that has hitherto met Mr. Washington has not always been of this broad character. In the South especially has he had to walk warily to avoid the harshest judgments,—and naturally so, for he is dealing with the one subject of deepest sensitiveness to that section. Twice —once when at the Chicago celebration of the Spanish-American War he alluded to the color-prejudice that is "eating away the vitals of the South," and once when he dined with President Roosevelt—has the resulting Southern criticism been violent enough to threaten seriously his popularity. In the North the feeling has several times forced itself into words, that Mr. Washington's counsels of submission overlooked certain elements of true manhood, and that his educational programme was unnecessarily narrow. Usually, however, such criticism has not found open expression, although, too, the spiritual sons of the Abolitionists have not been prepared to acknowledge that the schools founded before Tuskegee, by men of broad ideals and self-sacrificing spirit, were wholly failures or worthy of ridicule. While, then, criticism has not failed to follow Mr. Washington, yet the prevailing public opinion of the land has been but too willing to deliver the solution of a wearisome problem into his hands, and say, "If that is all you and your race ask, take it."

Among his own people, however, Mr. Washington has encountered the strongest and most lasting opposition, amounting at times to bitterness, and even to-day continuing strong and insistent even though largely silenced in outward expression by the public opinion of the nation. Some of this opposition is, of course, mere envy; the disappointment of displaced demagogues and the spite of narrow minds. But aside from this. there is among educated and thoughtful colored men in all parts of the land a feeling of deep regret, sorrow, and apprehension at the wide currency and ascendancy which some of Mr. Washington's theories have gained. These same men admire his sincerity of purpose, and are willing to forgive much to honest endeavor which is doing something worth the doing. They coöperate with Mr. Washington as far as they conscientiously can; and, indeed, it is no ordinary tribute to this man's tact and power that, steering as he must between so many diverse interests and opinions, he so largely retains the respect of all.

But the hushing of the criticism of honest opponents is a dangerous thing. It leads some of the best of the critics to unfortunate silence and paralysis of effort, and others to burst into speech so passionately and intemperately as to lose listeners. Honest and earnest criticism from those whose interests are most nearly touched,—criticism of writers by readers, of government by those governed, of leaders by those led,—this is the soul of democracy and the safeguard of modern society. If the best of the American Negroes receive by outer pressure a leader whom they had not recognized before, manifestly there is here a certain palpable gain. Yet there is also irreparable loss,—a loss of that peculiarly valuable education which a group receives when by search and criticism it finds and commissions its own leaders. The way in which this is done is at once the most elementary and the nicest problem of social growth. History is but the record of such group-leadership; and yet how infinitely changeful is its type and character! And of all types and kinds, what can be more instructive than the leadership of a group within a group?—that curious double movement where real progress may be negative and actual advance be relative retrogression. All this is the social student's inspiration and despair.

Now in the past the American Negro has had instructive experience in the choosing of group leaders, founding thus a peculiar dynasty which in the light of present conditions is worth while studying. When sticks and stones and beasts form the sole environment of a people, their attitude is largely one of determined opposition to and conquest of natural forces. But when to earth and brute is added an environment of men and ideas, then the attitude of the imprisoned group may take three main forms,—a feeling of revolt and revenge; an attempt to adjust all thought and action to the will of the greater group; or, finally, a determined effort at self-realization and self-development despite environing opinion. The influence of all of these attitudes at various times can be traced in the history of the American Negro, and in the evolution of his successive leaders.

Before 1750, while the fire of African freedom still burned in the veins of the slaves, there was in all leadership or attempted leadership but the one motive of revolt and revenge,—typified in the terrible Maroons, the Danish blacks, and Cato of Stono, and veiling all the Americas in fear of insurrection. The liberalizing tendencies of the latter half of the eighteenth century brought, along with kindlier relations between black and white, thoughts of ultimate adjustment and assimilation. Such aspiration was especially voiced in the earnest songs of Phyllis, in the martyrdom of Attucks, the fighting of Salem and Poor, the intellectual accomplishments of Banneker and Derham, and the political demands of the Cuffes.

Stern financial and social stress after the war cooled much of the previous humanitarian ardor. The disappointment and impatience of the Negroes at the persistence of slavery and serfdom voiced itself in two movements. The slaves in the South, aroused undoubtedly by vague rumors of the Haitian revolt, made three fierce attempts at insurrection,—in 1800 under Gabriel in Virginia, in 1822 under Vesey in Carolina, and in 1831 again in Virginia under the terrible Nat Turner. In the Free States, on the other hand, a new and curious attempt at self-development was made. In Philadelphia and New York color-prescription led to a withdrawal of Negro communicants from white churches and the formation of a peculiar socio-religious institution among the Negroes known as the African Church,—an organization still living and controlling in its various branches over a million of men.

Walker's wild appeal against the trend of the times showed how the world was changing after the coming of the cotton-gin. By 1830 slavery seemed hopelessly fastened on the South, and the slaves thoroughly cowed into submission. The free Negroes of the North, inspired by the mulatto immigrants from the West Indies, began to change the basis of their demands; they recognized the slavery of slaves, but insisted that they themselves were freemen, and sought assimilation and amalgamation with the nation on the same terms with other men. Thus, Forten and Purvis of Philadelphia, Shad of Wilmington, Du Bois of New Haven, Barbadoes of Boston, and others, strove singly and together as men, they said, not as slaves; as "people of color," not as "Negroes." The trend of the times, however, refused them recognition save in individual and exceptional cases, considered them as one with all the despised blacks, and they soon found themselves striving to keep even the rights they formerly had of voting and working and moving as freemen. Schemers of migration and colonization arose among them; but these they refused to entertain, and they eventually turned to the Abolition movement as a final refuge.

Here, led by Remond, Nell, Wells-Brown, and Douglass, a new period of self-assertion and self-development dawned. To be sure, ultimate freedom and assimilation was the ideal before the leaders, but the assertion of the manhood rights of the Negro by himself was the main reliance, and John Brown's raid was the extreme of its logic. After the war and emancipation, the great form of Frederick Douglass, the greatest of American Negro leaders, still led the host. Self-assertion, especially in political lines, was the main programme, and behind Douglass came Elliot, Bruce, and Langston, and the Reconstruction politicians, and, less conspicuous but of greater social significance Alexander Crummell and Bishop Daniel Payne.

Then came the Revolution of 1876, the suppression of the Negro votes, the changing and shifting of ideals, and the seeking of new lights in the great night. Douglass, in his old age, still bravely stood for the ideals of his early manhood,—ultimate assimilation *through* self-assertion, and on no other terms. For a time Price arose as a new leader, destined, it seemed, not to give up, but to re-state the old ideals in a form less repugnant to the white South. But he passed away in his prime. Then came the new leader. Nearly all the former ones had become leaders by the silent suffrage of their fellows, had sought to lead their own people alone, and were usually, save Douglass, little known outside their race. But Booker T. Washington arose as essentially the leader not of one race but of two,—a compromiser between the South, the North, and the Negro. Naturally the Negroes resented, at first bitterly, signs of compromise which surrendered their civil and political rights, even though this was to be exchanged for larger chances of economic development. The rich and dominating North, however, was not only weary of the race problem, but was investing largely in Southern enterprises, and welcomed any method of peaceful coöperation. Thus, by national opinion, the Negroes began to recognize Mr. Washington's leadership; and the voice of criticism was hushed.

Mr. Washington represents in Negro thought the old attitude of adjustment and submission; but adjustment at such a peculiar time as to make his programme unique. This is an age of unusual economic development, and Mr. Washington's programme naturally takes an economic cast, becoming a gospel of Work and Money to such an extent as apparently almost completely to overshadow the higher aims of life. Moreover, this is an age when the more advanced races are coming in closer contact with the less developed races, and the race-feeling is therefore intensified; and Mr. Washington's programme practically accepts the alleged inferiority of the Negro races. Again, in our own land, the reaction from the sentiment of war time has given impetus to race-prejudice against Negroes, and Mr. Washington withdraws many of the high demands of Negroes as men and American citizens. In other periods of intensified prejudice all the Negro's tendency to self-assertion has been called forth; at this period a policy of submission is advocated. In the history of nearly all other races and peoples the doctrine preached at such crises has been that manly self-respect is worth more than lands and houses, and that a people who voluntarily surrender such respect, or cease striving for it, are not worth civilizing.

In answer to this, it has been claimed that the Negro can survive only through submission. Mr. Washington distinctly asks that black people give up, at least for the present, three things,—

First, political power,

Second, insistence on civil rights,

Third, higher education of Negro youth,—

and concentrate all their energies on industrial education, the accumulation of wealth, and the conciliation of the South. This policy has been courageously and insistently advocated for over fifteen years, and has been triumphant for perhaps ten years. As a result of this tender of the palm-branch, what has been the return? In these years there have occurred:

1. The disfranchisement of the Negro.
2. The legal creation of a distinct status of civil inferiority for the Negro.
3. The steady withdrawal of aid from institutions for the higher training of the Negro.

These movements are not, to be sure, direct results of Mr. Washington's teachings; but his propaganda has, without a shadow of doubt, helped their speedier accomplishment. The question then comes: Is it possible, and probable, that nine millions of men can make effective progress in economic lines if they are deprived of political rights, made a servile caste, and allowed only the most meagre chance for developing their exceptional men? If history and reason give any distinct answer to these questions, it is an emphatic *No.* And Mr. Washington thus faces the triple paradox of his career:

1. He is striving nobly to make Negro artisans business men and property-owners; but it is utterly impossible, under modern competitive methods, for workingmen and property-owners to defend their rights and exist without the right of suffrage.
2. He insists on thrift and self-respect, but at the same time counsels a silent submission to civic inferiority such as is bound to sap the manhood of any race in the long run.
3. He advocates common-school and industrial training, and depreciates institutions of higher learning; but neither the Negro common-schools, nor Tuskegee itself, could remain open a day were it not for teachers trained in Negro colleges, or trained by their graduates.

This triple paradox in Mr. Washington's position is the object of criticism by two classes of colored Americans. One class is spiritually descended from Toussaint the Savior, through Gabriel, Vesey, and Turner, and they represent the attitude of revolt and revenge; they hate the white South blindly and distrust the white race generally, and so far as they agree on definite action, think that the Negro's only hope lies in emigration beyond the borders of the United States. And yet, by the irony of fate, nothing has more effectually made this programme seem hopeless than the recent course of the United States

toward weaker and darker peoples in the West Indies,
Hawaii, and the Philippines,—for where in the world may
we go and be safe from lying and brute force?

The other class of Negroes who cannot agree with Mr.
Washington has hitherto said little aloud. They deprecate
the sight of scattered counsels, of internal disagreement;
and especially they dislike making their just criticism of
a useful and earnest man an excuse for a general discharge of venom from small-minded opponents. Nevertheless, the questions involved are so fundamental and
serious that it is difficult to see how men like the Grimkes,
Kelly Miller, J. W. E. Bowen, and other representatives of
this group. can much longer be silent. Such men feel in
conscience bound to ask of this nation three things:

1. The right to vote.
2. Civic equality.
3. The education of youth according to ability.

They acknowledge Mr. Washington's invaluable service in
counselling patience and courtesy in such demands; they
do not ask that ignorant black men vote when ignorant
whites are debarred, or that any reasonable restrictions
in the suffrage should not be applied; they know that the
low social level of the mass of the race is responsible for
much discrimination against it, but they also know, and
the nation knows. that relentless color-prejudice is more
often a cause than a result of the Negro's degradation;
they seek the abatement of this relic of barbarism, and
not its systematic encouragement and pampering by all
agencies of social power from the Associated Press to
the Church of Christ. They advocate, with Mr. Washington, a broad system of Negro common schools supplemented by thorough industrial training; but they are
surprised that a man of Mr. Washington's insight cannot
see that no such educational system ever has rested or
can rest on any other basis than that of the well-equipped
college and university, and they insist that there is a demand for a few such institutions throughout the South
to train the best of the Negro youth as teachers, professional men, and leaders.

This group of men honor Mr. Washington for his attitude of conciliation toward the white South; they accept the "Atlanta Compromise" in its broadest interpretation; they recognize, with him, many signs of promise,
many men of high purpose and fair judgment, in this

section; they know that no easy task has been laid upon a region already tottering under heavy burdens. But, nevertheless, they insist that the way to truth and right lies in straightforward honesty, not in indiscriminate flattery; in praising those of the South who do well and criticising uncompromisingly those who do ill; in taking advantage of the opportunities at hand and urging their fellows to do the same, but at the same time in remembering that only a firm adherence to their higher ideals and aspirations will ever keep those ideals within the realm of possibility. They do not expect that the free right to vote, to enjoy civic rights, and to be educated, will come in a moment; they do not expect to see the bias and prejudices of years disappear at the blast of a trumpet; but they are absolutely certain that the way for a people to gain their reasonable rights is not by voluntarily throwing them away and insisting that they do not want them; that the way for a people to gain respect is not by continually belittling and ridiculing themselves; that, on the contrary, Negroes must insist continually, in season and out of season, that voting is necessary to modern manhood, that color discrimination is barbarism, and that black boys need education as well as white boys.

In failing thus to state plainly and unequivocally the legitimate demands of their people, even at the cost of opposing an honored leader, the thinking classes of American Negroes would shirk a heavy responsibility,—a responsibility to themselves, a responsibility to the struggling masses, a responsibility to the darker races of men whose future depends so largely on this American experiment, but especially a responsibility to this nation,—this common Fatherland. It is wrong to encourage a man or a people in evil-doing; it is wrong to aid and abet a national crime simply because it is unpopular not to do so. The growing spirit of kindliness and reconciliation between the North and South after the frightful difference of a generation ago ought to be a source of deep congratulation to all, and especially to those whose mistreatment caused the war; but if that reconciliation is to be marked by the industrial slavery and civic death of those same black men, with permanent legislation into a position of inferiority, then those black men, if they are really men, are called upon by every consideration of patriotism and loyalty to oppose such a course by all civilized methods, even though such opposition involves disagreement with Mr. Booker T. Washington. We have no right to sit silently by while the inevitable seeds are sown for a harvest of disaster to our children, black and white.

First, it is the duty of black men to judge the South discriminatingly. The present generation of Southerners are not responsible for the past, and they should not be blindly hated or blamed for it. Furthermore, to no class is the indiscriminate endorsement of the recent course of the South toward Negroes more nauseating than to the best thought of the South. The South is not "solid"; it is a land in the ferment of social change, wherein forces of all kinds are fighting for supremacy; and to praise the ill the South is to-day perpetrating is just as wrong as to condemn the good. Discriminating and broad-minded criticism is what the South needs,—needs it for the sake of her own white sons and daughters, and for the insurance of robust, healthy mental and moral development.

To-day even the attitude of the Southern whites toward the blacks is not, as so many assume, in all cases the same; the ignorant Southerner hates the Negro, the workingmen fear his competition, the money-makers wish to use him as a laborer, some of the educated see a menace in his upward development, while others—usually the sons of the masters—wish to help him to rise. National opinion has enabled this last class to maintain the Negro common schools, and to protect the Negro partially in property, life, and limb. Through the pressure of the money-makers, the Negro is in danger of being reduced to semi-slavery, especially in the country districts; the workingmen, and those of the educated who fear the Negro, have united to disfranchise him, and some have urged his deportation; while the passions of the ignorant are easily aroused to lynch and abuse any black man. To praise this intricate whirl of thought and prejudice is nonsense; to inveigh indiscriminately against "the South" is unjust; but to use the same breath in praising Governor Aycock, exposing Senator Morgan, arguing with Mr. Thomas Nelson Page, and denouncing Senator Ben Tillman, is not only sane, but the imperative duty of thinking black men.

It would be unjust to Mr. Washington not to acknowledge that in several instances he has opposed movements in the South which were unjust to the Negro; he sent memorials to the Louisiana and Alabama constitutional conventions, he has spoken against lynching, and in other ways has openly or silently set his influence against sinister schemes and unfortunate happenings. Notwithstanding this, it is equally true to assert that on the whole the distinct impression left by Mr. Washington's propaganda is, first, that the South is justified in its present attitude toward the Negro because of the Negro's degradation; secondly, that the prime cause of the Negro's failure to rise more quickly is his wrong education in the past; and, thirdly, that his future rise depends primarily on his own efforts. Each of these propositions is a dangerous half-truth. The supplementary truths must never be lost sight of: first, slavery and race-prejudice are potent if not sufficient causes of the Negro's position; second, industrial and common-school training were necessarily slow in planting because they had to await the black teachers trained by higher institutions,—it being extremely doubtful if any essentially different development was possible, and certainly a Tuskegee was unthinkable before 1880; and, third, while it is a great truth to say that the Negro must strive and strive mightily to help himself, it is equally true that unless his striving be not simply seconded, but rather aroused and encouraged, by the initiative of the richer and wiser environing group, he cannot hope for great success.

In his failure to realize and impress this last point, Mr. Washington is especially to be criticised. His doctrine has tended to make the whites, North and South, shift the burden of the Negro problem to the Negro's shoulders and stand aside as critical and rather pessimistic spectators; when in fact the burden belongs to the nation, and the hands of none of us are clean if we bend not our energies to righting these great wrongs.

The South ought to be led, by candid and honest criticism, to assert her better self and do her full duty to the race she has cruelly wronged and is still wronging. The North—her co-partner in guilt—cannot salve her conscience by plastering it with gold. We cannot settle this problem by diplomacy and suaveness, by "policy" alone. If worse comes to worst, can the moral fibre of this country survive the slow throttling and murder of nine millions of men?

The black men of America have a duty to perform, a
duty stern and delicate,—a forward movement to oppose
a part of the work of their greatest leader. So far as Mr.
Washington preaches Thrift, Patience, and Industrial Training for the masses, we must hold up his hands and
strive with him, rejoicing in his honors and glorying in
the strength of this Joshua called of God and of man to
lead the headless host. But so far as Mr. Washington
apologizes for injustice, North or South, does not rightly value the privilege and duty of voting, belittles the emas-
culating effects of caste distinctions, and opposes the
higher training and ambition of our brighter minds,—so
far as he, the South, or the Nation, does this,—we must
unceasingly and firmly oppose them. By every civilized and peaceful method we must strive for the rights
which the world accords to men, clinging unwaveringly
to those great words which the sons of the Fathers would
fain forget: "We hold these truths to be self-evident:
That all men are created equal; that they are endowed
by their Creator with certain unalienable rights; that among
these are life, liberty, and the pursuit of happiness."

IV

Of the Meaning of Progress

Willst Du Deine Macht verkünden,
Wähle sie die frei von Sünden,
Steh'n in Deinem ew'gen Haus!
Deine Geister sende ans!
Die Unsterblichen, die Reinen,
Die nicht fühlen, die nicht weinen!
Nicht die zarte Jungfrau wähle,
Nicht der Hirtin weiche Seele!

SCHILLER.

Once upon a time I taught school in the hills of Tennessee, where the broad dark vale of the Mississippi begins to roll and crumple to greet the Alleghenies. I was a Fisk student then, and all Fisk men thought that Tennessee—beyond the Veil—was theirs alone, and in vacation time they sallied forth in lusty bands to meet the county school-commissioners. Young and happy, I too went, and I shall not soon forget that summer, seventeen years ago.

First, there was a Teachers' Institute at the county-seat; and there distinguished guests of the superintendent taught the teachers fractions and spelling and other mysteries,—white teachers in the morning, Negroes at night. A picnic now and then, and a supper, and the rough world was softened by laughter and song. I remember how— But I wander.

There came a day when all the teachers left the Institute and began the hunt for schools. I learn from hearsay (for my mother was mortally afraid of firearms) that the hunting of ducks and bears and men is wonderfully interesting, but I am sure that the man who has never hunted a country school has something to learn of the pleasures of the chase. I see now the white, hot roads lazily rise and fall and wind before me under the burning July sun; I feel the deep weariness of heart and limb as ten, eight, six miles stretch relentlessly ahead; I feel my heart sink heavily as I hear again and again, "Got a teacher? Yes." So I walked on and on—horses were too expensive—until I had wandered beyond railways, beyond stage lines, to a land of "varmints" and rattlesnakes, where the coming of a stranger was an event, and men lived and died in the shadow of one blue hill.

Sprinkled over hill and dale lay cabins and farmhouses, shut out from the world by the forests and the rolling hills toward the east. There I found at last a little school. Josie told me of it; she was a thin, homely girl of twenty, with a dark-brown face and thick, hard hair. I had crossed the stream at Watertown, and rested under the great willows; then I had gone to the little cabin in the lot where Josie was resting on her way to town. The gaunt farmer made me welcome, and Josie, hearing my errand, told me anxiously that they wanted a school over the hill; that but once since the war had a teacher been there; that she herself longed to learn,—and thus she ran on, talking fast and loud, with much earnestness and energy.

Next morning I crossed the tall round hill, lingered to look at the blue and yellow mountains stretching toward the Carolinas, then plunged into the wood, and came out at Josie's home. It was a dull frame cottage with four rooms, perched just below the brow of the hill, amid peach-trees. The father was a quiet, simple soul, calmly ignorant, with no touch of vulgarity The mother was different,—strong, bustling, and energetic, with a quick, restless tongue, and an ambition to live "like folks." There was a crowd of children. Two boys had gone away. There remained two growing girls; a shy midget of eight; John, tall, awkward, and eighteen; Jim, younger, quicker, and better looking; and two babies of indefinite age. Then there was Josie herself. She seemed to be the centre of the family: always busy at service, or at home, or berry-picking, a little nervous and inclined to scold, like her mother, yet faithful, too, like her father. She had about her a certain fineness, the shadow of an unconscious moral heroism that would willingly give all of life to make life broader, deeper, and fuller for her and hers. I saw much of this family afterwards, and grew to love them for their honest efforts to be decent and comfortable, and for their knowledge of their own ignorance. There was with them no affectation. The mother would scold the father for being so "easy"; Josie would roundly berate the boys for carelessness; and all knew that it was a hard thing to dig a living out of a rocky side-hill.

I secured the school. I remember the day I rode horseback out to the commissioner's house with a pleasant young white fellow who wanted the white school. The road ran down the bed of a stream; the sun laughed and the water jingled, and we rode on. "Come in," said the commissioner,—"come in. Have a seat. Yes, that certificate will do. Stay to dinner. What do you want a month?" "Oh," thought I, "this is lucky"; but even then fell the awful shadow of the Veil, for they ate first, then I— alone.

The schoolhouse was a log hut, where Colonel Wheeler used to shelter his corn. It sat in a lot behind a rail fence and thorn bushes, near the sweetest of springs. There was an entrance where a door once was, and within, a massive rickety fireplace; great chinks between the logs served as windows. Furniture was scarce. A pale blackboard crouched in the corner. My desk was made of three boards, reinforced at critical points, and my chair, borrowed from the landlady, had to be returned every night. Seats for the children—these puzzled me much. I was haunted by a New England vision of neat little desks and chairs, but, alas! the reality was rough plank benches without backs, and at times without legs. They had the one virtue of making naps dangerous,—possibly fatal, for the floor was not to be trusted.

It was a hot morning late in July when the school opened. I trembled when I heard the patter of little feet down the dusty road, and saw the growing row of dark solemn faces and bright eager eyes facing me. First came Josie and her brothers and sisters. The longing to know, to be a student in the great school at Nashville, hovered like a star above this child-woman amid her work and worry, and she studied doggedly. There were the Dowells from their farm over toward Alexandria,—Fanny, with her smooth black face and wondering eyes; Martha, brown and dull; the pretty girl-wife of a brother, and the younger brood.

There were the Burkes,—two brown and yellow lads, and a tiny haughty-eyed girl. Fat Reuben's little chubby girl came, with golden face and old-gold hair, faithful and solemn. 'Thenie was on hand early,—a jolly, ugly, good-hearted girl, who slyly dipped snuff and looked after her little bow-legged brother. When her mother could spare her, 'Tildy came,—a midnight beauty, with starry eyes and tapering limbs; and her brother, correspondingly homely. And then the big boys,—the hulking Lawrences; the lazy Neills, unfathered sons of mother and daughter; Hickman, with a stoop in his shoulders; and the rest.

There they sat, nearly thirty of them, on the rough benches, their faces shading from a pale cream to a deep brown, the little feet bare and swinging, the eyes full of expectation, with here and there a twinkle of mischief, and the hands grasping Webster's blue-back spelling-book. I loved my school, and the fine faith the children had in the wisdom of their teacher was truly marvellous. We read and spelled together, wrote a little, picked flowers, sang, and listened to stories of the world beyond the hill. At times the school would dwindle away, and I would start out. I would visit Mun Eddings, who lived in two very dirty rooms, and ask why little Lugene, whose flaming face seemed ever ablaze with the dark-red hair uncombed, was absent all last week, or why I missed so often the inimitable rags of Mack and Ed. Then the father, who worked Colonel Wheeler's farm on shares, would tell me how the crops needed the boys; and the thin, slovenly mother, whose face was pretty when washed, assured me that Lugene must mind the baby. "But we'll start them again next week." When the Lawrences stopped, I knew that the doubts of the old folks about book-learning had conquered again, and so, toiling up the hill, and getting as far into the cabin as possible, I put Cicero "pro Archia Poeta" into the simplest English with local applications, and usually convinced them—for a week or so.

On Friday nights I often went home with some of the children,—sometimes to Doc Burke's farm. He was a great, loud, thin Black, ever working, and trying to buy the seventy-five acres of hill and dale where he lived; but people said that he would surely fail, and the "white folks would get it all." His wife was a magnificent Amazon, with saffron face and shining hair, uncorseted and barefooted, and the children were strong and beautiful. They lived in a one-and-a-half-room cabin in the hollow of the farm, near the spring. The front room was full of great fat white beds, scrupulously neat; and there were bad chromos on the walls, and a tired centre-table. In the tiny back kitchen I was often invited to "take out and help" myself to fried chicken and wheat biscuit, "meat" and corn pone, string-beans and berries. At first I used to be a little alarmed at the approach of bedtime in the one lone bedroom, but embarrassment was very deftly avoided. First, all the children nodded and slept, and were stowed away in one great pile of goose feathers; next, the mother and the father discreetly slipped away to the kitchen while I went to bed; then, blowing out the dim light, they retired in the dark. In the morning all were up and away before I thought of awaking. Across the road, where fat Reuben lived, they all went outdoors while the teacher retired, because they did not boast the luxury of a kitchen.

I liked to stay with the Dowells, for they had four rooms and plenty of good country fare. Uncle Bird had a small, rough farm, all woods and hills, miles from the

big road; but he was full of tales,—he preached now and
then,—and with his children, berries, horses, and wheat
he was happy and prosperous. Often, to keep the peace,
I must go where life was less lovely; for instance, 'Tildy's
mother was incorrigibly dirty, Reuben's larder was limited
seriously, and herds of untamed insects wandered over
the Eddingses' beds. Best of all I loved to go to Josie's, and
sit on the porch, eating peaches, while the mother bustled
and talked: how Josie had bought the sewing-machine;
how Josie worked at service in winter, but that four dollars a month was "mighty little" wages; how Josie longed
to go away to school, but that it "looked like" they never
could get far enough ahead to let her; how the crops
failed and the well was yet unfinished; and, finally, how
"mean" some of the white folks were.

For two summers I lived in this little world; it was dull
and humdrum. The girls looked at the hill in wistful longing, and the boys fretted and haunted Alexandria. Alexandria was "town,"—a straggling, lazy village of houses,
churches, and shops, and an aristocracy of Toms, Dicks,
and Captains. Cuddled on the hill to the north was the
village of the colored folks who lived in three- or four-
room unpainted cottages, some neat and homelike, and
some dirty. The dwellings were scattered rather aimlessly,
but they centred about the twin temples of the hamlet,
the Methodist, and the Hard-Shell Baptist churches. These,
in turn, leaned gingerly on a sad-colored schoolhouse.
Hither my little world wended its crooked way on Sunday
to meet other worlds, and gossip, and wonder, and make
the weekly sacrifice with frenzied priest at the altar of the
"old-time religion." Then the soft melody and mighty
cadences of Negro song fluttered and thundered.

I have called my tiny community a world, and so its
isolation made it; and yet there was among us but a half-
awakened common consciousness, sprung from common
joy and grief, at burial, birth, or wedding; from a common hardship in poverty, poor land, and low wages; and,
above all, from the sight of the Veil that hung between us
and Opportunity. All this caused us to think some thoughts
together; but these, when ripe for speech, were spoken in
various languages. Those whose eyes twenty-five and more
years before had seen "the glory of the coming of the
Lord," saw in every present hindrance or help a dark
fatalism bound to bring all things right in His own good

time. The mass of those to whom slavery was a dim recollection of childhood found the world a puzzling thing: it
asked little of them, and they answered with little, and
yet it ridiculed their offering. Such a paradox they could
not understand, and therefore sank into listless indifference, or shiftlessness, or reckless bravado. There were,
however, some—such as Josie, Jim, and Ben—to whom
War, Hell, and Slavery were but childhood tales, whose
young appetites had been whetted to an edge by school
and story and half-awakened thought. Ill could they be
content, born without and beyond the World. And their
weak wings beat against their barriers,—barriers of caste,
of youth, of life; at last, in dangerous moments, against
everything that opposed even a whim.

The ten years that follow youth, the years when first
the realization comes that life is leading somewhere,—
these were the years that passed after I left my little
school. When they were past, I came by chance once
more to the walls of Fisk University, to the halls of the
chapel of melody. As I lingered there in the joy and
pain of meeting old school-friends, there swept over me a
sudden longing to pass again beyond the blue hill, and to
see the homes and the school of other days, and to learn
how life had gone with my school-children; and I went.

Josie was dead, and the gray-haired mother said simply, "We've had a heap of trouble since you've been away."
I had feared for Jim. With a cultured parentage and a
social caste to uphold him, he might have made a venture-
some merchant or a West Point cadet. But here he was,
angry with life and reckless; and when Farmer Durham
charged him with stealing wheat, the old man had to ride
fast to escape the stones which the furious fool hurled
after him. They told Jim to run away; but he would not
run, and the constable came that afternoon. It grieved
Josie, and great awkward John walked nine miles every
day to see his little brother through the bars of Lebanon
jail. At last the two came back together in the dark night.
The mother cooked supper, and Josie emptied her purse,
and the boys stole away. Josie grew thin and silent, yet
worked the more. The hill became steep for the quiet old
father, and with the boys away there was little to do in
the valley. Josie helped them to sell the old farm, and they
moved nearer town. Brother Dennis, the carpenter, built a

new house with six rooms; Josie toiled a year in Nashville, and brought back ninety dollars to furnish the house and change it to a home.

When the spring came, and the birds twittered, and the stream ran proud and full, little sister Lizzie, bold and thoughtless, flushed with the passion of youth, bestowed herself on the tempter, and brought home a nameless child. Josie shivered and worked on, with the vision of schooldays all fled, with a face wan and tired,—worked until, on a summer's day, some one married another; then Josie crept to her mother like a hurt child, and slept—and sleeps.

I paused to scent the breeze as I entered the valley. The Lawrences have gone,—father and son forever,—and the other son lazily digs in the earth to live. A new young widow rents out their cabin to fat Reuben. Reuben is a Baptist preacher now, but I fear as lazy as ever, though his cabin has three rooms; and little Ella has grown into a bouncing woman, and is ploughing corn on the hot hillside. There are babies a-plenty, and one half-witted girl. Across the valley is a house I did not know before, and there I found, rocking one baby and expecting another, one of my schoolgirls, a daughter of Uncle Bird Dowell. She looked somewhat worried with her new duties, but soon bristled into pride over her neat cabin and the tale of her thrifty husband, and the horse and cow, and the farm they were planning to buy.

My log schoolhouse was gone. In its place stood Progress; and Progress, I understand, is necessarily ugly. The crazy foundation stones still marked the former site of my poor little cabin, and not far away, on six weary boulders, perched a jaunty board house, perhaps twenty by thirty feet, with three windows and a door that locked. Some of the window-glass was broken, and part of an old iron stove lay mournfully under the house. I peeped through the window half reverently, and found things that were more familiar. The blackboard had grown by about two feet, and the seats were still without backs. The county owns the lot now, I hear, and every year there is a session of school. As I sat by the spring and looked on the Old and the New I felt glad, very glad, and yet—

After two long drinks I started on. There was the great double log-house on the corner. I remembered the broken, blighted family that used to live there. The strong, hard

face of the mother, with its wilderness of hair, rose before
me. She had driven her husband away, and while I taught
school a strange man lived there, big and jovial, and
people talked. I felt sure that Ben and 'Tildy would come
to naught from such a home. But this is an odd world;
for Ben is a busy farmer in Smith County, "doing well,
too," they say, and he had cared for little 'Tildy until
last spring, when a lover married her. A hard life the lad
had led, toiling for meat, and laughed at because he was
homely and crooked. There was Sam Carlon, an impudent
old skinflint, who had definite notions about "niggers," and
hired Ben a summer and would not pay him. Then the
hungry boy gathered his sacks together, and in broad daylight went into Carlon's corn; and when the hard-fisted
farmer set upon him, the angry boy flew at him like a
beast. Doc Burke saved a murder and a lynching that day.

The story reminded me again of the Burkes, and an
impatience seized me to know who won in the battle, Doc
or the seventy-five acres. For it is a hard thing to make a
farm out of nothing, even in fifteen years. So I hurried on,
thinking of the Burkes. They used to have a certain mag-
nificient barbarism about them that I liked. They were
never vulgar, never immoral, but rather rough and primitive, with an unconventionality that spent itself in loud
guffaws, slaps on the back, and naps in the corner. I
hurried by the cottage of the misborn Neill boys. It was
empty, and they were grown into fat, lazy farm-hands.
I saw the home of the Hickmans, but Albert, with his
stooping shoulders, had passed from the world. Then I
came to the Burkes' gate and peered through; the in-
closure looked rough and untrimmed, and yet there were
the same fences around the old farm save to the left, where
lay twenty-five other acres. And lo! the cabin in the hollow
had climbed the hill and swollen to a half-finished six-
room cottage.

The Burkes held a hundred acres, but they were still
in debt. Indeed, the gaunt father who toiled night and day
would scarcely be happy out of debt, being so used to it.
Some day he must stop, for his massive frame is showing
decline. The mother wore shoes, but the lion-like
physique of other days was broken. The children had
grown up. Rob, the image of his father, was loud and
rough with laughter. Birdie, my school baby of six, had
grown to a picture of maiden beauty, tall and tawny.

"Edgar is gone," said the mother, with head half bowed, —"gone to work in Nashville; he and his father couldn't agree."

Little Doc, the boy born since the time of my school, took me horseback down the creek next morning toward Farmer Dowell's. The road and the stream were battling for mastery, and the stream had the better of it. We splashed and waded, and the merry boy, perched behind me, chattered and laughed. He showed me where Simon Thompson had bought a bit of ground and a home; but his daughter Lana, a plump, brown, slow girl, was not there. She had married a man and a farm twenty miles away. We wound on down the stream till we came to a gate that I did not recognize, but the boy insisted that it was "Uncle Bird's." The farm was fat with the growing crop. In that little valley was a strange stillness as I rode up; for death and marriage had stolen youth and left age and childhood there. We sat and talked that night after the chores were done. Uncle Bird was grayer, and his eyes did not see so well, but he was still jovial. We talked of the acres bought,—one hundred and twenty-five,—of the new guest-chamber added, of Martha's marrying. Then we talked of death: Fanny and Fred were gone; a shadow hung over the other daughter, and when it lifted she was to go to Nashville to school. At last we spoke of the neighbors, and as night fell, Uncle Bird told me how, on a night like that, 'Thenie came wandering back to her home over yonder, to escape the blows of her husband. And next morning she died in the home that her little bow-legged brother, working and saving, had bought for their widowed mother.

My journey was done, and behind me lay hill and dale, and Life and Death. How shall man measure Progress there where the dark-faced Josie lies? How many heartfuls of sorrow shall balance a bushel of wheat? How hard a thing is life to the lowly, and yet how human and real! And all this life and love and strife and failure,—is it the twilight of nightfall or the flush of some faint-dawning day?

Thus sadly musing, I rode to Nashville in the Jim Crow car.

V

Of the Wings of Atalanta

O black boy of Atlanta!
  But half was spoken;
The slave's chains and the master's
  Alike are broken;
The one curse of the races
  Held both in tether;
They are rising—all are rising—
  The black and white together.

WHITTIER.

South of the North, yet north of the South, lies the City
of a Hundred Hills, peering out from the shadows of the
past into the promise of the future. I have seen her in
the morning, when the first flush of day had half-roused
her; she lay gray and still on the crimson soil of Georgia;
then the blue smoke began to curl from her chimneys, the
tinkle of bell and scream of whistle broke the silence, the
rattle and roar of busy life slowly gathered and swelled,
until the seething whirl of the city seemed a strange thing
in a sleepy land.

Once, they say, even Atlanta slept dull and drowsy at
the foot-hills of the Alleghenies, until the iron baptism of
war awakened her with its sullen waters, aroused and mad-
dened her, and left her listening to the sea. And the sea
cried to the hills and the hills answered the sea, till the
city rose like a widow and cast away her weeds, and toiled
for her daily bread; toiled steadily, toiled cunningly,—
perhaps with some bitterness, with a touch of *réclame*,
and yet with real earnestness, and real sweat.

It is a hard thing to live haunted by the ghost of an
untrue dream; to see the wide vision of empire fade into
real ashes and dirt; to feel the pang of the conquered, and

yet know that with all the Bad that fell on one black day, something was vanquished that deserved to live, something killed that in justice had not dared to die; to know that with the Right that triumphed, triumphed something of Wrong, something sordid and mean, something less than the broadest and best. All this is bitter hard; and many a man and city and people have found in it excuse for sulking, and brooding, and listless waiting.

Such are not men of the sturdier make; they of Atlanta turned resolutely toward the future; and that future held aloft vistas of purple and gold:—Atlanta, Queen of the cotton kingdom; Atlanta, Gateway to the Land of the Sun; Atlanta, the new Lachesis, spinner of web and woof for the world. So the city crowned her hundred hills with factories, and stored her shops with cunning handiwork, and stretched long iron ways to greet the busy Mercury in his coming. And the Nation talked of her striving.

Perhaps Atlanta was not christened for the winged maiden of dull Bœotia; you know the tale,—how swarthy Atalanta, tall and wild, would marry only him who outraced her; and how the wily Hippomenes laid three apples of gold in the way. She fled like a shadow, paused, startled over the first apple, but even as he stretched his hand, fled again; hovered over the second, then, slipping from his hot grasp, flew over river, vale, and hill; but as she lingered over the third, his arms fell round her, and looking on each other, the blazing passion of their love profaned the sanctuary of Love, and they were cursed. If Atlanta be not named for Atalanta, she ought to have been.

Atalanta is not the first or the last maiden whom greed of gold has led to defile the temple of Love; and not maids alone, but men in the race of life, sink from the high and generous ideals of youth to the gambler's code of the Bourse; and in all our Nation's striving is not the Gospel of Work befouled by the Gospel of Pay? So common is this that one-half think it normal; so unquestioned, that we almost fear to question if the end of racing is not gold, if the aim of man is not rightly to be rich. And if this is the fault of America, how dire a danger lies before a new land and a new city, lest Atlanta, stooping for mere gold, shall find that gold accursed!

It was no maiden's idle whim that started this hard

racing; a fearful wilderness lay about the feet of that city
after the War,—feudalism, poverty, the rise of the Third
Estate, serfdom, the re-birth of Law and Order, and above
and between all, the Veil of Race. How heavy a journey
for weary feet! what wings must Atalanta have to flit over
all this hollow and hill, through sour wood and sullen
water, and by the red waste of sun-baked clay! How fleet
must Atalanta be if she will not be tempted by gold to
profane the Sanctuary!

The Sanctuary of our fathers has, to be sure, few Gods,
—some sneer, "all too few." There is the thrifty Mercury
of New England, Pluto of the North, and Ceres of the
West; and there, too, is the half-forgotten Apollo of the
South, under whose ægis the maiden ran,—and as she
ran she forgot him, even as there in Bœotia Venus was
forgot. She forgot the old ideal of the Southern gentleman,—that new-world heir of the grace and courtliness of
patrician, knight, and noble; forgot his honor with his
foibles, his kindliness with his carelessness, and stooped
to apples of gold,—to men busier and sharper, thriftier
and more unscrupulous. Golden apples are beautiful—I
remember the lawless days of boyhood, when orchards in
crimson and gold tempted me over fence and field—and,
too, the merchant who has dethroned the planter is no
despicable *parvenu*. Work and wealth are the mighty levers
to lift this old new land; thrift and toil and saving are
the highways to new hopes and new possibilities; and yet
the warning is needed lest the wily Hippomenes tempt
Atalanta to thinking that golden apples are the goal of
racing, and not mere incidents by the way.

Atlanta must not lead the South to dream of material
prosperity as the touchstone of all success; already the
fatal might of this idea is beginning to spread; it is re-
placing the finer type of Southerner with vulgar money-
getters; it is burying the sweeter beauties of Southern life
beneath pretence and ostentation. For every social ill the
panacea of Wealth has been urged,—wealth to overthrow
the remains of the slave feudalism; wealth to raise the
"cracker" Third Estate; wealth to employ the black serfs,
and the prospect of wealth to keep them working; wealth
as the end and aim of politics, and as the legal tender for
law and order; and, finally, instead of Truth, Beauty, and
Goodness, wealth as the ideal of the Public School.

Not only is this true in the world which Atlanta typifies,

but it is threatening to be true of a world beneath and beyond that world,—the Black World beyond the Veil. To-day it makes little difference to Atlanta, to the South, what the Negro thinks or dreams or wills. In the soul-life of the land he is to-day, and naturally will long remain, unthought of, half forgotten; and yet when he does come to think and will and do for himself,—and let no man dream that day will never come,—then the part he plays will not be one of sudden learning, but words and thoughts he has been taught to lisp in his race-childhood. To-day the ferment of his striving toward self-realization is to the strife of the white world like a wheel within a wheel: beyond the Veil are smaller but like problems of ideals, of leaders and the led, of serfdom, of poverty, of order and subordination, and, through all, the Veil of Race. Few know of these problems, few who know notice them; and yet there they are, awaiting student, artist, and seer,—a field for somebody sometime to discover. Hither has the temptation of Hippomenes penetrated; already in this smaller world, which now indirectly and anon directly must influence the larger for good or ill, the habit is forming of interpreting the world in dollars. The old leaders of Negro opinion, in the little groups where there is a Negro social consciousness, are being replaced by new; neither the black preacher nor the black teacher leads as he did two decades ago. Into their places are pushing the farmers and gardeners, the well-paid porters and artisans, the business-men,—all those with property and money. And with all this change, so curiously parallel to that of the Other-world, goes too the same inevitable change in ideals. The South laments to-day the slow, steady disappearance of a certain type of Negro,—the faithful, courteous slave of other days, with his incorruptible honesty and dignified humility. He is passing away just as surely as the old type of Southern gentleman is passing, and from not dissimilar causes,—the sudden transformation of a fair far-off ideal of Freedom into the hard reality of bread-winning and the consequent deification of Bread.

In the Black World, the Preacher and Teacher embodied once the ideals of this people,—the strife for another and a juster world, the vague dream of righteousness, the mystery of knowing; but to-day the danger is that these ideals, with their simple beauty and weird inspiration, will suddenly sink to a question of cash and a

lust for gold. Here stands this black young Atalanta, gird-
ing herself for the race that must be run; and if her eyes
be still toward the hills and sky as in the days of old,
then we may look for noble running; but what if some
ruthless or wily or even thoughtless Hippomenes lay
golden apples before her? What if the Negro people be
wooed from a strife for righteousness, from a love of
knowing, to regard dollars as the be-all and end-all of life?
What if to the Mammonism of America be added the rising
Mammonism of the re-born South, and the Mammonism
of this South be reinforced by the budding Mammonism
of its half-wakened black millions? Whither, then, is the
new-world quest of Goodness and Beauty and Truth gone
glimmering? Must this, and that fair flower of Freedom
which, despite the jeers of latter-day striplings, sprung
from our fathers' blood, must that too degenerate into a
dusty quest of gold,—into lawless lust with Hippomenes?

The hundred hills of Atlanta are not all crowned with
factories. On one, toward the west, the setting sun throws
three buildings in bold relief against the sky. The beauty
of the group lies in its simple unity:—a broad lawn of
green rising from the red street with mingled roses and
peaches; north and south, two plain and stately halls;
and in the midst, half hidden in ivy, a larger building,
boldly graceful, sparingly decorated, and with one low
spire. It is a restful group,—one never looks for more; it
is all here, all intelligible. There I live, and there I hear
from day to day the low hum of restful life. In winter's
twilight, when the red sun glows, I can see the dark
figures pass between the halls to the music of the night-
bell. In the morning, when the sun is golden, the clang
of the day-bell brings the hurry and laughter of three
hundred young hearts from hall and street, and from the
busy city below,—children all dark and heavy-haired,—
to join their clear young voices in the music of the morning sacrifice. In a half-dozen class-rooms they gather then,
—here to follow the love-song of Dido, here to listen to
the tale of Troy divine; there to wander among the stars,
there to wander among men and nations,—and elsewhere other well-worn ways of knowing this queer world.
Nothing new, no time-saving devices,—simply old time-
glorified methods of delving for Truth, and searching out
the hidden beauties of life, and learning the good of living.

The riddle of existence is the college curriculum that was laid before the Pharaohs, that was taught in the groves by Plato, that formed the *trivium* and *quadrivium*, and is to-day laid before the freedman's sons by Atlanta University. And this course of study will not change; its methods will grow more deft and effectual, its content richer by toil of scholar and sight of seer; but the true college will ever have one goal,—not to earn meat, but to know the end and aim of that life which meat nourishes.

The vision of life that rises before these dark eyes has in it nothing mean or selfish. Not at Oxford or at Leipsig, not at Yale or Columbia, is there an air of higher resolve or more unfettered striving; the determination to realize for men, both black and white, the broadest possibilities of life, to seek the better and the best, to spread with their own hands the Gospel of Sacrifice,—all this is the burden of their talk and dream. Here, amid a wide desert of caste and proscription, amid the heart-hurting slights and jars and vagaries of a deep race-dislike, lies this green oasis, where hot anger cools, and the bitterness of disappointment is sweetened by the springs and breezes of Parnassus; and here men may lie and listen, and learn of a future fuller than the past, and hear the voice of Time:

"Entbehren sollst du, sollst entbehren."

They made their mistakes, those who planted Fisk and Howard and Atlanta before the smoke of battle had lifted; they made their mistakes, but those mistakes were not the things at which we lately laughed somewhat uproariously. They were right when they sought to found a new educational system upon the University: where, forsooth, shall we ground knowledge save on the broadest and deepest knowledge? The roots of the tree, rather than the leaves, are the sources of its life; and from the dawn of history, from Academus to Cambridge, the culture of the University has been the broad foundation-stone on which is built the kindergarten's A B C.

But these builders did make a mistake in minimizing the gravity of the problem before them; in thinking it a matter of years and decades; in therefore building quick-ly and laying their foundation carelessly, and lowering the standard of knowing, until they had scattered haphazard through the South some dozen poorly equipped high schools and miscalled them universities. They forgot, too, just as their successors are forgetting, the rule of inequality:—that of the million black youth, some were fitted to know and some to dig; that some had the talent and capacity of university men, and some the talent and capacity of blacksmiths; and that true training meant neither that all should be college men nor all artisans, but that the one should be made a missionary of culture to an untaught people, and the other a free workman among serfs. And to seek to make the blacksmith a scholar is almost as silly as the more modern scheme of making the scholar a blacksmith; almost, but not quite.

The function of the university is not simply to teach bread-winning, or to furnish teachers for the public schools or to be a centre of polite society; it is, above all, to be the organ of that fine adjustment between real life and the growing knowledge of life, an adjustment which forms the secret of civilization. Such an institution the South of to-day sorely needs. She has religion, earnest, bigoted:—religion that on both sides of the Veil often omits the sixth, seventh, and eighth commandments, but substitutes a dozen supplementary ones. She has, as Atlanta shows, growing thrift and love of toil; but she lacks that broad knowledge of what the world knows and knew of human living and doing, which she may apply to the thousand problems of real life to-day confronting her. The need of the South is knowledge and culture,—not in dainty limited quantity, as before the war, but in broad busy abundance in the world of work; and until she has this, not all the Apples of Hesperides, be they golden and bejewelled, can save her from the curse of the Bœotian lovers.

The Wings of Atalanta are the coming universities of the South. They alone can bear the maiden past the temtation of golden fruit. They will not guide her flying feet away from the cotton and gold; for—ah, thoughtful Hippomenes!—do not the apples lie in the very Way of Life? But they will guide her over and beyond them, and leave her kneeling in the Sanctuary of Truth and Freedom and broad Humanity, virgin and undefiled. Sadly did the Old South err in human education, despising the education of the masses, and niggardly in the support of colleges. Her ancient university foundations dwindled and withered under the foul breath of slavery; and even since the war they have fought a failing fight for life in the tainted air of social unrest and commercial selfishness, stunted by the death of criticism, and starving for lack of broadly cultured men. And if this is the white South's need and danger, how much heavier the danger and need of the freedmen's sons! how pressing here the need of broad ideals and true culture, the conservation of soul from sordid aims and petty passions! Let us build the Southern university—William and Mary, Trinity, Georgia, Texas, Tulane, Vanderbilt, and the others—fit to live; let us build, too, the Negro universities:—Fisk, whose foundation was ever broad; Howard, at the heart of the Nation; Atlanta at Atlanta, whose ideal of scholarship has been held above the temptation of numbers. Why not here, and perhaps elsewhere, plant deeply and for all time centres of learning and living, colleges that yearly would send into the life of the South a few white men and a few black men of broad culture, catholic tolerance, and trained ability, joining their hands to other hands, and giving to this squabble of the Races a decent and dignified peace?

Patience, Humility, Manners, and Taste, common schools and kindergartens, industrial and technical schools, literature and tolerance,—all these spring from knowledge and culture, the children of the university. So must men and nations build, not otherwise, not upside down.

Teach workers to work,—a wise saying; wise when applied to German boys and American girls; wiser when said of Negro boys, for they have less knowledge of working and none to teach them. Teach thinkers to think,—a needed knowledge in a day of loose and careless logic; and they whose lot is gravest must have the carefulest training to think aright. If these things are so, how foolish to ask what is the best education for one or seven or sixty million souls! shall we teach them trades, or train them in liberal arts? Neither and both: teach the workers to work and the thinkers to think; make carpenters of carpenters, and philosophers of philosophers, and fops of fools. Nor can we pause here. We are training not isolated men but a living group of men,—nay, a group within a group. And the final product of our training must be neither a psychologist nor a brickmason, but a man. And to make men, we must have ideals, broad, pure, and inspiring ends of living,—not sordid money-getting, not apples of gold. The worker must work for the glory of his handiwork, not simply for pay; the thinker must think for truth, not for fame. And all this is gained only by human strife and longing; by ceaseless training and education; by founding Right on righteousness and Truth on the unhampered search for Truth; by founding the common school on the university, and the industrial school on the common school; and weaving thus a system, not a distortion, and bringing a birth, not an abortion.

When night falls on the City of a Hundred Hills, a wind gathers itself from the seas and comes murmuring westward. And at its bidding, the smoke of the drowsy factories sweeps down upon the mighty city and covers it like a pall, while yonder at the University the stars twinkle above Stone Hall. And they say that yon gray mist is the tunic of Atalanta pausing over her golden apples. Fly, my maiden, fly, for yonder comes Hippomenes!

VI

Of the Training of Black Men

Why, if the Soul can fling the Dust aside,
And naked on the Air of Heaven ride,
    Were't not a Shame—were't not a Shame for him
In this clay carcase crippled to abide?
                            OMAR KHAYYÁM (FITZGERALD).

From the shimmering swirl of waters where many, many thoughts ago the slave-ship first saw the square tower of Jamestown, have flowed down to our day three streams of thinking: one swollen from the larger world here and over-seas, saying, the multiplying of human wants in culture-lands calls for the world-wide coöperation of men in satisfying them. Hence arises a new human unity, pulling the ends of earth nearer, and all men, black, yellow, and white. The larger humanity strives to feel in this contact of living Nations and sleeping hordes a thrill of new life in the world, crying, "If the contact of Life and Sleep be Death, shame on such Life." To be sure, behind this thought lurks the afterthought of force and dominion,—the making of brown men to delve when the temptation of beads and red calico cloys.

The second thought streaming from the death-ship and the curving river is the thought of the older South,—the sincere and passionate belief that somewhere between men and cattle, God created a *tertium quid,* and called it a Negro,—a clownish, simple creature, at times even lovable within its limitations, but straitly foreordained to walk within the Veil. To be sure, behind the thought lurks the afterthought,—some of them with favoring chance might become men, but in sheer self-defence we dare not let them, and we build about them walls so high, and hang between them and the light a veil so thick, that they shall not even think of breaking through.

And last of all there trickles down that third and darker thought,—the thought of the things themselves, the confused, half-conscious mutter of men who are black and whitened, crying "Liberty, Freedom, Opportunity— vouchsafe to us, O boastful World, the chance of living men!" To be sure, behind the thought lurks the afterthought,—suppose, after all, the World is right and we are less than men? Suppose this mad impulse within is all wrong, some mock mirage from the untrue?

So here we stand among thoughts of human unity, even through conquest and slavery; the inferiority of black men, even if forced by fraud; a shriek in the night for the freedom of men who themselves are not yet sure of their right to demand it. This is the tangle of thought and afterthought wherein we are called to solve the problem of training men for life.

Behind all its curiousness, so attractive alike to sage and *dilettante,* lie its dim dangers, throwing across us shadows at once grotesque and awful. Plain it is to us that what the world seeks through desert and wild we have

within our threshold,—a stalwart laboring force, suited
to the semi-tropics; if, deaf to the voice of the Zeitgeist,
we refuse to use and develop these men, we risk poverty
and loss. If, on the other hand, seized by the brutal afterthought, we debauch the race thus caught in our talons,
selfishly sucking their blood and brains in the future as in
the past, what shall save us from national decadence?
Only that saner selfishness, which Education teaches, can
find the rights of all in the whirl of work.

Again, we may decry the color-prejudice of the South,
yet it remains a heavy fact. Such curious kinks of the
human mind exist and must be reckoned with soberly.
They cannot be laughed away, nor always successfully
stormed at, nor easily abolished by act of legislature. And
yet they must not be encouraged by being let alone. They
must be recognized as facts, but unpleasant facts; things
that stand in the way of cililization and religion and
common decency. They can be met in but one way,—by
the breadth and broadening of human reason, by catholic-
ity of taste and culture. And so, too, the native ambition
and aspiration of men, even though they be black, backward, and ungraceful, must not lightly be dealt with. To
stimulate wildly weak and untrained minds is to play with
mighty fires; to flout their striving idly is to welcome a
harvest of brutish crime and shameless lethargy in our
very laps. The guiding of thought and the deft coordina-
tion of deed is at once the path of honor and humanity.

And so, in this great question of reconciling three vast
and partially contradictory streams of thought, the one
panacea of Education leaps to the lips of all:—such human
training as will best use the labor of all men without
enslaving or brutalizing; such training as will give us
poise to encourage the prejudices that bulwark society,
and to stamp out those that in sheer barbarity deafen us to
the wail of prisoned souls within the Veil, and the mount-
ing fury of shackled men.

But when we have vaguely said that Education will
set this tangle straight, what have we uttered but a truism?
Training for life teaches living; but what training for the
profitable living together of black men and white? A
hundred and fifty years ago our task would have seemed
easier. Then Dr. Johnson blandly assured us that education
was needful solely for the embellishments of life, and
was useless for ordinary vermin. To-day we have climbed

to heights where we would open at least the outer courts of knowledge to all, display its treasures to many, and select the few to whom its mystery of Truth is revealed, not wholly by birth or the accidents of the stock market, but at least in part according to deftness and aim, talent and character. This programme, however, we are sorely puzzled in carrying out through that part of the land where the blight of slavery fell hardest, and where we are dealing with two backward peoples. To make here in human education that ever necessary combination of the permanent and the contingent—of the ideal and the practical in workable equilibrium—has been there, as it ever must be in every age and place, a matter of infinite experiment and frequent mistakes.

In rough approximation we may point out four varying decades of work in Southern education since the Civil War. From the close of the war until 1876, was the period of uncertain groping and temporary relief. There were army schools, mission schools, and schools of the Freedman's Bureau in chaotic disarrangement seeking system and coöperation. Then followed ten years of constructive definite effort toward the building of complete school systems in the South. Normal schools and colleges were founded for the freedmen, and teachers trained there to man the public schools. There was the inevitable tendency of war to underestimate the prejudices of the master and the ignorance of the slave, and all seemed clear sailing out of the wreckage of the storm. Meantime, starting in this decade yet especially developing from 1885 to 1895, began the industrial revolution of the South. The land saw glimpses of a new destiny and the stirring of new ideals. The educational system striving to complete itself saw new obstacles and a field of work ever broader and deeper. The Negro colleges, hurriedly founded, were inadequately equipped, illogically distributed, and of varying efficiency and grade; the normal and high schools were doing little more than common-school work, and the common schools were training but a third of the children who ought to be in them, and training these too often poorly. At the same time the white South, by reason of its sudden conversion from the slavery ideal, by so much the more became set and strengthened in its racial prejudice, and crystallized it into harsh law and harsher custom; while the marvellous pushing forward of the poor white daily threatened to take even bread and butter from the mouths of the heavily handicapped sons of the freedmen. In the midst, then, of the larger problem of Negro education sprang up the more practical question of work, the inevitable economic quandary that faces a people in the transition from slavery to freedom, and especially those who make that change amid hate and prejudice, lawlessness and ruthless competition.

The industrial school springing to notice in this decade, but coming to full recognition in the decade beginning with 1895, was the proffered answer to this combined educational and economic crisis, and an answer of singular wisdom and timeliness. From the very first in nearly all the schools some attention had been given to training in handiwork, but now was this training first raised to a dignity that brought it in direct touch with the South's magnificent industrial development, and given an emphasis which reminded black folk that before the Temple of Knowledge swing the Gates of Toil.

Yet after all they are but gates, and when turning our eyes from the temporary and the contingent in the Negro problem to the broader question of the permanent uplifting and civilization of black men in America, we have a right to inquire, as this enthusiasm for material advancement mounts to its height, if after all the industrial school is the final and sufficient answer in the training of the Negro race; and to ask gently, but in all sincerity, the ever-recurring query of the ages, Is not life more than meat, and the body more than raiment? And men ask this to-day all the more eagerly because of sinister signs in recent educational movements. The tendency is here, born of slavery and quickened to renewed life by the crazy imperialism of the day, to regard human beings as among the material resources of a land to be trained with an eye single to future dividends. Race-prejudices, which keep brown and black men in their "places," we are coming to regard as useful allies with such a theory, no matter how much they may dull the ambition and sicken the hearts of struggling human beings. And above all, we daily hear that an education that encourages aspiration, that sets the loftiest of ideals and seeks as an end culture and character rather than bread-winning, is the privilege of white men and the danger and delusion of black.

Especially has criticism been directed against the former

educational efforts to aid the Negro. In the four periods
I have mentioned, we find first, boundless, planless enthusiasm and sacrifice; then the preparation of teachers
for a vast public-school system; then the launching and
expansion of that school system amid increasing difficulties; and finally the training of workmen for the new and
growing industries. This development has been sharply
ridiculed as a logical anomaly and flat reversal of nature.
Soothly we have been told that first industrial and manual
training should have taught the Negro to work, then
simple schools should have taught him to read and write,
and finally, after years, high and normal schools could
have completed the system, as intelligence and wealth
demanded.

That a system logically so complete was historically
impossible, it needs but a little thought to prove. Progress
in human affairs is more often a pull than a push, a surg-
ing forward of the exceptional man, and the lifting of his
duller brethren slowly and painfully to his vantage-ground.
Thus it was no accident that gave birth to universities
centuries before the common schools, that made fair Harvard the first flower of our wilderness. So in the South:
the mass of the freedmen at the end of the war lacked
the intelligence so necessary to modern working-men.
They must first have the common school to teach them
to read, write, and cipher; and they must have higher
schools to teach teachers for the common schools. The
white teachers who flocked South went to establish such
a common-school system. Few held the idea of founding
colleges; most of them at first would have laughed at the
idea. But they faced, as all men since them have faced,
that central paradox of the South,—the social separation
of the races. At that time it was the sudden volcanic
rupture of nearly all relations between black and white,
in work and government and family life. Since then a
new adjustment of relations in economic and political
affairs has grown up,—an adjustment subtle and difficult
to grasp, yet singularly ingenious, which leaves still that
frightful chasm at the color-line across which men pass
at their peril. Thus, then and now, there stand in the South
two separate worlds; and separate not simply in the higher
realms of social intercourse, but also in church and
school, on railway and street-car, in hotels and theatres,
in streets and city sections, in books and newspapers,

in asylums and jails, in hospitals and graveyards. There is still enough of contact for large economic and group coöperation, but the separation is so thorough and deep that it absolutely precludes for the present between the races anything like that sympathetic and effective group-training and leadership of the one by the other, such as the American Negro and all backward peoples must have for effectual progress.

This the missionaries of '68 soon saw; and if effective industrial and trade schools were impracticable before the establishment of a common-school system, just as certainly no adequate common schools could be founded until there were teachers to teach them. Southern whites would not teach them; Northern whites in sufficient numbers could not be had. If the Negro was to learn, he must teach himself, and the most effective help that could be given him was the establishment of schools to train Negro teachers. This conclusion was slowly but surely reached by every student of the situation until simultaneously, in widely separated regions, without consultation or systematic plan, there arose a series of institutions designed to furnish teachers for the untaught. Above the sneers of critics at the obvious defects of this procedure must ever stand its one crushing rejoinder: in a single generation they put thirty thousand black teachers in the South; they wiped out the illiteracy of the majority of the black people of the land, and they made Tuskegee possible.

Such higher training-schools tended naturally to deepen broader development: at first they were common and grammar schools, then some became high schools. And finally, by 1900, some thirty-four had one year or more of studies of college grade. This development was reached with different degrees of speed in different institutions: Hampton is still a high school, while Fisk University started her college in 1871, and Spelman Seminary about 1896. In all cases the aim was identical,—to maintain the standards of the lower training by giving teachers and leaders the best practicable training; and above all, to furnish the black world with adequate standards of human culture and lofty ideals of life. It was not enough that the teachers of teachers should be trained in technical normal methods; they must also, so far as possible, be broad-minded, cultured men and women, to scatter civilization among a people whose ignorance was not simply of letters, but of life itself.

It can thus be seen that the work of education in the South began with higher institutions of training, which threw off as their foliage common schools, and later industrial schools, and at the same time strove to shoot their roots ever deeper toward college and university training. That this was an inevitable and necessary development, sooner or later, goes without saying; but there has been, and still is, a question in many minds if the natural growth was not forced, and if the higher training was not either overdone or done with cheap and unsound methods. Among white Southerners this feeling is widespread and positive. A prominent Southern journal voiced this in a recent editorial.

"The experiment that has been made to give the colored students classical training has not been satisfactory. Even though many were able to pursue the course, most of them did so in a parrot-like way, learning what was taught, but not seeming to appropriate the truth and import of their instruction, and graduating without sensible aim or valuable occupation for their future. The whole scheme has proved a waste of time, effort, and the money of the state."

While most fair-minded men would recognize this as extreme and overdrawn, still without doubt many are asking, Are there a sufficient number of Negroes ready for college training to warrant the undertaking? Are not too many students prematurely forced into this work? Does it not have the effect of dissatisfying the young Negro with his environment? And do these graduates succeed in real life? Such natural questions cannot be evaded, nor on the other hand must a Nation naturally skeptical as to Negro ability assume an unfavorable answer without careful inquiry and patient openness to conviction. We must not forget that most Americans answer all queries regarding the Negro *a priori*, and that the least that human courtesy can do is to listen to evidence.

The advocates of the higher education of the Negro would be the last to deny the incompleteness and glaring defects of the present system: too many institutions have attempted to do college work, the work in some cases has not been thoroughly done, and quantity rather than quality has sometimes been sought. But all this can be said of higher education throughout the land; it is the almost inevitable incident of educational growth, and leaves the deeper question of the legitimate demand for the higher training of Negroes untouched. And this latter question can be settled in but one way,—by a first-hand study of the facts. If we leave out of view all institutions which have not actually graduated students from a course higher than that of a New England high school, even though they be called colleges; if then we take the thirty-four remaining institutions, we may clear up many misapprehensions by asking searchingly, What kind of institutions are they? what do they teach? and what sort of men do they graduate?

And first we may say that this type of college, including Atlanta, Fisk, and Howard, Wilberforce and Claflin, Shaw, and the rest, is peculiar, almost unique. Through the shining trees that whisper before me as I write, I catch glimpses of a boulder of New England granite, covering a grave, which graduates of Atlanta University have placed there,—

"GRATEFUL MEMORY OF THEIR
FORMER TEACHER AND FRIEND
AND OF THE UNSELFISH LIFE HE
LIVED, AND THE NOBLE WORK HE
WROUGHT; THAT THEY, THEIR
CHILDREN, AND THEIR CHIL-
DREN'S CHILDREN MIGHT BE
BLESSED."

This was the gift of New England to the freed Negro: not alms, but a friend; not cash, but character. It was not and is not money these seething millions want, but love and sympathy, the pulse of hearts beating with red blood; —a gift which to-day only their own kindred and race can bring to the masses, but which once saintly souls brought to their favored children in the crusade of the sixties, that finest thing in American history, and one of the few things untainted by sordid greed and cheap vainglory. The teachers in these institutions came not to keep the Negroes in their place, but to raise them out of the defilement of the places where slavery had wallowed them. The colleges they founded were social settlements; homes where the best of the sons of the freedmen came in close and sympathetic touch with the best traditions of New England. They lived and ate together, studied and worked, hoped and harkened in the dawning light. In actual formal content their curriculum was doubtless old-fashioned, but in educational power it was supreme, for it was the contact of living souls.

From such schools about two thousand Negroes have gone forth with the bachelor's degree. The number in itself is enough to put at rest the argument that too large a proportion of Negroes are receiving higher training. If the ratio to population of all Negro students throughout the land, in both college and secondary training, be counted, Commissioner Harris assures us "it must be increased to five times its present average" to equal the average of the land.

Fifty years ago the ability of Negro students in any appreciable numbers to master a modern college course would have been difficult to prove. To-day it is proved by the fact that four hundred Negroes, many of whom have been reported as brilliant students, have received the bachelor's degree from Harvard, Yale, Oberlin, and seventy other leading colleges. Here we have, then, nearly twenty-five hundred Negro graduates, of whom the crucial query must be made. How far did their training fit them for life? It is of course extremely difficult to collect satisfactory data on such a point,—difficult to reach the men, to get trustworthy testimony, and to gauge that testimony by any generally acceptable criterion of success. In 1900, the Conference at Atlanta University undertook to study these graduates, and published the results. First they sought to know what these graduates were doing, and succeeded in getting answers from nearly two-thirds of the living. The direct testimony was in almost all cases corroborated by the reports of the colleges where they graduated, so that in the main the reports were worthy of credence. Fifty-three per cent of these graduates were teachers,- presidents of institutions, heads of normal schools, principals of city school-systems, and the like. Seventeen per cent were clergymen; another seventeen per cent were in the professions, chiefly as physicians. Over six per cent were merchants, farmers, and artisans, and four per cent were in the government civil-service. Granting even that a considerable proportion of the third unheard from are unsuccessful, this is a record of usefulness. Personally I know many hundreds of these graduates, and have corresponded with more than a thousand; through

others I have followed carefully the life-work of scores;
I have taught some of them and some of the pupils whom
they have taught, lived in homes which they have builded,
and looked at life through their eyes. Comparing them as
a class with my fellow students in New England and in
Europe, I cannot hesitate in saying that nowhere have I
met men and women with a broader spirit of helpfulness,
with deeper devotion to their life-work, or with more
consecrated determination to succeed in the face of bitter
difficulties than among Negro college-bred men. They have,
to be sure, their proportion of ne'er-do-wells, their pedants
and lettered fools, but they have a surprisingly small proportion of them; they have not that culture of manner
which we instinctively associate with university men, forgetting that in reality it is the heritage from cultured
homes, and that no people a generation removed from
slavery can escape a certain unpleasant rawness and
gaucherie, despite the best of training.

With all their larger vision and deeper sensibility, these
men have usually been conservative, careful leaders. They
have seldom been agitators, have withstood the temptation
to head the mob, and have worked steadily and faithfully
in a thousand communities in the South. As teachers, they
have given the South a commendable system of city
schools and large numbers of private normal-schools and
academies. Colored college-bred men have worked side
by side with white college graduates at Hampton; almost
from the beginning the backbone of Tuskegee's teaching
force has been formed of graduates from Fisk and Atlanta.
And to-day the institute is filled with college graduates,
from the energetic wife of the principal down to the
teacher of agriculture, including nearly half of the executive council and a majority of the heads of departments.
In the professions, college men are slowly but surely
leavening the Negro church, are healing and preventing
the devastations of disease, and beginning to furnish legal
protection for the liberty and property of the toiling
masses. All this is needful work. Who would do it if
Negroes did not? How could Negroes do it if they were not
trained carefully for it? If white people need colleges to
furnish teachers, ministers, lawyers, and doctors, do black
people need nothing of the sort?

If it is true that there are an appreciable number of
Negro youths in the land capable by character and talent

to receive that higher training, the end of which is culture, and if the two and a half thousand who have had something of this training in the past have in the main proved themselves useful to their race and generation, the question then comes, What place in the future development of the South ought the Negro college and college-bred man to occupy? That the present social separation and acute race-sensitiveness must eventually yield to the influences of culture, as the South grows civilized, is clear. But such transformation calls for singular wisdom and patience. If, while the healing of this vast sore is progressing, the races are to live for many years side by side, united in economic effort, obeying a common government, sensitive to mutual thought and feeling, yet subtly and silently separate in many matters of deeper human intimacy,—if this unusual and dangerous development is to progress amid peace and order, mutual respect and growing intelligence, it will call for social surgery at once the delicatest and nicest in modern history. It will demand broadminded, upright men, both white and black, and in its final accomplishment American civilization will triumph. So far as white men are concerned, this fact is to-day being recognized in the South, and a happy renaissance of university education seems imminent. But the very voices that cry hail to this good work are, strange to relate, largely silent or antagonistic to the higher education of the Negro.

Strange to relate! for this is certain, no secure civilization can be built in the South with the Negro as an ignorant, turbulent proletariat. Suppose we seek to remedy this by making them laborers and nothing more: they are not fools, they have tasted of the Tree of Life, and they will not cease to think, will not cease attempting to read the riddle of the world. By taking away their best equipped teachers and leaders, by slamming the door of opportunity in the faces of their bolder and brighter minds, will you make them satisfied with their lot? or will you not rather transfer their leading from the hands of men taught to think to the hands of untrained demagogues? We ought not to forget that despite the pressure of poverty, and despite the active discouragement and even ridicule of friends, the demand for higher training steadily increases among Negro youth: there were, in the years from 1875 to 1880, 22 Negro graduates from Northern colleges; from 1885 to 1890 there were 43, and from 1895 to 1900, nearly 100 graduates. From Southern Negro colleges there were, in the same three periods, 143, 413, and over 500 graduates. Here, then, is the plain thirst for training; by refusing to give this Talented Tenth the key to knowledge, can any sane man imagine that they will lightly lay aside their yearning and contentedly become hewers of wood and drawers of water?

No. The dangerously clear logic of the Negro's position will more and more loudly assert itself in that day when increasing wealth and more intricate social organization preclude the South from being, as it so largely is, simply an armed camp for intimidating black folk. Such waste of energy cannot be spared if the South is to catch up with civilization. And as the black third of the land grows in thrift and skill, unless skilfully guided in its larger philosophy, it must more and more brood over the red past and the creeping, crooked present, until it grasps a gospel of revolt and revenge and throws its new-found energies athwart the current of advance. Even to-day the masses of the Negroes see all too clearly the anomalies of their position and the moral crookedness of yours. You may marshal strong indictments against them, but their counter-cries, lacking though they be in formal logic, have burning truths within them which you may not wholly ignore, O Southern Gentlemen! If you deplore their presence here, they ask, Who brought us? When you cry, Deliver us from the vision of intermarriage, they answer that legal marriage is infinitely better than systematic concubinage and prostitution. And if in just fury you accuse their vagabonds of violating women, they also in fury quite as just may reply: The rape which your gentlemen have done against helpless black women in defiance of your own laws is written on the foreheads of two millions of mulattoes, and written in ineffaceable blood. And finally, when you fasten crime upon this race as its peculiar trait, they answer that slavery was the arch-crime, and lynching and lawlessness its twin abortion; that color and race are not crimes, and yet it is they which in this land receive most unceasing condemnation, North, East, South, and West.

I will not say such arguments are wholly justified,—I will not insist that there is no other side to the shield; but I do say that of the nine millions of Negroes in this nation, there is scarcely one out of the cradle to whom these arguments do not daily present themselves in the guise of terrible truth. I insist that the question of the future is how best to keep these millions from brooding over the wrongs of the past and the difficulties of the present, so that all their energies may be bent toward a cheerful striving and coöperation with their white neighbors toward a larger, juster, and fuller future. That one wise method of doing this lies in the closer knitting of the Negro to the great industrial possibilities of the South is a great truth. And this the common schools and the manual training and trade schools are working to accomplish. But these alone are not enough. The foundations of knowledge in this race, as in others, must be sunk deep in the college and university if we would build a solid, permanent structure. Internal problems of social advance must inevitably come,—problems of work and wages, of families and homes, of morals and the true valuing of the things of life; and all these and other inevitable problems of civilization the Negro must meet and solve largely for himself, by reason of his isolation; and can there be any possible solution other than by study and thought and an appeal to the rich experience of the past? Is there not, with such a group and in such a crisis, infinitely more danger to be apprehended from half-trained minds and shallow thinking than from over-education and over-refinement? Surely we have wit enough to found a Negro college so manned and equipped as to steer successfully between the *dilettante* and the fool. We shall hardly induce black men to believe that if their stomachs be full, it matters little about their brains. They already dimly perceive that the paths of peace winding between honest toil and dignified manhood call for the guidance of skilled thinkers, the loving, reverent comradeship between the black lowly and the black men emancipated by training and culture.

The function of the Negro college, then, is clear: it must maintain the standards of popular education, it must seek the social regeneration of the Negro, and it must help in the solution of problems of race contact and coöperation. And finally, beyond all this, it must develop men. Above our modern socialism, and out of the worship of the mass, must persist and evolve that higher individualism which the centres of culture protect; there must come

a loftier respect for the sovereign human soul that seeks
to know itself and the world about it; that seeks a freedom for expansion and self-development; that will love
and hate and labor in its own way, untrammeled alike
by old and new. Such souls aforetime have inspired and
guided worlds, and if we be not wholly bewitched by
our Rhinegold, they shall again. Herein the longing of
black men must have respect: the rich and bitter depth
of their experience, the unknown treasures of their inner
life, the strange rendings of nature they have seen, may
give the world new points of view and make their loving,
living, and doing precious to all human hearts. And to
themselves in these the days that try their souls, the chance
to soar in the dim blue air above the smoke is to their
finer spirits boon and guerdon for what they lose on earth
by being black.

I sit with Shakespeare and he winces not. Across the
color line I move arm in arm with Balzac and Dumas,
where smiling men and welcoming women glide in gilded
halls. From out the caves of evening that swing between
the strong-limbed earth and the tracery of the stars, I
summon Aristotle and Aurelius and what soul I will,
and they come all graciously with no scorn nor condescension. So, wed with Truth, I dwell above the Veil.
Is this the life you grudge us, O knightly America? Is
this the life you long to change into the dull red
hideousness of Georgia? Are you so afraid lest peering
from this high Pisgah, between Philistine and Amalekite,
we sight the Promised Land?

VII

Of the Black Belt

I am black but comely, O ye daughters of Jerusalem,
As the tents of Kedar, as the curtains of Solomon.
Look not upon me, because I am black,
Because the sun hath looked upon me:
My mother's children were angry with me;
They made me the keeper of the vineyards;
But mine own vineyard have I not kept.

THE SONG OF SOLOMON.

Out of the North the train thundered, and we woke to see the crimson soil of Georgia stretching away bare and monotonous right and left. Here and there lay straggling, unlovely villages, and lean men loafed leisurely at the depots; then again came the stretch of pines and clay. Yet we did not nod, nor weary of the scene; for this is historic ground. Right across our track, three hundred and sixty years ago, wandered the cavalcade of Hernando de Soto, looking for gold and the Great Sea; and he and his foot-sore captives disappeared yonder in the grim forests to the west. Here sits Atlanta, the city of a hundred hills, with something Western, something Southern, and something quite its own, in its busy life. Just this side of Atlanta is the land of the Cherokees and to the southwest, not far from where Sam Hose was crucified, you may stand on a spot which is to-day the centre of the Negro problem,—the centre of those nine million men who are America's dark heritage from slavery and the slave-trade.

Not only is Georgia thus the geographical focus of our Negro population, but in many other respects, both now and yesterday, the Negro problems have seemed to be centered in this State. No other State in the Union can count a million Negroes among its citizens,—a population as large as the slave population of the whole Union in 1800; no other State fought so long and strenuously to gather this host of Africans. Oglethorpe thought slavery against law and gospel; but the circumstances which gave Georgia its first inhabitants were not calculated to furnish citizens over-nice in their ideas about rum and slaves. De-

spite the prohibitions of the trustees, these Georgians, like
some of their descendants, proceeded to take the law into
their own hands; and so pliant were the judges, and so
flagrant the smuggling, and so earnest were the prayers
of Whitefield, that by the middle of the eighteenth century all restrictions were swept away, and the slave-
trade went merrily on for fifty years and more.

Down in Darien, where the Delegal riots took place
some summers ago, there used to come a strong protest
against slavery from the Scotch Highlanders; and the
Moravians of Ebenezer did not like the system. But not
till the Haytian Terror of Toussaint was the trade in men
even checked; while the national statute of 1808 did not
suffice to stop it. How the Africans poured in—fifty thousand between 1790 and 1810, and then, from Virginia
and from smugglers, two thousand a year for many years
more. So the thirty thousand Negroes of Georgia in 1790
doubled in a decade,—were over a hundred thousand in
1810, had reached two hundred thousand in 1820, and
half a million at the time of the war. Thus like a snake
the black population writhed upward.

But we must hasten on our journey. This that we pass
as we near Atlanta is the ancient land of the Cherokees,—
that brave Indian nation which strove so long for its
fatherland, until Fate and the United States Government
drove them beyond the Mississippi. If you wish to ride
with me you must come into the "Jim Crow Car." There
will be no objection,—already four other white men, and a
little white girl with her nurse, are in there. Usually
the races are mixed in there; but the white coach is all
white. Of course this car is not so good as the other, but
it is fairly clean and comfortable. The discomfort lies
chiefly in the hearts of those four black men yonder—
and in mine.

We rumble south in quite a business-like way. The bare
red clay and pines of Northern Georgia begin to disappear, and in their place appears a rich rolling land,
luxuriant, and here and there well tilled. This is the land
of the Creek Indians; and a hard time the Georgians had
to seize it. The towns grow more frequent and more interesting, and brand-new cotton mills rise on every side.
Below Macon the world grows darker; for now we approach the Black Belt,—that strange land of shadows, at
which even slaves paled in the past, and whence come

now only faint and half-intelligible murmurs to the world beyond. The "Jim Crow Car" grows larger and a shade better; three rough field-hands and two or three white loafers accompany us, and the newsboy still spreads his wares at one end. The sun is setting, but we can see the great cotton country as we enter it,—the soil now dark and fertile, now thin and gray, with fruit-trees and dilapidated buildings,—all the way to Albany.

At Albany, in the heart of the Black Belt, we stop. Two hundred miles south of Atlanta, two hundred miles west of the Atlantic, and one hundred miles north of the Great Gulf lies Dougherty County, with ten thousand Negroes and two thousand whites. The Flint River winds down from Andersonville, and, turning suddenly at Albany, the county-seat, hurries on to join the Chattahoochee and the sea. Andrew Jackson knew the Flint well, and marched across it once to avenge the Indian Massacre at Fort Mims. That was in 1814, not long before the battle of New Orleans; and by the Creek treaty that followed this campaign, all Dougherty County, and much other rich land, was ceded to Georgia. Still, settlers fought shy of this land, for the Indians were all about, and they were unpleasant neighbors in those days. The panic of 1837, which Jackson bequeathed to Van Buren, turned the planters from the impoverished lands of Virginia, the Carolinas, and east Georgia, toward the West. The Indians were removed to Indian Territory, and settlers poured into these coveted lands to retrieve their broken fortunes. For a radius of a hundred miles about Albany, stretched a great fertile land, luxuriant with forests of pine, oak, ash, hickory, and poplar; hot with the sun and damp with the rich black swamp-land; and here the cornerstone of the Cotton Kingdom was laid.

Albany is to-day a wide-streeted, placid, Southern town, with a broad sweep of stores and saloons, and flanking rows of homes,—whites usually to the north, and blacks to the south. Six days in the week the town looks decidedly too small for itself, and takes frequent and prolonged naps. But on Saturday suddenly the whole county disgorges itself upon the place, and a perfect flood of black peasantry pours through the streets, fills the stores, blocks the sidewalks, chokes the thoroughfares, and takes full possession of the town. They are black, sturdy, uncouth country folk, good-natured and simple, talkative to a de-

gree, and yet far more silent and brooding than the crowds
of the Rhine-pfalz, or Naples, or Cracow. They drink considerable quantities of whiskey, but do not get very drunk;
they talk and laugh loudly at times, but seldom quarrel
or fight. They walk up and down the streets, meet and
gossip with friends, stare at the shop windows, buy coffee,
cheap candy, and clothes, and at dusk drive home—happy?
well no, not exactly happy, but much happier than as
though they had not come.

Thus Albany is a real capital,—a typical Southern
county town, the centre of the life of ten thousand souls;
their point of contact with the outer world, their centre
of news and gossip, their market for buying and selling,
borrowing and lending, their fountain of justice and law.
Once upon a time we knew country life so well and city
life so little, that we illustrated city life as that of a closely
crowded country district. Now the world has well-nigh
forgotten what the country is, and we must imagine a little
city of black people scattered far and wide over three
hundred lonesome square miles of land, without train or
trolley, in the midst of cotton and corn, and wide patches
of sand and gloomy soil.

It gets pretty hot in Southern Georgia in July,—a sort
of dull, determined heat that seems quite independent of
the sun; so it took us some days to muster courage
enough to leave the porch and venture out on the long
country roads, that we might see this unknown world.
Finally we started. It was about ten in the morning, bright
with a faint breeze, and we jogged leisurely southward
in the valley of the Flint. We passed the scattered box-
like cabins of the brick-yard hands, and the long tenement-
row facetiously called "The Ark," and were soon in the
open country, and on the confines of the great plantations of other days. There is the "Joe Fields place"; a rough
old fellow was he, and had killed many a "nigger" in his
day. Twelve miles his plantation used to run,—a regular
barony. It is nearly all gone now; only straggling bits
belong to the family, and the rest has passed to Jews
and Negroes. Even the bits which are left are heavily
mortgaged, and, like the rest of the land, tilled by tenants.
Here is one of them now,—a tall brown man, a hard
worker and a hard drinker, illiterate, but versed in farm-
lore, as his nodding crops declare. This distressingly new

board house is his, and he has just moved out of yonder moss-grown cabin with its one square room.

From the curtains in Benton's house, down the road, a dark comely face is staring at the strangers; for passing carriages are not every-day occurrences here. Benton is an intelligent yellow man with a good-sized family, and manages a plantation blasted by the war and now the broken staff of the widow. He might be well-to-do, they say; but he carouses too much in Albany. And the half-desolate spirit of neglect born of the very soil seems to have settled on these acres. In times past there were cotton-gins and machinery here; but they have rotted away.

The whole land seems forlorn and forsaken. Here are the remnants of the vast plantations of the Sheldons, the Pellots, and the Rensons; but the souls of them are passed. The houses lie in half ruin, or have wholly disappeared; the fences have flown, and the families are wandering in the world. Strange vicissitudes have met these whilom masters. Yonder stretch the wide acres of Bildad Reasor; he died in war-time, but the upstart overseer hastened to wed the widow. Then he went, and his neighbors too, and now only the black tenant remains; but the shadow-hand of the master's grand-nephew or cousin or creditor stretches out of the gray distance to collect the rack-rent remorselessly, and so the land is uncared-for and poor. Only black tenants can stand such a system, and they only because they must. Ten miles we have ridden to-day and have seen no white face.

A resistless feeling of depression falls slowly upon us, despite the gaudy sunshine and the green cottonfields. This, then, is the Cotton Kingdom,—the shadow of a marvellous dream. And where is the King? Perhaps this is he,—the sweating ploughman, tilling his eighty acres with two lean mules, and fighting a hard battle with debt. So we sit musing, until, as we turn a corner on the sandy road, there comes a fairer scene suddenly in view,—a neat cottage snugly ensconced by the road, and near it a little store. A tall bronzed man rises from the porch as we hail him, and comes out to our carriage. He is six feet in height, with a sober face that smiles gravely. He walks too straight to be a tenant,—yes, he owns two hundred and forty acres. "The land is run down since the boom-days of eighteen hundred and fifty," he explained, and cotton is low. Three black tenants live on his place, and in his little store he keeps a small stock of tobacco, snuff, soap, and soda, for the neighborhood. Here is his gin-house with new machinery just installed. Three hundred bales of cotton went through it last year. Two children he has sent away to school. Yes, he says sadly, he is getting on, but cotton is down to four cents; I know how Debt sits staring at him.

Wherever the King may be, the parks and palaces of the Cotton Kingdom have not wholly disappeared. We plunge even now into great groves of oak and towering pine, with an undergrowth of myrtle and shrubbery. This was the "home-house" of the Thompsons,—slave-barons who drove their coach and four in the merry past. All is silence now, and ashes, and tangled weeds. The owner put his whole fortune into the rising cotton industry of the fifties, and with the falling prices of the eighties he packed up and stole away. Yonder is another grove, with unkempt lawn, great magnolias, and grass-grown paths. The Big House stands in half-ruin, its great front door staring blankly at the street, and the back part grotesquely restored for its black tenant. A shabby, well-built Negro he is, unlucky and irresolute. He digs hard to pay rent to the white girl who owns the remnant of the place. She married a policeman, and lives in Savannah.

Now and again we come to churches. Here is one now, —Shepherd's, they call it,—a great whitewashed barn of a thing, perched on stilts of stone, and looking for all the world as though it were just resting here a moment and might be expected to waddle off down the road at almost any time. And yet it is the centre of a hundred cabin homes; and sometimes, of a Sunday, five hundred persons from far and near gather here and talk and eat and sing. There is a school-house near,—a very airy, empty shed; but even this is an improvement, for usually the school is held in the church. The churches vary from log-huts to those like Shepherd's, and the schools from nothing to this little house that sits demurely on the county line. It is a tiny plank-house, perhaps ten by twenty, and has within a double row of rough unplaned benches, resting mostly on legs, sometimes on boxes. Opposite the door is a square home-made desk. In one corner are the ruins of a stove, and in the other a dim blackboard. It is the cheerfulest schoolhouse I have seen in Dougherty, save in town. Back of the schoolhouse is a lodgehouse two stories high and not quite finished. Societies meet there,—societies "to care for the sick and bury the dead"; and these societies grow and flourish.

We had come to the boundaries of Dougherty, and were about to turn west along the county-line, when all these sights were pointed out to us by a kindly old man, black, white-haired, and seventy. Forty-five years he had lived here, and now supports himself and his old wife by the help of the steer tethered yonder and the charity of his black neighbors. He shows us the farm of the Hills just across the county line in Baker,—a widow and two strapping sons, who raised ten bales (one need not add "cotton" down here) last year. There are fences and pigs and cows, and the soft-voiced, velvet-skinned young Memnon, who sauntered half-bashfully over to greet the strangers, is proud of his home. We turn now to the west along the county line. Great dismantled trunks of pines tower above the green cotton-fields, cracking their naked gnarled fingers toward the border of living forest beyond. There is little beauty in this region, only a sort of crude abandon that suggests power,—a naked grandeur, as it were. The houses are bare and straight; there are no hammocks or easy-chairs, and few flowers. So when, as here at Rawdon's, one sees a vine clinging to a little porch, and home-like windows peeping over the fences, one takes a long breath. I think I never before quite realized the place of the Fence in civilization. This is the Land of the Unfenced, where crouch on either hand scores of ugly one-room cabins, cheerless and dirty. Here lies the Negro problem in its naked dirt and penury. And here are no fences. But now and then the criss-cross rails or straight palings break into view, and then we know a touch of culture is near. Of course Harrison Gohagen,—a quiet yellow man, young, smooth-faced, and diligent,—of course he is lord of some hundred acres, and we expect to see a vision of well-kept rooms and fat beds and laughing children. For has he not fine fences? And those over yonder, why should they build fences on the rack-rented land? It will only increase their rent.

On we wind, through sand and pines and glimpses of old plantations, till there creeps into sight a cluster of buildings,—wood and brick, mills and houses, and scattered cabins. It seemed quite a village. As it came nearer and nearer, however, the aspect changed: the buildings

were rotten, the bricks were falling out, the mills were
silent, and the store was closed. Only in the cabins appeared now and then a bit of lazy life. I could imagine
the place under some weird spell, and was half-minded
to search out the princess. An old ragged black man,
honest, simple, and improvident, told us the tale. The
Wizard of the North—the Capitalist—had rushed down
in the seventies to woo this coy dark soil. He bought a
square mile or more, and for a time the field-hands sang,
the gins groaned, and the mills buzzed. Then came a
change. The agent's son embezzled the funds and ran off
with them. Then the agent himself disappeared. Finally
the new agent stole even the books, and the company in
wrath closed its business and its houses, refused to sell,
and let houses and furniture and machinery rust and rot.
So the Waters-Loring plantation was stilled by the spell
of dishonesty, and stands like some gaunt rebuke to a
scarred land.

Somehow that plantation ended our day's journey; for
I could not shake off the influence of that silent scene.
Back toward town we glided, past the straight and thread-
like pines, past a dark tree-dotted pond where the air was
heavy with a dead sweet perfume. White slender-legged
curlews flitted by us, and the garnet blooms of the cotton
looked gay against the green and purple stalks. A peasant
girl was hoeing in the field, white-turbaned and black-
limbed. All this we saw, but the spell still lay upon us.

How curious a land is this,—how full of untold story,
of tragedy and laughter, and the rich legacy of human
life; shadowed with a tragic past, and big with future
promise! This is the Black Belt of Georgia. Dougherty
County is the west end of the Black Belt, and men once
called it the Egypt of the Confederacy. It is full of historic interest. First there is the Swamp, to the west, where
the Chickasawhatchee flows sullenly southward. The shadow of an old plantation lies at its edge, forlorn and dark.
Then comes the pool; pendent gray moss and brackish
waters appear, and forests filled with wildfowl. In one
place the wood is on fire, smouldering in dull red anger;
but nobody minds. Then the swamp grows beautiful; a
raised road, built by chained Negro convicts, dips down
into it, and forms a way walled and almost covered in
living green. Spreading trees spring from a prodigal luxuriance of undergrowth; great dark green shadows fade

into the black background, until all is one mass of tangled semi-tropical foliage, marvellous in its weird savage splendor. Once we crossed a black silent stream, where the sad trees and writhing creepers, all glinting fiery yellow and green, seemed like some vast cathedral,—some green Milan builded of wildwood. And as I crossed, I seemed to see again that fierce tragedy of seventy years ago. Osceola, the Indian-Negro chieftain, had risen in the swamps of Florida, vowing vengeance. His war-cry reached the red Creeks of Dougherty, and their war-cry rang from the Chattahoochee to the sea. Men and women and children fled and fell before them as they swept into Dougherty. In yonder shadows a dark and hideously painted warrior glided stealthily on,—another and another, until three hundred had crept into the treacherous swamp. Then the false slime closing about them called the white men from the east. Waist-deep, they fought beneath the tall trees, until the war-cry was hushed and the Indians glided back into the west. Small wonder the wood is red.

Then came the black slaves. Day after day the clank of chained feet marching from Virginia and Carolina to Georgia was heard in these rich swamp lands. Day after day the songs of the callous, the wail of the motherless, and the muttered curses of the wretched echoed from the Flint to the Chickasawhatchee, until by 1860 there had risen in West Dougherty perhaps the richest slave kingdom the modern world ever knew. A hundred and fifty barons commanded the labor of nearly six thousand Negroes, held sway over farms with ninety thousand acres of tilled land, valued even in times of cheap soil at three millions of dollars. Twenty thousand bales of ginned cotton went yearly to England, New and Old; and men that came there bankrupt made money and grew rich. In a single decade the cotton output increased four-fold and the value of lands was tripled. It was the heyday of the *nouveau riche*, and a life of careless extravagance among the masters. Four and six bob-tailed thoroughbreds rolled their coaches to town; open hospitality and gay entertainment were the rule. Parks and groves were laid out, rich with flower and vine, and in the midst stood the low wide-halled "big house," with its porch and columns and great fire-places.

And yet with all this there was something sordid, something forced,—a certain feverish unrest and recklessness;

for was not all this show and tinsel built upon a groan? "This land was a little Hell," said a ragged, brown, and grave-faced man to me. We were seated near a roadside blacksmith-shop, and behind was the bare ruin of some master's home. "I've seen niggers drop dead in the furrow, but they were kicked aside, and the plough never stopped. Down in the guard-house, there's where the blood ran."

With such foundations a kingdom must in time sway and fall. The masters moved to Macon and Augusta, and left only the irresponsible overseers on the land. And the result is such ruin as this, the Lloyd "home-place":— great waving oaks, a spread of lawn, myrtles and chestnuts, all ragged and wild; a solitary gate-post standing where once was a castle entrance; an old rusty anvil lying amid rotting bellows and wood in the ruins of a blacksmith shop; a wide rambling old mansion, brown and dingy, filled now with the grandchildren of the slaves who once waited on its tables; while the family of the master has dwindled to two lone women, who live in Macon and feed hungrily off the remnants of an earldom. So we ride on, past phantom gates and falling homes,—past the once flourishing farms of the Smiths, the Gandys, and the Lagores,—and find all dilapidated and half ruined, even there where a solitary white woman, a relic of other days, sits alone in state among miles of Negroes and rides to town in her ancient coach each day.

This was indeed the Egypt of the Confederacy,—the rich granary whence potatoes and corn and cotton poured out to the famished and ragged Confederate troops as they battled for a cause lost long before 1861. Sheltered and secure, it became the place of refuge for families, wealth, and slaves. Yet even then the hard ruthless rape of the land began to tell. The red-clay sub-soil already had begun to peer above the loam. The harder the slaves were driven the more careless and fatal was their farming. Then came the revolution of war and Emancipation, the bewilderment of Reconstruction,—and now, what is the Egypt of the Confederacy, and what meaning has it for the nation's weal or woe?

It is a land of rapid contrasts and of curiously mingled hope and pain. Here sits a pretty blue-eyed quadroon hiding her bare feet; she was married only last week, and yonder in the field is her dark young husband, hoeing to

support her, at thirty cents a day without board. Across
the way is Gatesby, brown and tall, lord of two thousand
acres shrewdly won and held. There is a store conducted
by his black son, a blacksmith shop, and a ginnery. Five
miles below here is a town owned and controlled by one
white New Englander. He owns almost a Rhode Island
county, with thousands of acres and hundreds of black
laborers. Their cabins look better than most, and the
farm, with machinery and fertilizers, is much more
business-like than any in the county, although the manager drives hard bargains in wages. When now we turn
and look five miles above, there on the edge of town are
five houses of prostitutes,—two of blacks and three of
whites; and in one of the houses of the whites a worthless black boy was harbored too openly two years ago;
so he was hanged for rape. And here, too, is the high
whitewashed fence of the "stockade," as the county prison
is called; the white folks say it is ever full of black
criminals,—the black folks say that only colored boys are
sent to jail, and they not because they are guilty, but because the State needs criminals to eke out its income by
their forced labor.

Immigrants are heirs of the slave baron in Dougherty;
and as we ride westward, by wide stretching cornfields
and stubby orchards of peach and pear, we see on all
sides within the circle of dark forest a Land of Canaan.
Here and there are tales of projects for money-getting,
born in the swift days of Reconstruction,—"improvement" companies, wine companies, mills and factories;
most failed, and foreigners fell heir. It is a beautiful land,
this Dougherty, west of the Flint. The forests are wonderful, the solemn pines have disappeared, and this is the
"Oakey Woods," with its wealth of hickories, beeches,
oaks and palmettos. But a pall of debt hangs over the
beautiful land; the merchants are in debt to the whole-
salers, the planters are in debt to the merchants, the
tenants owe the planters, and laborers bow and bend beneath the burden of it all. Here and there a man has
raised his head above these murky waters. We passed
one fenced stock-farm with grass and grazing cattle, that
looked very homelike after endless corn and cotton. Here
and there are black free-holders: there is the gaunt dull-
black Jackson, with his hundred acres. "I says, 'Look
up! If you don't look up you can't get up,' " remarks Jack-

son, philosophically. And he's gotten up. Dark Carter's
neat barns would do credit to New England. His master
helped him to get a start, but when the black man died
last fall the master's sons immediately laid claim to the
estate. "And them white folks will get it, too," said my
yellow gossip.

I turn from these well-tended acres with a comfortable
feeling that the Negro is rising. Even then, however, the
fields, as we proceed, begin to redden and the trees disappear. Rows of old cabins appear filled with renters and
laborers,—cheerless, bare, and dirty, for the most part,
although here and there the very age and decay makes
the scene pictuesque. A young black fellow greets us.
He is twenty-two, and just married. Until last year he
had good luck renting; then cotton fell, and the sheriff
seized and sold all he had. So he moved here, where the
rent is higher, the land poorer, and the owner inflexible;
he rents a forty-dollar mule for twenty dollars a year.
Poor lad!—a slave at twenty-two. This plantation, owned
now by a foreigner, was a part of the famous Bolton
estate. After the war it was for many years worked by
gangs of Negro convicts,—and black convicts then were
even more plentiful than now; it was a way of making
Negroes work, and the question of guilt was a minor
one. Hard tales of cruelty and mistreatment of the chained
freemen are told, but the county authorities were deaf
until the free-labor market was nearly ruined by wholesale migration. Then they took the convicts from the plantations, but not until one of the fairest regions of the
"Oakey Woods" had been ruined and ravished into a red
waste, out of which only a Yankee or an immigrant could
squeeze more blood from debt-cursed tenants.

No wonder that Luke Black, slow, dull, and discouraged, shuffles to our carriage and talks hopelessly. Why
should he strive? Every year finds him deeper in debt.
How strange that Georgia, the world-heralded refuge of
poor debtors, should bind her own to sloth and misfortune as ruthlessly as ever England did! The poor land
groans with its birth-pains, and brings forth scarcely a
hundred pounds of cotton to the acre, where fifty years
ago it yielded eight times as much. Of this meagre
yield the tenant pays from a quarter to a third in rent,
and most of the rest in interest on food and supplies
bought on credit. Twenty years yonder a sunken-cheeked,

old black man has labored under that system, and now, turned day-laborer, is supporting his wife and boarding himself on his wages of a dollar and a half a week, received only part of the year.

The Bolton convict farm formerly included the neighboring plantation. Here it was that the convicts were lodged in the great log prison still standing. A dismal place it still remains, with rows of ugly huts filled with surly ignorant tenants. "What rent do you pay here?" I inquired. "I don't know,—what is it, Sam?" "All we make," answered Sam. It is a depressing place,—bare, unshaded, with no charm of past association, only a memory of forced human toil,—now, then, and before the war. They are not happy, these black men whom we meet throughout this region. There is little of the joyous abandon and playfulness which we are wont to associate with the plantation Negro. At best, the natural good-nature is edged with complaint or has changed into sullenness and gloom. And now and then it blazes forth in veiled but hot anger. I remember one big red-eyed black whom we met by the roadside. Forty-five years he had labored on this farm, beginning with nothing, and still having nothing. To be sure, he had given four children a common-school training, and perhaps if the new fence-law had not allowed unfenced crops in West Dougherty he might have raised a little stock and kept ahead. As it is, he is hopelessly in debt, disappointed, and embittered. He stopped us to inquire after the black boy in Albany, whom it was said a policeman had shot and killed for loud talking on the sidewalk. And then he said slowly: "Let a white man touch me, and he dies; I don't boast this,—I don't say it around loud, or before the children, —but I mean it. I've seen them whip my father and my old mother in them cotton-rows till the blood ran; by—" and we passed on.

Now Sears, whom we met next lolling under the chubby oak-trees, was of quite different fibre. Happy?— Well, yes; he laughed and flipped pebbles, and thought the world was as it was. He had worked here twelve years and has nothing but a mortgaged mule. Children? Yes, seven; but they hadn't been to school this year,—couldn't afford books and clothes, and couldn't spare their work. There go part of them to the fields now,—three big boys astride mules, and a strapping girl with bare brown legs.

Careless ignorance and laziness here, fierce hate and vindictiveness there;—these are the extremes of the Negro problem which we met that day, and we scarce knew which we preferred.

Here and there we meet distinct characters quite out of the ordinary. One came out of a piece of newly cleared ground, making a wide detour to avoid the snakes. He was an old, hollow-cheeked man, with a drawn and characterful brown face. He had a sort of self-contained quaintness and rough humor impossible to describe; a certain cynical earnestness that puzzled one. "The niggers were jealous of me over on the other place," he said, "and so me and the old woman begged this piece of woods, and I cleared it up myself. Made nothing for two years, but I reckon I've got a crop now." The cotton looked tall and rich, and we praised it. He curtsied low, and then bowed almost to the ground, with an imperturbable gravity that seemed almost suspicious. Then he continued, "My mule died last week,"—a calamity in this land equal to a devastating fire in town,—"but a white man loaned me another." Then he added, eyeing us, "Oh, I gets along with white folks." We turned the conversation. "Bears? deer?" he answered, "well, I should say there were," and he let fly a string of brave oaths, as he told hunting-tales of the swamp. We left him standing still in the middle of the road looking after us, and yet apparently not noticing us.

The Whistle place, which includes his bit of land, was bought soon after the war by an English syndicate, the "Dixie Cotton and Corn Company." A marvellous deal of style their factor put on, with his servants, and coach-and-six; so much so that the concern soon landed in inextricable bankruptcy. Nobody lives in the old house now, but a man comes each winter out of the North and collects his high rents. I know not which are the more touching,—such old empty houses, or the homes of the masters' sons. Sad and bitter tales lie hidden back of those white doors,—tales of poverty, of struggle, of disappointment. A revolution such as that of '63 is a terrible thing; they that rose rich in the morning often slept in paupers' beds. Beggars and vulgar speculators rose to rule over them, and their children went astray. See yonder sad-colored house, with its cabins and fences and glad crops! It is not glad within; last month the prodigal son of the struggling father wrote home from the city for money. Money! Where was it to come from? And so the son rose in the night and killed his baby, and killed his wife, and shot himself dead. And the world passed on.

I remember wheeling around a bend in the road beside a graceful bit of forest and a singing brook. A long low house faced us, with porch and flying pillars, great oaken door, and a broad lawn shining in the evening sun. But the window-panes were gone, the pillars were worm-eaten, and the moss-grown roof was falling in. Half curiously I peered through the unhinged door, and saw where, on the wall across the hall, was written in once gay letters a faded "Welcome."

Quite a contrast to the southwestern part of Dougherty County is the northwest. Soberly timbered in oak and pine, it has none of that half-tropical luxuriance of the southwest. Then, too, there are fewer signs of a romantic past, and more of systematic modern land-grabbing and money-getting. White people are more in evidence here, and farmer and hired labor replace to some extent the absentee landlord and rack-rented tenant. The crops have neither the luxuriance of the richer land nor the signs of neglect so often seen, and there were fences and meadows here and there. Most of this land was poor, and beneath the notice of the slave-baron, before the war. Since then his poor relations and foreign immigrants have seized it. The returns of the farmer are too small to allow much for wages, and yet he will not sell off small farms. There is the Negro Sanford; he has worked fourteen years as overseer on the Ladson place, and "paid out enough for fertilizers to have bought a farm," but the owner will not sell off a few acres.

Two children—a boy and a girl—are hoeing sturdily in the fields on the farm where Corliss works. He is smooth-faced and brown, and is fencing up his pigs. He used to run a successful cotton-gin, but the Cotton Seed Oil Trust has forced the price of ginning so low that he says it hardly pays him. He points out a stately old house over the way as the home of "Pa Willis." We eagerly ride over, for "Pa Willis" was the tall and powerful black Moses who led the Negroes for a generation, and led them well. He was a Baptist preacher, and when he died, two thousand black people followed him to the grave; and now they preach his funeral sermon each year. His widow lives here,—a weazened, sharp-featured little woman, who curtsied quaintly as we greeted her. Further on lives Jack Delson, the most prosperous Negro farmer in the county. It is a joy to meet him,—a great broad-shouldered, handsome black man, intelligent and jovial. Six hundred and fifty acres he owns, and has eleven black tenants. A neat and tidy home nestled in a flower-garden, and a little store stands beside it.

We pass the Munson place, where a plucky white widow is renting and struggling; and the eleven hundred acres of the Sennet plantation, with its Negro overseer. Then the character of the farms begins to change. Nearly all the lands belong to Russian Jews; the overseers are white, and the cabins are bare board-houses scattered here and there. The rents are high, and day-laborers and "contract" hands abound. It is a keen, hard struggle for living here, and few have time to talk. Tired with the long ride, we gladly drive into Gillonsville. It is a silent cluster of farm-houses standing on the crossroads, with one of its stores closed and the other kept by a Negro preacher. They tell great tales of busy times at Gillonsville before all the railroads came to Albany; now it is chiefly a memory. Riding down the street, we stop at the preacher's and seat ourselves before the door. It was one of those scenes one cannot soon forget:—a wide, low, little house, whose motherly roof reached over and sheltered a snug little porch. There we sat, after the long hot drive, drinking cool water,—the talkative little storekeeper who is my daily companion; the silent old black woman patching pantaloons and saying never a word; the ragged picture of helpless misfortune who called in just to see the preacher; and finally the neat matronly preacher's wife, plump, yellow, and intelligent. "Own land?" said the wife; "well, only this house." Then she added quietly, "we did buy seven hundred acres up yonder, and paid for it; but they cheated us out of it. Sells was the owner." "Sells!" echoed the ragged misfortune, who was leaning against the balustrade and listening, "he's a regular cheat. I worked for him thirty-seven days this spring, and he paid me in cardboard checks which were to be cashed at the end of the month. But he never cashed them,—kept putting me off. Then the sheriff came and took my mule

and corn and furniture—" "Furniture? But furniture is
exempt from seizure by law." "Well, he took it just the
same," said the hard-faced man.

VIII

Of the Quest of The Golden Fleece

But the Brute said in his breast, "Till the mills I grind
    have ceased,
The riches shall be dust of dust, dry ashes be the feast!

"On the strong and cunning few
    Cynic favors I will strew;
I will stuff their maw with overplus until their spirit dies;
    From the patient and the low
    I will take the joys they know;
    They shall hunger after vanities and still an-hungered
        go.
Madness shall be on the people, ghastly jealousies arise;
Brother's blood shall cry on brother up the dead and
    empty skies."

WILLIAM VAUGHN MOODY.

Have you even seen a cotton-field white with the harvest,—its golden fleece hovering above the black earth
like a silvery cloud edged with dark green, its bold white
signals waving like the foam of billows from Carolina to
Texas across that Black and human Sea? I have sometimes half suspected that here the winged ram Chrysomal-
lus left that Fleece after which Jason and his Argonauts
went vaguely wandering into the shadowy East three

thousand years ago; and certainly one might frame a
pretty and not far-fetched analogy of witchery and drag-
on's teeth, and blood and armed men, between the ancient and the modern quest of the Golden Fleece in the
Black Sea.

And now the golden fleece is found; not only found,
but, in its birthplace, woven. For the hum of the cotton-
mills is the newest and most significant thing in the New
South to-day. All through the Carolinas and Georgia,
away down to Mexico, rise these gaunt red buildings, bare
and homely, and yet so busy and noisy withal that they
scarce seem to belong to the slow and sleepy land. Perhaps they sprang from dragons' teeth. So the Cotton
Kingdom still lives; the world still bows beneath her
sceptre. Even the markets that once defied the *parvenu*
have crept one by one across the seas, and then slowly
and reluctantly, but surely, have started toward the Black
Belt.

To be sure, there are those who wag their heads
knowingly and tell us that the capital of the Cotton
Kingdom has moved from the Black to the White Belt,—
that the Negro of to-day raises not more than half of the
cotton crop. Such men forget that the cotton crop has
doubled, and more than doubled, since the era of slavery,
and that, even granting their contention, the Negro is still
supreme in a Cotton Kingdom larger than that on which
the Confederacy builded its hopes. So the Negro forms
to-day one of the chief figures in a great world-industry;
and this, for its own sake, and in the light of historic
interest, makes the field-hands of the cotton country
worth studying.

We seldom study the condition of the Negro to-day
honestly and carefully. It is so much easier to assume that
we know it all. Or perhaps, having already reached conclusions in our own minds, we are loth to have them
disturbed by facts. And yet how little we really know of
these millions,—of their daily lives and longings, of their
homely joys and sorrows, of their real shortcomings and
the meaning of their crimes! All this we can only learn
by intimate contact with the masses, and not by wholesale arguments covering millions separate in time and
space, and differing widely in training and culture. To-day,
then, my reader, let us turn our faces to the Black Belt

of Georgia and seek simply to know the condition of the
black farm-laborers of one county there.

Here in 1890 lived ten thousand Negroes and two
thousand whites. The country is rich, yet the people are
poor. The keynote of the Black Belt is debt; not commercial credit, but debt in the sense of continued inability
on the part of the mass of the population to make income
cover expense. This is the direct heritage of the South
from the wasteful economies of the slave *régime;* but it
was emphasized and brought to a crisis by the Emancipation of the slaves. In 1860, Dougherty County had six
thousand slaves, worth at least two and a half millions
of dollars; its farms were estimated at three millions,—
making five and a half millions of property, the value of
which depended largely on the slave system, and on the
speculative demand for land once marvellously rich but
already partially devitalized by careless and exhaustive
culture. The war then meant a financial crash; in place
of the five and a half millions of 1860, there remained in
1870 only farms valued at less than two millions. With
this came increased competition in cotton culture from
the rich lands of Texas; a steady fall in the normal price
of cotton followed, from about fourteen cents a pound
in 1860 until it reached four cents in 1898. Such a financial revolution was it that involved the owners of the
cotton-belt in debt. And if things went ill with the master,
how fared it with the man?

The plantations of Dougherty County in slavery days
were not as imposing and aristocratic as those of Virginia.
The Big House was smaller and usually one-storied, and
sat very near the slave cabins. Sometimes these cabins
stretched off on either side like wings; sometimes only
on one side, forming a double row, or edging the road
that turned into the plantation from the main thorough-
fare. The form and disposition of the laborers' cabins
throughout the Black Belt is to-day the same as in slavery
days. Some live in the self-same cabins, others in cabins
rebuilt on the sites of the old. All are sprinkled in little
groups over the face of the land, centering about some
dilapidated Big House where the head-tenant or agent
lives. The general character and arrangement of these
dwellings remains on the whole unaltered. There were in
the county, outside the corporate town of Albany, about
fifteen hundred Negro families in 1898. Out of all these,

only a single family occupies a house with seven rooms; only fourteen have five rooms or more. The mass live in one- and two-room homes.

The size and arrangements of a people's homes are no unfair index of their condition. If, then, we inquire more carefully into these Negro homes, we find much that is unsatisfactory. All over the face of the land is the one-room cabin,—now standing in the shadow of the Big House, now staring at the dusty road, now rising dark and sombre amid the green of the cotton-fields. It is nearly always old and bare, built of rough boards, and neither plastered nor ceiled. Light and ventilation are supplied by the single door and by the square hole in the wall with its wooden shutter. There is no glass, porch, or ornamentation without. Within is a fireplace, black and smoky, and usually unsteady with age. A bed or two, a table, a wooden chest, and a few chairs compose the furniture; while a stray show-bill or a newspaper makes up the decorations for the walls. Now and then one may find such a cabin kept scrupulously neat, with merry steaming fireplace and hospitable door; but the majority are dirty and dilapidated, smelling of eating and sleeping, poorly ventilated, and anything but homes.

Above all, the cabins are crowded. We have come to associate crowding with homes in cities almost exclusively. This is primarily because we have so little accurate knowledge of country life. Here in Dougherty County one may find families of eight and ten occupying one or two rooms, and for every ten rooms of house accommodations for the Negroes there are twenty-five persons. The worst tenement abominations of New York do not have above twenty-two persons for every ten rooms. Of course, one small, close room in a city, without a yard, is in many respects worse than the larger single country room. In other respects it is better; it has glass windows, a decent chimney, and a trustworthy floor. The single great advantage of the Negro peasant is that he may spend most of his life outside his hovel, in the open fields.

There are four chief causes of these wretched homes: First, long custom born of slavery has assigned such homes to Negroes; white laborers would be offered better accommodations, and might, for that and similar reasons, give better work. Secondly, the Negroes, used to such accommodations, do not as a rule demand better; they do not know what better houses mean. Thirdly, the landlords as a class have not yet come to realize that it is a good business investment to raise the standard of living among labor by slow and judicious methods; that a Negro laborer who demands three rooms and fifty cents a day would give more efficient work and leave a larger profit than a discouraged toiler herding his family in one room and working for thirty cents. Lastly, among such conditions of life there are few incentives to make the laborer become a better farmer. If he is ambitious, he moves to town or tries other labor; as a tenant-farmer his outlook is almost hopeless, and following it as a makeshift, he takes the house that is given him without protest.

In such homes, then, these Negro peasants live. The families are both small and large; there are many single tenants,—widows and bachelors, and remnants of broken groups. The system of labor and the size of the houses both tend to the breaking up of family groups: the grown children go away as contract hands or migrate to town, the sister goes into service; and so one finds many families with hosts of babies, and many newly married couples, but comparatively few families with half-grown and grown sons and daughters. The average size of Negro families has undoubtedly decreased since the war, primarily from economic stress. In Russia over a third of the bridegrooms and over half the brides are under twenty; the same was true of the ante-bellum Negroes. To-day, however, very few of the boys and less than a fifth of the Negro girls under twenty are married. The young men marry between the ages of twenty-five and thirty-five; the young women between twenty and thirty. Such postponement is due to the difficulty of earning sufficient to rear and support a family; and it undoubtedly leads, in the country districts, to sexual immorality. The form of this immorality, however, is very seldom that of prostitution, and less frequently that of illegitimacy than one would imagine. Rather, it takes the form of separation and desertion after a family group has been formed. The number of separated persons is thirty-five to the thousand,—a very large number. It would of course be unfair to compare this number with divorce statistics, for many of these separated women are in reality widowed, were

the truth known, and in other cases the separation is
not permanent. Nevertheless, here lies the seat of greatest
moral danger. There is little or no prostitution among
these Negroes, and over three-fourths of the families, as
found by house-to-house investigation, deserve to be
classed as decent people with considerable regard for female chastity. To be sure, the ideas of the mass would
not suit New England, and there are many loose habits
and notions. Yet the rate of illegitimacy is undoubtedly
lower than in Austria or Italy, and the women as a class
are modest. The plague-spot in sexual relations is easy marriage and easy separation. This is no sudden development,
nor the fruit of Emancipation. It is the plain heritage
from slavery. In those days Sam, with his master's consent, "took up" with Mary. No ceremony was necessary,
and in the busy life of the great plantations of the
Black Belt it was usually dispensed with. If now the
master needed Sam's work in another plantation or in another part of the same plantation, or if he took a notion
to sell the slave, Sam's married life with Mary was usually
unceremoniously broken, and then it was clearly to the
master's interest to have both of them take new mates.
This widespread custom of two centuries has not been
eradicated in thirty years. To-day Sam's grandson "takes
up" with a woman without license or ceremony; they live
together decently and honestly, and are, to all intents and
purposes, man and wife. Sometimes these unions are
never broken until death; but in too many cases family
quarrels, a roving spirit, a rival suitor, or perhaps more
frequently the hopeless battle to support a family, lead
to separation, and a broken household is the result. The
Negro church has done much to stop this practice, and
now most marriage ceremonies are performed by the pas-
tors. Nevertheless, the evil is still deep seated, and only
a general raising of the standard of living will finally
cure it.

Looking now at the county black population as a whole,
it is fair to characterize it as poor and ignorant. Perhaps
ten per cent compose the well-to-do and the best of the
laborers, while at least nine per cent are thoroughly lewd
and vicious. The rest, over eighty per cent, are poor and
ignorant, fairly honest and well meaning, plodding, and
to a degree shiftless, with some but not great sexual
looseness. Such class lines are by no means fixed; they

vary, one might almost say, with the price of cotton. The degree of ignorance cannot easily be expressed. We may say, for instance, that nearly two-thirds of them cannot read or write. This but partially expresses the fact. They are ignorant of the world about them, of modern economic organization, of the function of government, of individual worth and possibilities,—of nearly all those things which slavery in self-defence had to keep them from learning. Much that the white boy inbibes from his earliest social atmosphere forms the puzzling problems of the black boy's mature years. America is not another world for Opportunity to *all* her sons.

It is easy for us to lose ourselves in details in endeavoring to grasp and comprehend the real condition of a mass of human beings. We often forget that each unit in the mass is a throbbing human soul. Ignorant it may be, and poverty stricken, black and curious in limb and ways and thought; and yet it loves and hates, it toils and tires, it laughs and weeps its bitter tears, and looks in vague and awful longing at the grim horizon of its life,—all this, even as you and I. These black thousands are not in reality lazy; they are improvident and careless; they insist on breaking the monotony of toil with a glimpse at the great town-world on Saturday; they have their loafers and their rascals; but the great mass of them work continuously and faithfully for a return, and under circumstances that would call forth equal voluntary effort from few if any other modern laboring class. Over eighty-eight per cent of them—men, women, and children—are farmers. Indeed, this is almost the only industry. Most of the children get their schooling after the "crops are laid by," and very few there are that stay in school after the spring work has begun. Child-labor is to be found here in some of its worst phases, as fostering ignorance and stunting physical development. With the grown men of the county there is little variety in work: thirteen hundred are farmers, and two hundred are laborers, teamsters, etc., including twenty-four artisans, ten merchants, twenty-one preachers, and four teachers. This narrowness of life reaches its maximum among the women: thirteen hundred and fifty of these are farm laborers, one hundred are servants and washerwomen, leaving sixty-five housewives, eight teachers, and six seamstresses.

Among this people there is no leisure class. We often

forget that in the United States over half the youth and
adults are not in the world earning incomes, but are making homes, learning of the world, or resting after the heat
of the strife. But here ninety-six per cent are toiling; no
one with leisure to turn the bare and cheerless cabin into
a home, no old folks to sit beside the fire and hand down
traditions of the past; little of careless happy childhood
and dreaming youth. The dull monotony of daily toil is
broken only by the gayety of the thoughtless and the
Saturday trip to town. The toil, like all farm toil, is monotonous, and here there are little machinery and few
tools to relieve its burdensome drudgery. But with all this,
it is work in the pure open air, and this is something in a
day when fresh air is scarce.

The land on the whole is still fertile, despite long abuse.
For nine or ten months in succession the crops will come
if asked: garden vegetables in April, grain in May, melons
in June and July, hay in August, sweet potatoes in September, and cotton from then to Christmas. And yet on
two-thirds of the land there is but one crop, and that
leaves the toilers in debt. Why is this?

Away down the Baysan road, where the broad flat fields
are flanked by great oak forests, is a plantation; many
thousands of acres it used to run, here and there, and
beyond the great wood. Thirteen hundred human beings
here obeyed the call of one,—were his in body, and largely
in soul. One of them lives there yet,—a short, stocky man,
his dull-brown face seamed and drawn, and his tightly
curled hair gray-white. The crops? Just tolerable, he said;
just tolerable. Getting on? No—he wasn't getting on at
all. Smith of Albany "furnishes" him, and his rent is
eight hundred pounds of cotton. Can't make anything at
that. Why didn't he buy land! Humph! Takes money to buy
land. And he turns away. Free! The most piteous thing
amid all the black ruin of war-time, amid the broken
fortunes of the masters, the blighted hopes of mothers and
maidens, and the fall of an empire,—the most piteous thing
amid all this was the black freedman who threw down his
hoe because the world called him free. What did such a
mockery of freedom mean? Not a cent of money, not an
inch of land, not a mouthful of victuals,—not even ownership of the rags on his back. Free! On Saturday, once or
twice a month, the old master, before the war, used to
dole out bacon and meal to his Negroes. And after the

first flush of freedom wore off, and his true helplessness dawned on the freedman, he came back and picked up his hoe, and old master still doled out his bacon and meal. The legal form of service was theoretically far different; in practice, task-work or "cropping" was substituted for daily toil in gangs; and the slave gradually became a metayer, or tenant on shares, in name, but a laborer with indeterminate wages in fact.

Still the price of cotton fell, and gradually the landlords deserted their plantations, and the reign of the merchant began. The merchant of the Black Belt is a contractor, and part despot. His store, which used most frequently to stand at the cross-roads and become the centre of a weekly village, has now moved to town; and thither the Negro tenant follows him. The merchant keeps everything,—clothes and shoes, coffee and sugar, pork and meal, canned and dried goods, wagons and ploughs, seed and fertilizer,—and what he has not in stock he can give you an order for at the store across the way. Here then, comes the tenant, Sam Scott, after he has contracted with some absent landlord's agent for hiring forty acres of land; he fingers his hat nervously until the merchant finishes his morning chat with Colonel Saunders, and calls out, "Well, Sam, what do you want?" Sam wants him to "furnish" him,—*i.e.*, to advance him food and clothing for the year, and perhaps seed and tools, until his crop is raised and sold. If Sam seems a favorable subject, he and the merchant go to a lawyer, and Sam executes a chattel mortgage on his mule and wagon in return for seed and a week's rations. As soon as the green cotton-leaves appear above the ground, another mortgage is given on the "crop." Every Saturday, or at longer intervals, Sam calls upon the merchant for his "rations"; a family of five usually gets about thirty pounds of fat side-pork and a couple of bushels of cornmeal a month. Besides this, clothing and shoes must be furnished; if Sam or his family is sick, there are orders on the druggist and doctor; if the mule wants shoeing, an order on the blacksmith, etc. If Sam is a hard worker and crops promise well, he is often encouraged to buy more,—sugar, extra clothes, perhaps a buggy. But he is seldom encouraged to save. When cotton rose to ten cents last fall, the shrewd merchants of Dougherty County sold a thousand buggies in one season, mostly to black men.

The security offered for such transactions—a crop and chattel mortgage—may at first seem slight. And, indeed, the merchants tell many a true tale of shiftlessness and cheating; of cotton picked at night, mules disappearing, and tenants absconding. But on the whole the merchant of the Black Belt is the most prosperous man in the section. So skilfully and so closely has he drawn the bonds of the law about the tenant, that the black man has often simply to choose between pauperism and crime; he "waives" all homestead exemptions in his contract; he cannot touch his own mortgaged crop, which the laws put almost in the full control of the land-owner and of the merchant. When the crop is growing the merchant watches it like a hawk; as soon as it is ready for market he takes possession of it, sells it, pays the land-owner his rent, subtracts his bill for supplies, and if, as sometimes happens, there is anything left, he hands it over to the black serf for his Christmas celebration.

The direct result of this system is an all-cotton scheme of agriculture and the continued bankruptcy of the tenant. The currency of the Black Belt is cotton. It is a crop always salable for ready money, not usually subject to great yearly fluctations in price, and one which the Negroes know how to raise. The landlord therefore demands his rent in cotton, and the merchant will accept mortgages on no other crop. There is no use asking the black tenant, then, to diversify his crops,—he cannot under this system. Moreover, the system is bound to bankrupt the tenant. I remember once meeting a little one-mule wagon on the River road. A young black fellow sat in it driving listlessly, his elbows on his knees. His dark-faced wife sat beside him, stolid, silent.

"Hello!" cried my driver,—he has a most impudent way of addressing these people, though they seem used to it,— "what have you got there?"

"Meat and meal," answered the man, stopping. The meat lay uncovered in the bottom of the wagon,—a great thin side of fat pork covered with salt; the meal was in a white bushel bag.

"What did you pay for that meat?"

"Ten cents a pound." It could have been bought for six or seven cents cash.

"And the meal?"

"Two dollars." One dollar and ten cents is the cash

price in town. Here was a man paying five dollars for goods which he could have bought for three dollars cash, and raised for one dollar or one dollar and a half.

Yet it is not wholly his fault. The Negro farmer started behind, started in debt. This was not his choosing, but the crime of this happy-go-lucky nation which goes blundering along with its Reconstruction tragedies, its Spanish war interludes and Philippine matinees, just as though God really were dead. Once in debt, it is no easy matter for a whole race to emerge.

In the year of low-priced cotton, 1898, out of three hundred tenant families one hundred and seventy-five ended their year's work in debt to the extent of fourteen thousand dollars; fifty cleared nothing, and the remaining seventy-five made a total profit of sixteen hundred dollars. The net indebtedness of the black tenant families of the whole county must have been at least sixty thousand dollars. In a more prosperous year the situation is far better; but on the average the majority of tenants end the year even, or in debt, which means that they work for board and clothes. Such an economic organization is radically wrong. Whose is the blame?

The underlying causes of this situation are complicated but discernible. And one of the chief, outside the carelessness of the nation in letting the slave start with nothing, is the widespread opinion among the merchants and employers of the Black Belt that only by the slavery of debt can the Negro be kept at work. Without doubt, some pressure was necessary at the beginning of the free-labor system to keep the listless and lazy at work; and even to-day the mass of the Negro laborers need stricter guardianship than most Northern laborers. Behind this honest and widespread opinion dishonesty and cheating of the ignorant laborers have a good chance to take refuge. And to all this must be added the obvious fact that a slave ancestry and a system of unrequited toil has not improved the efficiency or temper of the mass of black laborers. Nor is this peculiar to Sambo; it has in history been just as true of John and Hans, of Jacques and Pat, of all ground-down peasantries. Such is the situation of the mass of the Negroes in the Black Belt to-day; and they are thinking about it. Crime, and a cheap and dangerous socialism, are the inevitable results of this pondering. I see now that ragged black man sitting on a log, aimlessly whittling a stick. He muttered to me with the murmur of many ages, when he said: "White man sit down whole year; Nigger work day and night and make crop; Nigger hardly gits bread and meat; white man sittin' down gits all. *It's wrong*." And what do the better classes of Negroes do to improve their situation? One of two things: if any way possible, they buy land; if not, they migrate to town. Just as centuries ago it was no easy thing for the serf to escape into the freedom of town-life, even so to-day there are hindrances laid in the way of county laborers. In considerable parts of all the Gulf States, and especially in Mississippi, Louisiana, and Arkansas, the Negroes on the plantations in the back-country districts are still held at forced labor practically without wages. Especially is this true in districts where the farmers are composed of the more ignorant class of poor whites, and the Negroes are beyond the reach of schools and intercourse with their advancing fellows. If such a peon should run away, the sheriff, elected by white suffrage, can usually be depended on to catch the fugitive, return him, and ask no questions. If he escapes to another county, a charge of petty thieving, easily true, can be depended upon to secure his return. Even if some unduly officious person insist upon a trial, neighborly comity will probably make his conviction sure, and then the labor due the county can easily be bought by the master. Such a system is impossible in the more civilized parts of the south, or near the large towns and cities; but in those vast stretches of land beyond the telegraph and the newspaper the spirit of the Thirteenth Amendment is sadly broken. This represents the lowest economic depths of the black American peasant; and in a study of the rise and condition of the Negro freeholder we must trace his economic progress from the modern serfdom.

Even in the better-ordered country districts of the South the free movement of agricultural laborers is hindered by the migration-agent laws. The "Associated Press" recently informed the world of the arrest of a young white man in Southern Georgia who represented the "Atlantic Naval Supplies Company," and who "was caught in the act of enticing hands from the turpentine farm of Mr. John Greer." The crime for which this young man was arrested is taxed five hundred dollars for each county in which the employment agent proposes to gather laborers for work outside the State. Thus the Negroes' ignorance of the labor-market outside his own vicinity is increased rather than diminished by the laws of nearly every Southern State.

Similar to such measures is the unwritten law of the back districts and small towns of the South, that the character of all Negroes unknown to the mass of the community must be vouched for by some white man. This is really a revival of the old Roman idea of the patron under whose protection the new-made freedman was put. In many instances this system has been of great good to the Negro, and very often under the protection and guidance of the former master's family, or other white friends, the freedman progressed in wealth and morality. But the same system has in other cases resulted in the refusal of whole communities to recognize the right of a Negro to change his habitation and to be master of his own fortunes. A black stranger in Baker County, Georgia, for instance, is liable to be stopped anywhere on the public highway and made to state his business to the satisfaction of any white interrogator. If he fails to give a suitable answer, or seems too independent or "sassy," he may be arrested or summarily driven away.

Thus it is that in the country districts of the South, by written or unwritten law, peonage, hindrances to the migration of labor, and a system of white patronage exists over large areas. Besides this, the chance for lawless oppression and illegal exactions is vastly greater in the country than in the city, and nearly all the more serious race disturbances of the last decade have arisen from disputes in the county between master and man,—as, for instance, the Sam Hose affair. As a result of such a situation, there arose, first, the Black Belt; and, second, the Migration to Town. The Black Belt was not, as many assumed, a movement toward fields of labor under more genial climatic conditions; it was primarily a huddling for self-protection,—a massing of the black population for mutual defence in order to secure the peace and tranquility necessary to economic advance. This movement took place between Emancipation and 1880, and only partially accomplished the desired results. The rush to town since 1880 is the counter-movement of men disappointed in the economic opportunities of the Black Belt.

In Dougherty County, Georgia, one can see easily the

results of this experiment in huddling for protection. Only ten per cent of the adult population was born in the county, and yet the blacks outnumber the whites four or five to one. There is undoubtedly a security to the blacks in their very numbers,—a personal freedom from arbitrary treatment, which makes hundreds of laborers cling to Dougherty in spite of low wages and economic distress. But a change is coming, and slowly but surely even here the agricultural laborers are drifting to town and leaving the broad acres behind. Why is this? Why do not the Negroes become land-owners, and build up the black landed peasantry, which has for a generation and more been the dream of philanthropist and statesman?

To the car-window sociologist, to the man who seeks to understand and know the South by devoting the few leisure hours of a holiday trip to unravelling the snarl of centuries,—to such men very often the whole trouble with the black field-hand may be summed up by Aunt Ophelia's word, "Shiftless!" They have noted repeatedly scenes like one I saw last summer. We were riding along the highroad to town at the close of a long hot day. A couple of young black fellows passed us in a muleteam, with several bushels of loose corn in the ear. One was driving, listlessly bent forward, his elbows on his knees, —a happy-go-lucky, careless picture of irresponsibility. The other was fast asleep in the bottom of the wagon. As we passed we noticed an ear of corn fall from the wagon. They never saw it,—not they. A rod farther on we noted another ear on the ground; and between that creeping mule and town we counted twenty-six ears of corn. Shiftless? Yes, the personification of shiftlessness. And yet follow those boys: they are not lazy; to-morrow morning they'll be up with the sun; they work hard when they do work, and they work willingly. They have no sordid, selfish, money-getting ways, but rather a fine disdain for mere cash. They'll loaf before your face and work behind your back with good-natured honesty. They'll steal a watermelon, and hand you back your lost purse intact. Their great defect as laborers lies in their lack of incentive to work beyond the mere pleasure of physical exertion. They are careless because they have not found that it pays to be careful; they are improvident because the improvident ones of their acquaintance get on about as well as the provident. Above all, they cannot see why they should take unusual pains to make the white man's land better, or to fatten his mule, or save his corn. On the other hand, the white land-owner argues that any attempt to improve these laborers by increased responsibility, or higher wages, or better homes, or land of their own, would be sure to result in failure. He shows his Northern visitor the scarred and wretched land; the ruined mansions, the worn-out soil and mortgaged acres, and says, this is Negro freedom!

Now it happens that both master and man have just enough argument on their respective sides to make it difficult for them to understand each other. The Negro dimly personifies in the white man all his ills and misfortunes; if he is poor, it is because the white man seizes the fruit of his toil; if he is ignorant, it is because the white man gives him neither time nor facilities to learn; and, indeed, if any misfortune happens to him, it is because of some hidden machinations of "white folks." On the other hand, the masters and the masters' sons have never been able to see why the Negro, instead of settling down to be day-laborers for bread and clothes, are infected with a silly desire to rise in the world, and why they are sulky, dissatisfied, and careless, where their fathers were happy and dumb and faithful. "Why, you niggers have an easier time than I do," said a puzzled Albany merchant to his black customer. "Yes," he replied, "and so does yo' hogs."

Taking, then, the dissatisfied and shiftless field-hand as a starting-point, let us inquire how the black thousands of Dougherty have struggled from him up toward their ideal, and what that ideal is. All social struggle is evidenced by the rise, first of economic, then of social classes, among a homogeneous population. To-day the following economic classes are plainly differentiated among these Negroes.

A "submerged tenth" of croppers, with a few paupers; forty per cent who are metayers and thirty-nine per cent of semi-metayers and wage-laborers. There are left five per cent of money-renters and six per cent of freeholders, —the "Upper Ten" of the land. The croppers are entirely without capital, even in the limited sense of food or money to keep them from seed-time to harvest. All they furnish is their labor; the land-owner furnishes land, stock, tools, seed, and house; and at the end of the year the laborer gets from a third to a half of the crop. Out of his share, however, comes pay and interest for food and clothing advanced him during the year. Thus we have a laborer without capital and without wages, and an employer whose capital is largely his employees' wages. It is an unsatisfactory arrangement, both for hirer and hired, and is usually in vogue on poor land with hard-pressed owners.

Above the croppers come the great mass of the black population who work the land on their own responsibility, paying rent in cotton and supported by the crop-mortgage system. After the war this system was attractive to the freedmen on account of its larger freedom and its possibility for making a surplus. But with the carrying out of the crop-lien system, the deterioration of the land, and the slavery of debt, the position of the metayers has sunk to a dead level of practically unrewarded toil. Formerly all tenants had some capital, and often considerable; but absentee landlordism, rising rack-rent, and falling cotton have stripped them well-nigh of all, and probably not over half of them to-day own their mules. The change from cropper to tenant was accomplished by fixing the rent. If, now, the rent fixed was reasonable, this was an incentive to the tenant to strive. On the other hand, if the rent was too high, or if the land deteriorated, the result was to discourage and check the efforts of the black peasantry. There is no doubt that the latter case is true; that in Dougherty County every economic advantage of the price of cotton in market and of the strivings of the tenant has been taken advantage of by the landlords and merchants, and swallowed up in rent and interest. If cotton rose in price, the rent rose even higher; if cotton fell, the rent remained or followed reluctantly. If the tenant worked hard and raised a large crop, his rent was raised the next year; if that year the crop failed, his corn was confiscated and his mule sold for debt. There were, of course, exceptions to this,—cases of personal kindness and forbearance; but in the vast majority of cases the rule was to extract the uttermost farthing from the mass of the black farm laborers.

The average metayer pays from twenty to thirty per cent of his crop in rent. The result of such rack-rent can only be evil,—abuse and neglect of the soil, deterioration in the character of the laborers, and a widespread sense of injustice. "Wherever the country is poor," cried Arthur Young, "it is in the hands of metayers," and "their condition is more wretched than that of day-laborers." He was talking of Italy a century ago; but he might have been talking of Dougherty County to-day. And especially is that true to-day which he declares was true in France before the Revolution: "The metayers are considered as little better than menial servants, removable at pleasure, and obliged to conform in all things to the will of the landlords." On this low plane half the black population of Dougherty County—perhaps more than half the black millions of this land—are to-day struggling.

A degree above these we may place those laborers who receive money wages for their work. Some receive a house with perhaps a garden-spot; then supplies of food and clothing are advanced, and certain fixed wages are given at the end of the year, varying from thirty to sixty dollars, out of which the supplies must be paid for, with interest. About eighteen per cent of the population belong to this class of semi-metayers, while twenty-two per cent are laborers paid by the month or year, and are either "furnished" by their own savings or perhaps more usually by some merchant who takes his chances of payment. Such laborers receive from thirty-five to fifty cents a day during the working season. They are usually young unmarried persons, some being women; and when they marry they sink to the class of metayers, or, more seldom, become renters.

The renters for fixed money rentals are the first of the emerging classes, and form five per cent of the families. The sole advantage of this small class is their freedom to choose their crops, and the increased responsibility which comes through having money transactions. While some of the renters differ little in condition from the metayers, yet on the whole they are more intelligent and responsible persons, and are the ones who eventually become land-owners. Their better character and greater shrewdness enable them to gain, perhaps to demand, better terms in rents; rented farms, varying from forty to a hundred acres, bear an average rental of about fifty-four dollars a year. The men who conduct such farms do not long remain renters; either they sink to metayers, or with a successful series of harvests rise to be land-owners.

In 1870 the tax-books of Dougherty report no Negroes as landholders. If there were any such at that time,— and there may have been a few,—their land was probably held in the name of some white patron,—a method not uncommon during slavery. In 1875 ownership of land had begun with seven hundred and fifty acres; ten years later this had increased to over sixty-five hundred acres, to nine thousand acres in 1890 and ten thousand in 1900. The total assessed property has in this same period risen from eighty thousand dollars in 1875 to two hundred and forty thousand dollars in 1900.

Two circumstances complicate this development and make it in some respects difficult to be sure of the real tendencies; they are the panic of 1893, and the low price of cotton in 1898. Besides this, the system of assessing property in the country districts of Georgia is somewhat antiquated and of uncertain statistical value; there are no assessors, and each man makes a sworn return to a tax-receiver. Thus public opinion plays a large part, and the returns vary strangely from year to year. Certainly these figures show the small amount of accumulated capital among the Negroes, and the consequent large dependence of their property on temporary prosperity. They have little to tide over a few years of economic depression, and are at the mercy of the cotton-market far more than the whites. And thus the land-owners, despite their marvellous efforts, are really a transient class, continually being depleted by those who fall back into the class of renters or metayers, and augmented by newcomers from the masses. Of the one hundred land-owners in 1898, half had bought their land since 1893, a fourth between 1890 and 1893, a fifth between 1884 and 1890, and the rest between 1870 and 1884. In all, one hundred and eighty-five Negroes have owned land in this county since 1875.

If all the black land-owners who had ever held land here had kept it or left it in the hands of black men, the Negroes would have owned nearer thirty thousand acres than the fifteen thousand they now hold. And yet these fifteen thousand acres are a creditable showing,—a proof of no little weight of the worth and ability of the Negro people. If they had been given an economic start at Emancipation, if they had been in an enlightened and rich community which really desired their best good, then we might perhaps call such a result small or even insignificant. But for a few thousand poor ignorant field-hands, in the face of poverty, a falling market, and social stress, to save and capitalize two hundred thousand dollars in a generation has meant a tremendous effort. The rise of a nation, the pressing forward of a social class, means a bitter struggle, a hard and soul-sickening battle with the world such as few of the more favored classes know or appreciate.

Out of the hard economic conditions of this portion of the Black Belt, only six per cent of the population have succeeded in emerging into peasant proprietorship; and these are not all firmly fixed, but grow and shrink in number with the wavering of the cotton-market. Fully ninety-four per cent have struggled for land and failed, and half of them sit in hopeless serfdom. For these there is one other avenue of escape toward which they have turned in increasing numbers, namely, migration to town. A glance at the distribution of land among the black owners curiously reveals this fact. In 1898 the holdings were as follows: Under forty acres, forty-nine families; forty to two hundred and fifty acres, seventeen families, two hundred and fifty to one thousand acres, thirteen families; one thousand or more acres, two families. Now in 1890 there were forty-four holdings, but only nine of these were under forty acres. The great increase of holdings, then, has come in the buying of small homesteads near town, where their owners really share in the town life; this is a part of the rush to town. And for every landowner who has thus hurried away from the narrow and hard conditions of country life, how many field-hands, how many tenants, how many ruined renters, have joined that long procession? Is it not strange compensation? The sin of the country districts is visited on the town, and the social sores of city life to-day may, here in Dougherty County, and perhaps in many places near and far, look for their final healing without the city walls.

IX

Of the Sons of Master and Man

Life treads on life, and heart on heart;
We press too close in church and mart
To keep a dream or grave apart.

MRS. BROWNING.

The world-old phenomenon of the contact of diverse races of men is to have new exemplification during the new century. Indeed, the characteristic of our age is the contact of European civilization with the world's undeveloped peoples. Whatever we may say of the results of such contact in the past, it certainly forms a chapter in human action not pleasant to look back upon. War, murder, slavery, extermination, and debauchery,—this has again and again been the result of carrying civilization and the blessed gospel to the isles of the sea and the heathen without the law. Nor does it altogether satisfy the conscience of the modern world to be told complacently that all this has been right and proper, the fated triumph of strength over weakness, of righteousness over evil, of superiors over inferiors. It would certainly be soothing if one could readily believe all this; and yet there are too many ugly facts for everything to be thus easily explained away. We feel and know that there are many delicate differences in race psychology, numberless changes that our crude social measurements are not yet able to follow minutely, which explain much of history and social development. At the same time, too, we know that these considerations have never adequately explained or excused the triumph of brute force and cunning over weakness and innocence.

It is, then, the strife of all honorable men of the twentieth century to see that in the future competition of races the survival of the fittest shall mean the triumph of the good, the beautiful, and the true; that we may be able to preserve for future civilization all that is really fine and noble and strong, and not continue to put a premium on greed and impudence and cruelty. To bring this hope to fruition, we are compelled daily to turn more and more to a conscientious study of the phenomena of race-contact,—to a study frank and fair, and not falsified and colored by our wishes or our fears. And we have in the South as fine a field for such a study as the world affords, —a field, to be sure, which the average American scientist deems somewhat beneath his dignity, and which the average man who is not a scientist knows all about, but nevertheless a line of study which by reason of the enormous race complications with which God seems about to punish this nation must increasingly claim our sober attention, study, and thought, we must ask, what are the actual relations of whites and blacks in the South? and we must be answered, not by apology or fault-finding, but by a plain, unvarnished tale.

In the civilized life of to-day the contact of men and their relations to each other falls in a few main lines of action and communication: there is, first, the physical proximity of homes and dwelling-places, the way in which neighborhoods group themselves, and the contiguity of neighborhoods. Secondly, and in our age chiefest, there are the economic relations,—the methods by which individuals coöperate for earning a living, for the mutual satisfaction of wants, for the production of wealth. Next, there are the political relations, the coöperation in social control, in group government, in laying and paying the burden of taxation. In the fourth place there are the less tangible but highly important forms of intellectual contact and commerce, the interchange of ideas through conversation and conference, through periodicals and libraries; and, above all, the gradual formation for each community of that curious *tertium quid* which we call public opinion. Closely allied with this come the various forms of social

contact in everyday life, in travel, in theatres, in house
gatherings, in marrying and giving in marriage. Finally,
there are the varying forms of religious enterprise, of moral
teaching and benevolent endeavor. These are the principle
ways in which men living in the same communities are
brought into contact with each other. It is my present task,
therefore, to indicate, from my point of view, how the
black race in the South meet and mingle with the whites
in these matters of everyday life.

First, as to physical dwelling. It is usually possible to
draw in nearly every Southern community a physical color-
line on the map, on the one side of which whites dwell
and on the other Negroes. The winding and intricacy of
the geographical color line varies, of course, in different
communities. I know some towns where a straight line
drawn through the middle of the main street separates
nine-tenths of the whites from nine-tenths of the blacks.
In other towns the older settlement of whites has been
encircled by a broad band of blacks; in still other cases
little settlements or nuclei of blacks have sprung up amid
surrounding whites. Usually in cities each street has its
distinctive color, and only now and then do the colors
meet in close proximity. Even in the country something of
this segregation is manifest in the smaller areas, and of
course in the larger phenomena of the Black Belt.

All this segregation by color is largely independent of
that natural clustering of social grades common to all
communities. A Negro slum may be in dangerous proximity to a white residence quarter, while it is quite common to find a white slum planted in the heart of a respectable Negro district. One thing, however, seldom occurs: the best of the whites and the best of the Negroes
almost never live in anything like close proximity. It thus
happens that in nearly every Southern town and city, both
whites and blacks see commonly the worst of each other.
This is a vast change from the situation in the past, when,
through the close contact of master and house-servant in
the patriarchal big house, one found the best of both races
in close contact and sympathy, while at the same time the
squalor and dull round of toil among the field-hands was
removed from the sight and hearing of the family. One
can easily see how a person who saw slavery thus from
his father's parlors, and sees freedom on the streets of a
great city, fails to grasp or comprehend the whole of the

new picture. On the other hand, the settled belief of the mass of the Negroes that the Southern white people do not have the black man's best interests at heart has been intensified in later years by this continual daily contact of the better class of blacks with the worst representatives of the white race.

Coming now to the economic relations of the races, we are on ground made familiar by study, much discussion, and no little philanthropic effort. And yet with all this there are many essential elements in the coöperation of Negroes and whites for work and wealth that are too readily overlooked or not thoroughly understood. The average American can easily conceive of a rich land awaiting development and filled with black laborers. To him the Southern problem is simply that of making efficient workingmen out of this material, by giving them the requisite technical skill and the help of invested capital. The problem, however, is by no means as simple as this, from the obvious fact that these workingmen have been trained for centuries as slaves. They exhibit, therefore, all the advantages and defects of such training; they are willing and good-natured, but not self-reliant, provident, or careful. If now the economic development of the South is to be pushed to the verge of exploitation, as seems probable, then we have a mass of workingmen thrown into relentless competition with the workingmen of the world, but handicapped by a training the very opposite to that of the modern self-reliant democratic laborer. What the black laborer needs is careful personal guidance, group leadership of men with hearts in their bosoms, to train them to foresight, carefulness, and honesty. Nor does it require any fine-spun theories of racial differences to prove the necessity of such group training after the brains of the race have been knocked out by two hundred and fifty years of assiduous education in submission, carelessness, and stealing. After Emancipation, it was the plain duty of some one to assume this group leadership and training of the Negro laborer. I will not stop here to inquire whose duty it was,—whether that of the white ex-master who had profited by unpaid toil, or the Northern philanthropist whose persistence brought on the crisis, or the National Government whose edict freed the bondsmen; I will not stop to ask whose duty it was, but I insist it was the duty of some one to see that these workingmen were not left alone and unguided, without capital, without land, without skill, without economic organization, without even the bald protection of law, order, and decency,—left in a great land, not to settle down to slow and careful internal development, but destined to be thrown almost immediately into relentless and sharp competition with the best of modern workingmen under an economic system where every participant is fighting for himself, and too often utterly regardless of the rights or welfare of his neighbor.

For we must never forget that the economic system of the South to-day which has succeeded the old regime is not the same system as that of the old industrial North, of England, or of France, with their trade-unions, their restrictive laws, their written and unwritten commercial customs, and their long experience. It is, rather, a copy of that England of the early nineteenth century, before the factory acts,—the England that wrung pity from thinkers and fired the wrath of Carlyle. The rod of empire that passed from the hands of Southern gentlemen in 1865, partly by force, partly by their own petulance, has never returned to them. Rather it has passed to those men who have come to take charge of the industrial exploitation of the New South,—the sons of poor whites fired with a new thirst for wealth and power, thrifty and avaricious Yankees, and unscrupulous immigrants. Into the hands of these men the Southern laborers, white and black, have fallen; and this to their sorrow. For the laborers as such, there is in these new captains of industry neither love nor hate, neither sympathy nor romance; it is a cold question of dollars and dividends. Under such a system all labor is bound to suffer. Even the white laborers are not yet intelligent, thrifty, and well trained enough to maintain themselves against the powerful inroads of organized capital. The results among them, even, are long hours of toil, low wages, child labor, and lack of protection against usury and cheating. But among the black laborers all this is aggravated, first, by a race prejudice which varies from a doubt and distrust among the best element of whites to a frenzied hatred among the worst; and, secondly, it is aggravated, as I have said before, by the wretched economic heritage of the freedmen from slavery. With this training it is difficult for the freedman to learn to grasp the opportunities already opened to him, and the new opportunities are seldom given him, but go by favor to the whites.

Left by the best elements of the South with little protection or oversight, he has been made in law and custom the victim of the worst and most unscrupulous men in each community. The crop-lien system which is depopulating the fields of the South is not simply the result of shiftlessness on the part of Negroes, but is also the result of cunningly devised laws as to mortgages, liens, and misdemeanors, which can be made by conscienceless men to entrap and snare the unwary until escape is impossible, further toil a farce, and protest a crime. I have seen, in the Black Belt of Georgia, an ignorant, honest Negro buy and pay for a farm in installments three separate times, and then in the face of law and decency the enterprising American who sold it to him pocketed money and deed and left the black man landless, to labor on his own land at thirty cents a day. I have seen a black farmer fall in debt to a white storekeeper, and that storekeeper go to his farm and strip it of every single marketable article,—mules, ploughs, stored crops, tools, furniture, bedding, clocks, looking-glass,—and all this without a sheriff or officer, in the face of the law for homestead exemptions, and without rendering to a single responsible person any account or reckoning. And such proceedings can happen, and will happen, in any community where a class of ignorant toilers are placed by custom and race-prejudice beyond the pale of sympathy and race-brotherhood. So long as the best elements of a community do not feel in duty bound to protect and train and care for the weaker members of their group, they leave them to be preyed upon by these swindlers and rascals.

This unfortunate economic situation does not mean the hindrance of all advance in the black South, or the absence of a class of black landlords and mechanics who, in spite of disadvantages, are accumulating property and making good citizens. But it does mean that this class is not nearly so large as a fairer economic system might easily make it, that those who survive in the competition are handicapped so as to accomplish much less than they deserve to, and that, above all, the *personnel* of the successful class is left to chance and accident, and not to any intelligent culling or reasonable methods of selection. As a remedy for this, there is but one possible procedure. We

must accept some of the race prejudice in the South as a
fact,—deplorable in its intensity, unfortunate in results,
and dangerous for the future, but nevertheless a hard fact
which only time can efface. We cannot hope, then, in this
generation, or for several generations, that the mass of
the whites can be brought to assume that close sympathetic
and self-sacrificing leadership of the blacks which their
present situation so eloquently demands. Such leadership,
such social teaching and example, must come from the
blacks themselves. For some time men doubted as to
whether the Negro could develop such leaders; but to-day
no one seriously disputes the capability of individual Negroes to assimilate the culture and common sense of
modern civilization, and to pass it on, to some extent at
least, to their fellows. If this is true, then here is the path
out of the economic situation, and here is the imperative
demand for trained Negro leaders of character and intelligence,—men of skill, men of light and leading, college-
bred men, black captains of industry, and missionaries of
culture; men who thoroughly comprehend and know modern civilization, and can take hold of Negro communities
and raise and train them by force of precept and example,
deep sympathy, and the inspiration of common blood and
ideals. But if such men are to be effective they must have
some power,—they must be backed by the best public
opinion of these communities, and able to wield for their
objects and aims such weapons as the experience of the
world has taught are indispensable to human progress.

Of such weapons the greatest, perhaps, in the modern
world is the power of the ballot; and this brings me to a
consideration of the third form of contact between whites
and blacks in the South,—political activity.

In the attitude of the American mind toward Negro
suffrage can be traced with unusual accuracy the prevalent
conceptions of government. In the fifties we were near
enough to the echoes of the French Revolution to believe
pretty thoroughly in universal suffrage. We argued, as we
thought then rather logically, that no social class was so
good, so true, and so disinterested as to be trusted wholly
with the political destiny of its neighbors; that in every
state the best arbiters of their own welfare are the persons
directly affected; consequently that it is only by arming
every hand with a ballot,—with the right to have a voice
in the policy of the state,—that the greatest good to the

greatest number could be attained. To be sure, there were objections to these arguments, but we thought we had answered them tersely and convincingly; if some one complained of the ignorance of voters, we answered, "Educate them." If another complained of their venality, we replied, "Disfranchise them or put them in jail." And, finally, to the men who feared demagogues and the natural perversity of some human beings we insisted that time and bitter experience would teach the most hardheaded. It was at this time that the question of Negro suffrage in the South was raised. Here was a defenceless people suddenly made free. How were they to be protected from those who did not believe in their freedom and were determined to thwart it? Not by force, said the North; not by government guardianship, said the South; then by the ballot, the sole and legitimate defence of a free people, said the Common Sense of the Nation. No one thought, at the time, that the ex-slaves could use the ballot intelligently or very effectively; but they did think that the possession of so great power by a great class in the nation would compel their fellows to educate this class to its intelligent use.

Meantime, new thoughts came to the nation: the inevitable period of moral retrogression and political trickery that ever follows in the wake of war overtook us. So flagrant became the political scandals that reputable men began to leave politics alone, and politics consequently became disreputable. Men began to pride themselves on having nothing to do with their own government, and to agree tacitly with those who regarded public office as a private perquisite. In this state of mind it became easy to wink at the suppression of the Negro vote in the South, and to advise self-respecting Negroes to leave politics entirely alone. The decent and reputable citizens of the North who neglected their own civic duties grew hilarious over the exaggerated importance with which the Negro regarded the franchise. Thus it easily happened that more and more the better class of Negroes followed the advice from abroad and the pressure from home, and took no further interest in politics, leaving to the careless and the venal of their race the exercise of their rights as voters. The black vote that still remained was not trained and educated, but further debauched by open and unblushing bribery, or force and fraud; until the Negro voter was thoroughly inoculated with the idea that politics was a method of private gain by disreputable means.

And finally, now, to-day, when we are awakening to the fact that the perpetuity of republican institutions on this continent depends on the purification of the ballot, the civic training of voters, and the raising of voting to the plane of a solemn duty which a patriotic citizen neglects to his peril and to the peril of his children's children,—in this day, when we are striving for a renaissance of civic virtue, what are we going to say to the black voter of the South? Are we going to tell him still that politics is a disreputable and useless form of human activity? Are we going to induce the best class of Negroes to take less and less interest in government, and to give up their right to take such an interest, without a protest? I am not saying a word against all legitimate efforts to purge the ballot of ignorance, pauperism, and crime. But few have pretended that the present movement for disfranchisement in the South is for such a purpose; it has been plainly and frankly declared in nearly every case that the object of the disfranchising laws is the elimination of the black man from politics.

Now, is this a minor matter which has no influence on the main question of the industrial and intellectual development of the Negro? Can we establish a mass of black laborers and artisans and landholders in the South who, by law and public opinion, have absolutely no voice in shaping the laws under which they live and work? Can the modern organization of industry, assuming as it does free democratic government and the power and ability of the laboring classes to compel respect for their welfare,—can this system be carried out in the South when half its laboring force is voiceless in the public councils and powerless in its own defence? To-day the black man of the South has almost nothing to say as to how much he shall be taxed, or how those taxes shall be expended; as to who shall execute the laws, and how they shall do it; as to who shall make the laws, and how they shall be made. It is pitiable that frantic efforts must be made at critical times to get law-makers in some States even to listen to the respectful presentation of the black man's side of a current controversy. Daily the Negro is coming more and more to look upon law and justice, not as protecting safeguards, but as sources of humiliation and oppression. The laws are made by men who have little interest in him; they are executed by men who have absolutely no motive for treating the black people with courtesy or consideration; and, finally, the accused lawbreaker is tried, not by his peers, but too often by men who would rather punish ten innocent Negroes than let one guilty one escape.

I should be the last one to deny the patent weaknesses and shortcomings of the Negro people; I should be the last to withhold sympathy from the white South in its efforts to solve its intricate social problems. I freely acknowledged that it is possible, and sometimes best, that a partially undeveloped people should be ruled by the best of their stronger and better neighbors for their own good, until such time as they can start and fight the world's battles alone. I have already pointed out how sorely in need of such economic and spiritual guidance the emancipated Negro was, and I am quite willing to admit that if the representatives of the best white Southern public opinion were the ruling and guiding powers in the South to-day the conditions indicated would be fairly well fulfilled. But the point I have insisted upon, and now emphasize again, is that the best opinion of the South to-day is not the ruling opinion. That to leave the Negro helpless and without a ballot to-day is to leave him, not to the guidance of the best, but rather to the exploitation and debauchment of the worst; that this is no truer of the South than of the North,—of the North than of Europe: in any land, in any country under modern free competition, to lay any class of weak and despised people, be they white, black, or blue, at the political mercy of their stronger, richer, and more resourceful fellows, is a temptation which human nature seldom has withstood and seldom will withstand.

Moreover, the political status of the Negro in the South is closely connected with the question of Negro crime. There can be no doubt that crime among Negroes has sensibly increased in the last thirty years, and that there has appeared in the slums of great cities a distinct criminal class among the blacks. In explaining this unfortunate development, we must note two things: (1) that the inevitable result of Emancipation was to increase crime and criminals, and (2) that the police system of the South was primarily designed to control slaves. As to the first point, we must not forget that under a strict slave system there

can scarcely be such a thing as crime. But when these
variously constituted human particles are suddenly thrown
broadcast on the sea of life, some swim, some sink, and
some hang suspended, to be forced up or down by the
chance currents of a busy hurrying world. So great an
economic and social revolution as swept the South in '63
meant a weeding out among the Negroes of the incom-
petents and vicious, the beginning of a differentiation of
social grades. Now a rising group of people are not lifted
bodily from the ground like an inert solid mass, but
rather stretch upward like a living plant with its roots
still clinging in the mould. The appearance, therefore, of
the Negro criminal was a phenomenon to be awaited;
and while it causes anxiety, it should not occasion surprise.

Here again the hope for the future depended peculiarly
on careful and delicate dealing with these criminals. Their
offences at first were those of laziness, carelessness, and
impulse, rather than of malignity or ungoverned viciousness. Such misdemeanors needed discriminating treatment, firm but reformatory, with no hint of injustice, and
full proof of guilt. For such dealing with criminals, white
or black, the South had no machinery, no adequate jails
or reformatories; its police system was arranged to deal
with blacks alone, and tacitly assumed that every white
man was *ipso facto* a member of that police. Thus grew
up a double system of justice, which erred on the white
side by undue leniency and the practical immunity of red-
handed criminals, and erred on the black side by undue
severity, injustice, and lack of discrimination. For, as I
have said, the police system of the South was originally
designed to keep track of all Negroes, not simply of criminals; and when the Negroes were freed and the whole
South was convinced of the impossibility of free Negro
labor, the first and almost universal device was to use
the courts as a means of reënslaving the blacks. It was
not then a question of crime, but rather one of color,
that settled a man's conviction on almost any charge.
Thus Negroes came to look upon courts as instruments
of injustice and oppression, and upon those convicted in
them as martyrs and victims.

When, now, the real Negro criminal appeared, and instead of petty stealing and vagrancy we began to have
highway robbery, burglary, murder, and rape, there was a

curious effect on both sides the color-line: the Negroes refused to believe the evidence of white witnesses or the fairness of white juries, so that the greatest deterrent to crime, the public opinion of one's own social caste, was lost, and the criminal was looked upon as crucified rather than hanged. On the other hand, the whites, used to being careless as to the guilt or innocence of accused Negroes, were swept in moments of passion beyond law, reason, and decency. Such a situation is bound to increase crime, and has increased it. To natural viciousness and vagrancy are being daily added motives of revolt and revenge which stir up all the latent savagery of both races and make peaceful attention to economic development often impossible.

But the chief problem in any community cursed with crime is not the punishment of the criminals, but the preventing of the young from being trained to crime. And here again the peculiar conditions of the South have prevented proper precautions. I have seen twelve-year-old boys working in chains on the public streets of Atlanta, directly in front of the schools, in company with old and hardened criminals; and this indiscriminate mingling of men and women and children makes the chain-gangs perfect schools of crime and debauchery. The struggle for reformatories, which has gone on in Virginia, Georgia, and other States, is the one encouraging sign of the awakening of some communities to the suicidal results of this policy.

It is the public schools, however, which can be made, outside the homes, the greatest means of training decent self-respecting citizens. We have been so hotly engaged recently in discussing trade-schools and the higher education that the pitiable plight of the public-school system in the South has almost dropped from view. Of every five dollars spent for public education in the State of Georgia, the white schools get four dollars and the Negro one dollar; and even then the white public-school system, save in the cities, is bad and cries for reform. If this is true of the whites, what of the blacks? I am becoming more and more convinced, as I look upon the system of common-school training in the South, that the national government must soon step in and aid popular education in some way. To-day it has been only by the most strenuous efforts on the part of the thinking men of the South that the Negro's share of the school fund has not been cut

down to a pittance in some half-dozen States; and that
movement not only is not dead, but in many communities
is gaining strength. What in the name of reason does
this nation expect of a people, poorly trained and hard
pressed in severe economic competition, without political
rights, and with ludicrously inadequate common-school facilities? What can it expect but crime and listlessness,
offset here and there by the dogged struggles of the fortunate and more determined who are themselves buoyed
by the hope that in due time the country will come to
its senses?

I have thus far sought to make clear the physical, economic, and political relations of the Negroes and whites
in the South, as I have conceived them, including, for the
reasons set forth, crime and education. But after all that
has been said on these more tangible matters of human
contact, there still remains a part essential to a proper
description of the South which it is difficult to describe
or fix in terms easily understood by strangers. It is, in
fine, the atmosphere of the land, the thought and feeling,
the thousand and one little actions which go to make up
life. In any community or nation it is these little things
which are most elusive to the grasp and yet most essential to any clear conception of the group life taken as a
whole. What is thus true of all communities is peculiarly
true of the South, where, outside of written history and
outside of printed law, there has been going on for a
generation as deep a storm and stress of human souls, as
intense a ferment of feeling, as intricate a writhing of
spirit, as ever a people experienced. Within and without
the sombre veil of color vast social forces have been at
work,—efforts for human betterment, movements toward
disintegration and despair, tragedies and comedies in social and economic life, and a swaying and lifting and
sinking of human hearts which have made this land a
land of mingled sorrow and joy, of change and excitement and unrest.

The centre of this spiritual turmoil has ever been the
millions of black freedmen and their sons, whose destiny
is so fatefully bound up with that of the nation. And yet
the casual observer visiting the South sees at first little
of this. He notes the growing frequency of dark faces as
he rides along,—but otherwise the days slip lazily on, the
sun shines, and this little world seems as happy and con-

tented as other worlds he has visited. Indeed, on the question of questions—the Negro problem—he hears so little that there almost seems to be a conspiracy of silence; the morning papers seldom mention it, and then usually in a far-fetched academic way, and indeed almost every one seems to forget and ignore the darker half of the land, until the astonished visitor is inclined to ask if after all there *is* any problem here. But if he lingers long enough there comes the awakening: perhaps in a sudden whirl of passion which leaves him gasping at its bitter intensity; more likely in a gradually dawning sense of things he had not at first noticed. Slowly but surely his eyes begin to catch the shadows of the color-line: here he meets crowds of Negroes and whites; then he is suddenly aware that he cannot discover a single dark face; or again at the close of a day's wandering he may find himself in some strange assembly, where all faces are tinged brown or black, and where he has the vague, uncomfortable feeling of the stranger. He realizes at last that silently, resistlessly, the world about flows by him in two great streams: they ripple on in the same sunshine, they approach and mingle their waters in seeming carelessness,— then they divide and flow wide apart. It is done quietly; no mistakes are made, or if one occurs, the swift arm of the law and of public opinion swings down for a moment, as when the other day a black man and a white woman were arrested for talking together on Whitehall Street in Atlanta.

Now if one notices carefully one will see that between these two worlds, despite much physical contact and daily intermingling, there is almost no community of intellectual life or point of transference where the thoughts and feelings of one race can come into direct contact and sympathy with the thoughts and feelings of the other. Before and directly after the war, when all the best of the Negroes were domestic servants in the best of the white families, there were bonds of intimacy, affection, and sometimes blood relationship, between the races. They lived in the same home, shared in the family life, often attended the same church, and talked and conversed with each other. But the increasing civilization of the Negro since then has naturally meant the development of higher classes: there are increasing numbers of ministers, teachers, physicians, merchants, mechanics, and independent farmers, who by nature and training are the aristocracy and leaders of the blacks. Between them, however, and the best element of the whites, there is little or no intellectual commerce. They go to separate churches, they live in separate sections, they are strictly separated in all public gatherings, they travel separately, and they are beginning to read different papers and books. To most libraries, lectures, concerts, and museums, Negroes are either not admitted at all, or on terms peculiarly galling to the pride of the very classes who might otherwise be attracted. The daily paper chronicles the doings of the black world from afar with no great regard for accuracy; and so on, throughout the category of means for intellectual communication,—schools, conferences, efforts for social betterment, and the like,—it is usually true that the very representatives of the two races, who for mutual benefit and the welfare of the land ought to be in complete understanding and sympathy, are so far strangers that one side thinks all whites are narrow and prejudiced, and the other thinks educated Negroes dangerous and insolent. Moreover, in a land where the tyranny of public opinion and the intolerance of criticism is for obvious historical reasons so strong as in the South, such a situation is extremely difficult to correct. The white man, as well as the Negro, is bound and barred by the color-line, and many a scheme of friendliness and philanthropy, of broadminded sympathy and generous fellowship between the two has dropped still-born because some busybody has forced the color-question to the front and brought the tremendous force of unwritten law against the innovators.

It is hardly necessary for me to add very much in regard to the social contact between the races. Nothing has come to replace that finer sympathy and love between some masters and house servants which the radical and more uncompromising drawing of the color-line in recent years has caused almost completely to disappear. In a world where it means so much to take a man by the hand and sit beside him, to look frankly into his eyes and feel his heart beating with red blood; in a world where a social cigar or a cup of tea together means more than legislative halls and magazine articles and speeches,—one can imagine the consequences of the almost utter absence of such social amenities between estranged races, whose separation extends even to parks and streetcars.

Here there can be none of that social going down to the people,—the opening of heart and hand of the best to the worst, in generous acknowledgment of a common humanity and a common destiny. On the other hand, in matters of simple almsgiving, where there can be no question of social contact, and in the succor of the aged and sick, the South, as if stirred by a feeling of its unfortunate limitations, is generous to a fault. The black beggar is never turned away without a good deal more than a crust, and a call for help for the unfortunate meets quick response. I remember, one cold winter, in Atlanta, when I refrained from contributing to a public relief fund lest Negroes should be discriminated against, I afterward inquired of a friend: "Were any black people receiving aid?" "Why," said he, "they were *all* black."

And yet this does not touch the kernel of the problem. Human advancement is not a mere question of almsgiving, but rather of sympathy and coöperation among classes who would scorn charity. And here is a land where, in the higher walks of life, in all the higher striving for the good and noble and true, the color-line comes to separate natural friends and co-workers; while at the bottom of the social group, in the saloon, the gambling-hell, and the brothel, that same line wavers and disappears.

I have sought to paint an average picture of real relations between the sons of master and man in the South. I have not glossed over matters for policy's sake, for I fear we have already gone too far in that sort of thing. On the other hand, I have sincerely sought to let no unfair exaggerations creep in. I do not doubt that in some Southern communities conditions are better than those I have indicated; while I am no less certain that in other communities they are far worse.

Nor does the paradox and danger of this situation fail to interest and perplex the best conscience of the South. Deeply religious and intensely democratic as are the mass of the whites, they feel acutely the false position in which the Negro problems place them. Such an essentially honest-hearted and generous people cannot cite the caste-levelling precepts of Christianity, or believe in equality of opportunity for all men, without coming to feel more and more with each generation that the present drawing of the color-line is a flat contradiction to their beliefs and professions. But just as often as they come to this point, the present social condition of the Negro stands as a menace and a portent before even the most openminded: if there were nothing to charge against the Negro but his blackness or other physical peculiarities, they argue, the problem would be comparatively simple; but what can we say to his ignorance, shiftlessness, poverty, and crime? can a self-respecting group hold anything but the least possible fellowship with such persons and survive? and shall we let a mawkish sentiment sweep away the culture of our fathers or the hope of our children? The argument so put is of great strength, but it is not a whit stronger than the argument of thinking Negroes: granted, they reply, that the condition of our masses is bad; there is certainly on the one hand adequate historical cause for this, and unmistakable evidence that no small number have, in spite of tremendous disadvantages, risen to the level of American civilization. And when, by proscription and prejudice, these same Negroes are classed with and treated like the lowest of their people, simply *because* they are Negroes, such a policy not only discourages thrift and intelligence among black men, but puts a direct premium on the very things you complain of,—inefficiency and crime. Draw lines of crime, of incompetency, of vice, as tightly and uncompromisingly as you will, for these things must be proscribed; but a color-line not only does not accomplish this purpose, but thwarts it.

In the face of two such arguments, the future of the South depends on the ability of the representatives of these opposing views to see and appreciate and sympathize with each other's position,—for the Negro to realize more deeply than he does at present the need of uplifting the masses of his people, for the white people to realize more vividly than they have yet done the deadening and disastrous effect of a color-prejudice that classes Phillis Wheatley and Sam Hose in the same despised class.

It is not enough for the Negroes to declare that color-prejudice is the sole cause of their social condition, nor for the white South to reply that their social condition is the main cause of prejudice. They both act as reciprocal cause and effect, and a change in neither alone will bring the desired effect. Both must change, or neither can improve to any great extent. The Negro cannot stand the present reactionary tendencies and unreasoning drawing of the color-line indefinitely without discouragement and retrogression. And the condition of the Negro is ever the excuse for further discrimination. Only by a union of intelligence and sympathy across the color-line in this critical period of the Republic shall justice and right triumph,

> "That mind and soul according well,
>   May make one music as before,
>   But vaster."

X

Of the Faith of the Fathers

Dim face of Beauty haunting all the world,
  Fair face of Beauty all too fair to see,
Where the lost stars adown the heavens are hurled,—
    There, there alone for thee
    May white peace be.

. . . .

Beauty, sad face of Beauty, Mystery, Wonder,
  What are these dreams to foolish babbling men
Who cry with little noises 'neath the thunder
    Of Ages ground to sand,
    To a little sand.

FIONA MACLEOD.

It was out in the country, far from home, far from my foster home, on a dark Sunday night. The road wandered from our rambling log-house up the stony bed of a creek, past wheat and corn, until we could hear dimly across the fields a rhythmic cadence of song,—soft, thrilling, powerful, that swelled and died sorrowfully in our ears. I was a country school-teacher then, fresh from the East, and had never seen a Southern Negro revival. To be sure, we in Berkshire were not perhaps as stiff and formal as they in Suffolk of olden time; yet we were very quiet and subdued, and I know not what would have happened those clear Sabbath mornings had some one punctuated the sermon with a wild scream, or interrupted the long prayer with a loud Amen! And so most striking to me, as I approached the village and the little plain church perched aloft, was the air of intense excitement that possessed that mass of black folk. A sort of suppressed terror hung in the air and seemed to seize us,—a pythian madness, a demoniac possession, that lent terrible reality to song and word. The black and massive form of the preacher swayed and quivered as the words crowded to his lips and flew at us in singular eloquence. The people moaned and fluttered, and then the gaunt-cheeked brown woman beside me suddenly leaped straight into the air and shrieked like a lost soul, while round about came wail and groan and outcry, and a scene of human passion such as I had never conceived before.

Those who have not thus witnessed the frenzy of a Negro revival in the untouched backwoods of the South can but dimly realize the religious feeling of the slave; as described, such scenes appear grotesque and funny, but as seen they are awful. Three things characterized this religion of the slave,—the Preacher, the Music, and the Frenzy. The Preacher is the most unique personality developed by the Negro on American soil. A leader, a politician, an orator, a "boss," an intriguer, an idealist, —all these he is, and ever, too, the centre of a group of men, now twenty, now a thousand in number. The combination of a certain adroitness with deep-seated earnestness, of tact with consummate ability, gave him his preëminence, and helps him maintain it. The type, of course, varies according to time and place, from the West Indies in the sixteenth century to New England in the nineteenth, and from the Mississippi bottoms to cities like New Orleans or New York.

The Music of Negro religion is that plaintive rhythmic melody, with its touching minor cadences, which, despite caricature and defilement, still remains the most original and beautiful expression of human life and longing yet born on American soil. Sprung from the African forests, where its counterpart can still be heard, it was adapted, changed, and intensified by the tragic soul-life of the slave, until, under the stress of law and whip, it became the one true expression of a people's sorrow, despair, and hope.

Finally the Frenzy or "Shouting," when the Spirit of the Lord passed by, and, seizing the devotee, made him mad with supernatural joy, was the last essential of Negro religion and the one more devoutly believed in than all the rest. It varied in expression from the silent rapt countenance or the low murmur and moan to the mad abandon of physical fervor,—the stamping, shrieking, and shouting, the rushing to and fro and wild waving of arms, the weeping and laughing, the vision and the trance. All this is nothing new in the world, but old as religion, a Delphi and Endor. And so firm a hold did it have on the Negro, that many generations firmly believed that without this visible manifestation of the God there could be no true communion with the Invisible.

These were the characteristics of Negro religious life as developed up to the time of Emancipation. Since under the peculiar circumstances of the black man's environment they were the one expression of his higher life, they are of deep interest to the student of his development, both socially and psychologically. Numerous are the attractive lines of inquiry that here group themselves. What did slavery mean to the African savage? What was his attitude toward the World and Life? What seemed to him good and evil,—God and Devil? Whither went his longings and strivings, and wherefore were his heartburnings and disappointments? Answers to such questions can come only from a study of Negro religion as a development, through its gradual changes from the heathenism of the Gold Coast to the institutional Negro church of Chicago.

Moreover, the religious growth of millions of men, even though they be slaves, cannot be without potent influence upon their contemporaries. The Methodists and Baptists of America owe much of their condition to the silent but potent influence of their millions of Negro converts. Especially is this noticeable in the South, where theology and religious philosophy are on this account a long way behind the North, and where the religion of the poor whites is a plain copy of Negro thought and methods. The mass of "gospel" hymns which has swept through American churches and well-nigh ruined our sense of song con-

sists largely of debased imitations of Negro melodies made
by ears that caught the jingle but not the music, the body
but not the soul, of the Jubilee songs. It is thus clear that
the study of Negro religion is not only a vital part of the
history of the Negro in America, but an interesting part
of American history.

The Negro church of to-day is the social centre of
Negro life in the United States, and the most characteristic
expression of African character. Take a typical church
in a small Virginia town: it is the "First Baptist"—a
roomy brick edifice seating five hundred or more persons,
tastefully finished in Georgia pine, with a carpet, a small
organ, and stained-glass windows. Underneath is a large
assembly room with benches. This building is the central
club-house of a community of a thousand or more Negroes. Various organizations meet here,—the church
proper, the Sunday-school, two or three insurance societies,
women's societies, secret societies, and mass meetings of
various kinds. Entertainments, suppers, and lectures are
held beside the five or six regular weekly religious services.
Considerable sums of money are collected and expended
here, employment is found for the idle, strangers are introduced, news is disseminated and charity distributed.
At the same time this social, intellectual, and economic
centre is a religious centre of great power. Depravity,
Sin, Redemption, Heaven, Hell, and Damnation are
preached twice a Sunday with much fervor, and revivals
take place every year after the crops are laid by; and
few indeed of the community have the hardihood to withstand conversion. Back of this more formal religion, the
Church often stands as a real conserver of morals, a
strengthener of family life, and the final authority on what
is Good and Right.

Thus one can see in the Negro church to-day, repro-
duced in microcosm, all that great world from which the
Negro is cut off by color-prejudice and social condition.
In the great city churches the same tendency is noticeable and in many respects emphasized. A great church
like the Bethel of Philadelphia has over eleven hundred
members, an edifice seating fifteen hundred persons and
valued at one hundred thousand dollars, an annual budget
of five thousand dollars, and a government consisting of
a pastor with several assisting local preachers, an executive and legislative board, financial boards and tax col-

lectors; general church meetings for making laws; sub-divided groups led by class leaders, a company of militia, and twenty-four auxiliary societies. The activity of a church like this is immense and far-reaching, and the bishops who preside over these organizations throughout the land are among the most powerful Negro rulers in the world.

Such churches are really governments of men, and con-sequently a little investigation reveals the curious fact that, in the South, at least, practically every American Negro is a church member. Some, to be sure, are not regularly enrolled, and a few do not habitually attend serv-ices; but, practically, a proscribed people must have a social centre, and that centre for this people is the Negro church. The census of 1890 showed nearly twenty-four thousand Negro churches in the country, with a total en-rolled membership of over two and a half millions, or ten actual church members to every twenty-eight persons, and in some Southern States one in every two persons. Besides these there is the large number who, while not enrolled as members, attend and take part in many of the activities of the church. There is an organized Negro church for every sixty black families in the nation, and in some States for every forty families, owning, on an average, a thousand dollars' worth of property each, or nearly twenty-six million dollars in all.

Such, then, is the large development of the Negro church since Emancipation. The question now is, What have been the successive steps of this social history and what are the present tendencies? First, we must realize that no such institution as the Negro church could rear itself without definite historical foundations. These founda-tions we can find if we remember that the social history of the Negro did not start in America. He was brought from a definite social environment,—the polygamous clan life under the headship of the chief and the potent influence of the priest. His religion was nature-worship, with profound belief in invisible surrounding influences, good and bad, and his worship was through incantation and sacrifice. The first rude change in this life was the slave ship and the West Indian sugar-fields. The planta-tion organization replaced the clan and tribe, and the white master replaced the chief with far greater and more despotic powers. Forced and long-continued toil became the rule of life, the old ties of blood relationship and kinship disappeared, and instead of the family appeared a new polygamy and polyandry, which, in some cases, almost reached promiscuity. It was a terrific social revolution, and yet some traces were retained of the former group life, and the chief remaining institution was the Priest or Medicine-man. He early appeared on the plantation and found his function as the healer of the sick, the interpreter of the Unknown, the comforter of the sorrowing, the supernatural avenger of wrong, and the one who rudely but picturesquely expressed the longing, disappointment, and resentment of a stolen and oppressed people. Thus, as bard, physician, judge, and priest, within the narrow limits allowed by the slave system, rose the Negro preacher, and under him the first Afro-American institution, the Negro church. This church was not at first by any means Christian nor definitely organized; rather it was an adaptation and mingling of heathen rites among the members of each plantation, and roughly designated as Voodooism. Association with the masters, missionary effort and motives of expediency gave these rites an early veneer of Christianity, and after the lapse of many generations the Negro church became Christian.

Two characteristic things must be noticed in regard to this church. First, it became almost entirely Baptist and Methodist in faith; secondly, as a social institution it antedated by many decades the monogamic Negro home. From the very circumstances of its beginning, the church was confined to the plantation, and consisted primarily of a series of disconnected units; although, later on, some freedom of movement was allowed, still this geographical limitation was always important and was one cause of the spread of the decentralized and democratic Baptist faith among the slaves. At the same time, the visible rite of baptism appealed strongly to their mystic temperament. To-day the Baptist Church is still largest in membership among Negroes, and has a million and a half communicants. Next in popularity came the churches organized in connection with the white neighboring churches, chiefly Baptist and Methodist, with a few Episcopalian and others. The Methodists still form the second greatest denomination, with nearly a million members. The faith of these two leading denominations was more suited to the slave church from the prominence they gave to religious feeling and fervor. The Negro membership in other denominations has always been small and relatively unimportant, although the Episcopalians and Presbyterians are gaining among the more intelligent classes to-day, and the Catholic Church is making headway in certain sections. After Emancipation, and still earlier in the North, the Negro churches largely severed such affiliations as they had had with the white churches, either by choice or by compulsion. The Baptist churches became independent, but the Methodists were compelled early to unite for purposes of episcopal government. This gave rise to the great African Methodist Church, the greatest Negro organization in the world, to the Zion Church and the Colored Methodist, and to the black conferences and churches in this and other denominations.

The second fact noted, namely, that the Negro church antedates the Negro home, leads to an explanation of much that is paradoxical in this communistic institution and in the morals of its members. But especially it leads us to regard this institution as peculiarly the expression of the inner ethical life of a people in a sense seldom true elsewhere. Let us turn, then, from the outer physical development of the church to the more important inner ethical life of the people who compose it. The Negro has already been pointed out many times as a religious animal,—a being of that deep emotional nature which turns instinctively toward the supernatural. Endowed with a rich tropical imagination and a keen, delicate appreciation of Nature, the transplanted African lived in a world animate with gods and devils, elves and witches; full of strange influences,—of Good to be implored, of Evil to be propitiated. Slavery, then, was to him the dark triumph of Evil over him. All the hateful powers of the Underworld were striving against him, and a spirit of revolt and revenge filled his heart. He called up all the resources of heathenism to aid,—exorcism and witchcraft, the mysterious Obi worship with its barbarous rites, spells, and blood-sacrifice even, now and then, of human victims. Weird midnight orgies and mystic conjurations were invoked, the witch-woman and the voodoo-priest became the centre of Negro group life, and that vein of vague superstition which characterizes the unlettered Negro even to-day was deepened and strengthened.

In spite, however, of such success as that of the fierce

Maroons, the Danish blacks, and others, the spirit of revolt gradually died away under the untiring energy and superior strength of the slave masters. By the middle of the eighteenth century the black slave had sunk, with hushed murmurs, to his place at the bottom of a new economic system, and was unconsciously ripe for a new philosophy of life. Nothing suited his condition then better than the doctrines of passive submission embodied in the newly learned Christianity. Slave masters early realized this, and cheerfully aided religious propaganda within certain bounds. The long system of repression and degradation of the Negro tended to emphasize the elements in his character which made him a valuable chattel: courtesy became humility, moral strength degenerated into submission, and the exquisite native appreciation of the beautiful became an infinite capacity for dumb suffering. The Negro, losing the joy of this world, eagerly seized upon the offered conceptions of the next; the avenging Spirit of the Lord enjoining patience in this world, under sorrow and tribulation until the Great Day when He should lead His dark children home,—this became his comforting dream. His preacher repeated the prophecy, and his bards sang,—

> "Children, we all shall be free
> When the Lord shall appear!"

This deep religious fatalism, painted so beautifully in "Uncle Tom," came soon to breed, as all fatalistic faiths will, the sensualist side by side with the martyr. Under the lax moral life of the plantation, where marriage was a farce, laziness a virtue, and property a theft, a religion of resignation and submission degenerated easily, in less strenuous minds, into a philosophy of indulgence and crime. Many of the worst characteristics of the Negro masses of to-day had their seed in this period of the slave's ethical growth. Here it was that the Home was ruined under the very shadow of the Church, white and black; here habits of shiftlessness took root, and sullen hopelessness replaced hopeful strife.

With the beginning of the abolition movement and the gradual growth of a class of free Negroes came a change. We often neglect the influence of the freedman before the war, because of the paucity of his numbers and the small weight he had in the history of the nation. But we must not forget that his chief influence was internal,—

was exerted on the black world; and that there he was the ethical and social leader. Huddled as he was in a few centres like Philadelphia, New York, and New Orleans, the masses of the freedmen sank into poverty and listlessness; but not all of them. The free Negro leader early arose and his chief characteristic was intense earnestness and deep feeling on the slavery question. Freedom became to him a real thing and not a dream. His religion became darker and more intense, and into his ethics crept a note of revenge, into his songs a day of reckoning close at hand. The "Coming of the Lord" swept this side of Death, and came to be a thing to be hoped for in this day. Through fugitive slaves and irrepressible discussion this desire for freedom seized the black millions still in bondage, and became their one ideal of life. The black bards caught new notes, and sometimes even dared to sing.—

> "O Freedom, O Freedom, O Freedom over me!
> Before I'll be a slave
> I'll be buried in my grave,
> And go home to my Lord
> And be free."

For fifty years Negro religion thus transformed itself and identified itself with the dream of Abolition, until that which was a radical fad in the white North and an anarchistic plot in the white South had become a religion to the black world. Thus, when Emancipation finally came, it seemed to the freedman a literal Coming of the Lord. His fervid imagination was stirred as never before, by the tramp of armies, the blood and dust of battle, and the wail and whirl of social upheaval. He stood dumb and motionless before the whirlwind: what had he to do with it? Was it not the Lord's doing, and marvellous in his eyes? Joyed and bewildered with what came, he stood awaiting new wonders till the inevitable Age of Reaction swept over the nation and brought the crisis of to-day.

It is difficult to explain clearly the present critical stage of Negro religion. First, we must remember that living as the blacks do in close contact with a great modern nation, and sharing, although imperfectly, the soul-life of that nation, they must necessarily be affected more or less directly by all the religious and ethical forces that are to-day moving the United States. These questions and movements are, however, overshadowed and dwarfed by the (to them) all-important question of their civil, political, and economic status. They must perpetually discuss the "Negro Problem,"—must live, move, and have their being in it, and interpret all else in its light or darkness. With this come, too, peculiar problems of their inner life,—of the status of women, the maintenance of Home, the training of children, the accumulation of wealth, and the prevention of crime. All this must mean a time of intense ethical ferment, of religious heart-searching and intellectual unrest. From the double life every American Negro must live, as a Negro and as an American, as swept on by the current of the nineteenth while yet struggling in the eddies of the fifteenth century,—from this must arise a painful self-consciousness, an almost morbid sense of personality and a moral hesitancy which is fatal to self-confidence. The worlds within and without the Veil of Color are changing, and changing rapidly, but not at the same rate, not in the same way; and this must produce a peculiar wrenching of the soul, a peculiar sense of doubt and bewilderment. Such a double life, with double thoughts, double duties, and double social classes, must give rise to double words and double ideals, and tempt the mind to pretence or revolt, to hypocrisy or radicalism.

In some such doubtful words and phrases can one perhaps most clearly picture the peculiar ethical paradox that faces the Negro of to-day and is tingeing and changing his religious life. Feeling that his rights and his dearest ideals are being trampled upon, that the public conscience is ever more deaf to his righteous appeal, and that all the reactionary forces of prejudice, greed, and revenge are daily gaining new strength and fresh allies, the Negro faces no enviable dilemma. Conscious of his impotence, and pessimistic; he often becomes bitter and vindictive; and his religion, instead of a worship, is a complaint and a curse, a wail rather than a hope, a sneer rather than a faith. On the other hand, another type of mind, shrewder and keener and more tortuous too, sees in the very strength of the anti-Negro movement its patent weaknesses, and with Jesuitic casuistry is deterred by no ethical considerations in the endeavor to turn this weakness to the black man's strength. Thus we have two great and hardly reconcilable streams of thought and ethical strivings; the danger of the one lies in anarchy, that of the other in hypocrisy. The one type of Negro stands almost ready to curse God and die, and the other is too often found a traitor to right and a coward before force; the one is wedded to ideals remote, whimsical, perhaps impossible of realization; the other forgets that life is more than meat and the body more than raiment. But, after all, is not this simply the writhing of the age translated into black,— the triumph of the Lie which to-day, with its false culture, faces the hideousness of the anarchist assassin?

To-day the two groups of Negroes, the one in the North, the other in the South, represent these divergent ethical tendencies, the first tending toward radicalism, the other toward hypocritical compromise. It is no idle regret with which the white South mourns the loss of the old-time Negro,—the frank, honest, simple old servant who stood for the earlier religious age of submission and humility. With all his laziness and lack of many elements of true manhood, he was at least open-hearted, faithful, and sincere. To-day he is gone, but who is to blame for his going? Is it not those very persons who mourn for him? Is it not the tendency, born of Reconstruction and Reaction, to found a society on lawlessness and deception, to tamper with the moral fibre of a naturally honest and straightforward people until the whites threaten to become ungovernable tyrants and the blacks criminal and hypocrites? Deception is the natural defence of the weak against the strong, and the South used it for many years against its conquerors; to-day it must be prepared to see its black proletariat turn that same two-edged weapon against itself. And how natural this is! The death of Denmark Vesey and Nat Turner proved long since to the Negro the present hopelessness of physical defence. Political defence is becoming less and less available, and economic defence is still only partially effective. But there is a patent defence at hand,—the defence of deception and flattery, of cajoling and lying. It is the same defence which peasants of the Middle Age used and which left its stamp on their character for centuries. To-day the young Negro of the South who would succeed cannot be frank and outspoken, honest and self-assertive, but rather he is daily tempted to be silence and wary, politic and sly; he must flatter and be pleasant, endure petty insults with a smile, shut his eyes to wrong; in too many cases he sees positive personal advantage in deception and lying. His real thoughts, his real aspirations, must be guarded in whispers; he must not criticise, he must not complain. Patience, humility, and

adroitness must, in these growing black youth, replace
impulse, manliness, and courage. With this sacrifice there
is an economic opening, and perhaps peace and some
prosperity. Without this there is riot, migration, or crime.
Nor is this situation peculiar to the Southern United
States, is it not rather the only method by which undeveloped races have gained the right to share modern
culture? The price of culture is a Lie.

On the other hand, in the North the tendency is to
emphasize the radicalism of the Negro. Driven from his
birthright in the South by a situation at which every fibre
of his more outspoken and assertive nature revolts, he
finds himself in a land where he can scarcely earn a decent living amid the harsh competition and the color discrimination. At the same time, through schools and periodicals, discussions and lectures, he is intellectually
quickened and awakened. The soul, long pent up and
dwarfed, suddenly expands in new-found freedom. What
wonder that every tendency is to excess,—radical complaint, radical remedies, bitter denunciation or angry silence. Some sink, some rise. The criminal and the sensualist leave the church for the gambling-hell and the
brothel, and fill the slums of Chicago and Baltimore; the
better classes segregate themselves from the group-life of
both white and black, and form an aristocracy, cultured
but pessimistic, whose bitter criticism stings while it points
out no way of escape. They despise the submission and
subserviency of the Southern Negroes, but offer no other
means by which a poor and oppressed minority can exist
side by side with its masters. Feeling deeply and keenly
the tendencies and opportunities of the age in which they
live, their souls are bitter at the fate which drops the
Veil between; and the very fact that this bitterness is
natural and justifiable only serves to intensify it and
make it more maddening.

Between the two extreme types of ethical attitude
which I have thus sought to make clear wavers the mass
of the millions of Negroes, North and South; and their
religious life and activity partake of this social conflict
within their ranks. Their churches are differentiating,—
now into groups of cold, fashionable devotees, in no way
distinguishable from similar white groups save in color of
skin; now into large social and business institutions cater-
ing to the desire for information and amusement of their

members, warily avoiding unpleasant questions both within
and without the black world, and preaching in effect if not
in word: *Dum vivimus, vivamus.*

But back of this still broods silently the deep religious
feeling of the real Negro heart, the stirring, unguided
might of powerful human souls who have lost the guiding
star of the past and seek in the great night a new religious,
ideal. Some day the Awakening will come, when the pent-
up vigor of ten million souls shall sweep irresistibly toward
the Goal, out of the Valley of the Shadow of Death, where
all that makes life worth living—Liberty, Justice, and
Right—is marked "For White People Only."

XI

Of the Passing of the First-Born

O sister, sister, thy first-begotten,
The hands that cling and the feet that follow,
The voice of the child's blood crying yet,
Who hath remembered me? who hath forgotten?
Thou hast forgotten, O summer swallow,
But the world shall end when I forget.

SWINBURNE.

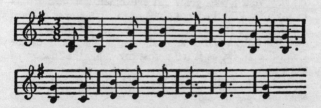

"Unto you a child is born," sang the bit of yellow paper
that fluttered into my room one brown October morning.
Then the fear of fatherhood mingled wildly with the joy
of creation; I wondered how it looked and how it felt,—
what were its eyes, and how its hair curled and crumpled
itself. And I thought in awe of her,—she who had slept
with Death to tear a man-child from underneath her heart,
while I was unconsciously wandering. I fled to my wife

and child, repeating the while to myself half wonderingly,
"Wife and child? Wife and child?"—fled fast and faster
than boat and steam-car, and yet must ever impatiently
await them; away from the hard-voiced city, away from
the flickering sea into my own Berkshire Hills that sit
all sadly guarding the gates of Massachusetts.

Up the stairs I ran to the wan mother and whimpering
babe, to the sanctuary on whose altar a life at my bidding
had offered itself to win a life, and won. What is this tiny
formless thing, this newborn wail from an unknown world,
—all head and voice? I handle it curiously, and watch
perplexed its winking, breathing, and sneezing. I did not
love it then; it seemed a ludicrous thing to love; but her I
loved, my girl-mother, she whom now I saw unfolding
like the glory of the morning—the transfigured woman.
Through her I came to love the wee thing, as it grew
strong; as its little soul unfolded itself in twitter and
cry and half-formed word, and as its eyes caught the
gleam and flash of life. How beautiful he was, with his
olive-tinted flesh and dark gold ringlets, his eyes of mingled
blue and brown, his perfect little limbs, and the soft
voluptuous roll which the blood of Africa had moulded
into his features! I held him in my arms, after we had
sped far away to our Southern home,—held him, and
glanced at the hot red soil of Georgia and the breathless
city of a hundred hills, and felt a vague unrest. Why was
his hair tinted with gold? An evil omen was golden hair
in my life. Why had not the brown of his eyes crushed
out and killed the blue?—for brown were his father's eyes,
and his father's father's. And thus in the Land of the
Color-line I saw, as it fell across my baby, the shadow of
the Veil.

Within the Veil was he born, said I; and there within
shall he live,—a Negro and a Negro's son. Holding in that
little head—ah, bitterly!—the unbowed pride of a hunted
race, clinging with that tiny dimpled hand—ah, wearily!
—to a hope not hopeless but unhopeful, and seeing with
those bright wondering eyes that peer into my soul a land
whose freedom is to us a mockery and whose liberty a
lie. I saw the shadow of the Veil as it passed over my
baby, I saw the cold city towering above the blood-red
land. I held my face beside his little cheek, showed him
the star-children and the twinkling lights as they began to

flash, and stilled with an even-song the unvoiced terror of my life.

So sturdy and masterful he grew, so filled with bubbling life, so tremulous with the unspoken wisdom of a life but eighteen months distant from the All-life,—we were not far from worshipping this revelation of the divine, my wife and I. Her own life builded and moulded itself upon the child; he tinged her every dream and idealized her every effort. No hands but hers must touch and garnish those little limbs; no dress or frill must touch them that had not wearied her fingers; no voice but hers could coax him off to Dreamland, and she and he together spoke some soft and unknown tongue and in it held communion. I too mused above his little white bed; saw the strength of my own arm stretched onward through the ages through the newer strength of his; saw the dream of my black fathers stagger a step onward in the wild phantasm of the world; heard in his baby voice the voice of the Prophet that was to rise within the Veil.

And so we dreamed and loved and planned by fall and winter, and the full flush of the long Southern spring, till the hot winds rolled from the fetid Gulf, till the roses shivered and the still stern sun quivered its awful light over the hills of Atlanta. And then one night the little feet pattered wearily to the wee white bed, and the tiny hands trembled; and a warm flushed face tossed on the pillow, and we knew baby was sick. Ten days he lay there,—a swift week and three endless days, wasting, wasting away. Cheerily the mother nursed him the first days, and laughed into the little eyes that smiled again. Tenderly then she hovered round him, till the smile fled away and Fear crouched beside the little bed.

Then the day ended not, and night was a dreamless terror, and joy and sleep slipped away. I hear now that Voice at midnight calling me from dull and dreamless trance,—crying, "The Shadow of Death! The Shadow of Death!" Out into the starlight I crept, to rouse the gray physician,—the Shadow of Death, the Shadow of Death. The hours trembled on; the night listened; the ghastly dawn glided like a tired thing across the lamplight. Then we two alone looked upon the child as he turned toward us with great eyes, and stretched his string-like hands, —the Shadow of Death! And we spoke no word, and turned away.

He died at eventide, when the sun lay like a brooding sorrow above the western hills, veiling its face; when the winds spoke not, and the trees, the great green trees he loved, stood motionless. I saw his breath beat quicker and quicker, pause, and then his little soul leapt like a star that travels in the night and left a world of darkness in its train. The day changed not; the same tall trees peeped in at the windows, the same green grass glinted in the setting sun. Only in the chamber of death writhed the world's most piteous thing—a childless mother.

I shirk not. I long for work. I pant for a life full of striving. I am no coward, to shrink before the rugged rush of the storm, nor even quail before the awful shadow of the Veil. But hearken, O Death! Is not this my life hard enough,—is not that dull land that stretches its sneering web about me cold enough,—is not all the world beyond these four little walls pitiless enough, but that thou must needs enter here,—thou, O Death? About my head the thundering storm beat like a heartless voice, and the crazy forest pulsed with the curses of the weak; but what cared I, within my home beside my wife and baby boy? Wast thou so jealous of one little coign of happiness that thou must needs enter there,—thou, O Death?

A perfect life was his, all joy and love, with tears to make it brighter,—sweet as a summer's day beside the Housatonic. The world loved him; the women kissed his curls, the men looked gravely into his wonderful eyes, and the children hovered and fluttered about him. I can see him now, changing like the sky from sparkling laughter to darkening frowns, and then to wondering thoughtfulness as he watched the world. He knew no color-line, poor dear,—and the Veil, though it shadowed him, had not yet darkened half his sun. He loved the white matron, he loved his black nurse; and in his little world walked souls alone, uncolored and unclothed. I—yea, all men—are larger and purer by the infinite breadth of that one little life. She who in simple clearness of vision sees beyond the stars said when he had flown. "He will be happy There; he ever loved beautiful things." And I, far more ignorant, and blind by the web of mine own weaving, sit alone winding words and muttering, "If still he be, and he be There, and there be a There, let him be happy, O Fate!"

Blithe was the morning of his burial, with bird and song

and sweet-smelling flowers. The trees whispered to the grass, but the children sat with hushed faces. And yet it seemed a ghostly unreal day,—the wraith of Life. We seemed to rumble down an unknown street behind a little white bundle of posies, with the shadow of a song in our ears. The busy city dinned about us; they did not say much, those pale-faced hurrying men and women; they did not say much,—they only glanced and said, "Niggers!"

We could not lay him in the ground there in Georgia, for the earth there is strangely red; so we bore him away to the northward, with his flowers and his little folded hands. In vain, in vain!—for where, O God! beneath thy broad blue sky shall my dark baby rest in peace,—where Reverence dwells, and Goodness, and a Freedom that is free?

All that day and all that night there sat an awful gladness in my heart,—nay, blame me not if I see the world thus darkly through the Veil,—and my soul whispers ever to me, saying, "Not dead, not dead, but escaped; not bound, but free." No bitter meanness now shall sicken his baby heart till it die a living death, no taunt shall madden his happy boyhood. Fool that I was to think or wish that this little soul should grow choked and deformed within the Veil! I might have known that yonder deep unworldly look that ever and anon floated past his eyes was peering far beyond this narrow Now. In the poise of his little curl-crowned head did there not sit all that wild pride of being which his father had hardly crushed in his own heart? For what, forsooth, shall a Negro want with pride amid the studied humiliations of fifty million fellows? Well sped, my boy, before the world had dubbed your ambition insolence, had held your ideals unattainable, and taught you to cringe and bow. Better far this nameless void that stops my life than a sea of sorrow for you.

Idle words; he might have borne his burden more bravely than we,—aye, and found it lighter too, some day; for surely, surely this is not the end. Surely there shall yet dawn some mighty morning to lift the Veil and set the prisoned free. Not for me,—I shall die in my bonds,—but for fresh young souls who have not known the night and waken to the morning; a morning when men ask of the workman, not "Is he white?" but "Can he work?" When men ask artists, not "Are they black?" but "Do they know?"

Some morning this may be, long, long years to come. But now there wails, on that dark shore within the Veil, the same deep voice, *Thou shalt forego!* And all have I foregone at that command, and with small complaint,—all save that fair young form that lies so coldly wed with death in the nest I had builded.

If one must have gone, why not I? Why may I not rest me from this restlessness and sleep from this wide waking? Was not the world's alembic, Time, in his young hands, and is not my time waning? Are there so many workers in the vineyard that the fair promise of this little body could lightly be tossed away? The wretched of my race that line the alleys of the nation sit fatherless and unmothered; but Love sat beside his cradle, and in his ear Wisdom waited to speak. Perhaps now he knows the All-love, and needs not to be wise. Sleep, then, child,—sleep till I sleep and waken to a baby voice and the ceaseless patter of little feet—above the Veil.

XII

Of Alexander Crummell

Then from the Dawn it seemed there came, but faint
As from beyond the limit of the world,
Like the last echo born of a great cry,
Sounds, as if some fair city were one voice
Around a king returning from his wars.

TENNYSON.

This is the history of a human heart,—the tale of a black boy who many years ago began to struggle with life that he might know the world and know himself. Three temptations he met on those dark dunes that lay gray and dismal before the wonder-eyes of the child: the temptation of Hate, that stood out against the red dawn; the temptation of Despair, that darkened noonday; and the temptation of Doubt, that ever steals along with twilight. Above all, you must hear of the vales he crossed,—the Valley of Humiliation and the Valley of the Shadow of Death.

I saw Alexander Crummell first at a Wilberforce commencement season, admid its bustle and crush. Tall, frail, and black he stood, with simple dignity and an unmistakable air of good breeding. I talked with him apart, where the storming of the lusty young orators could not harm us. I spoke to him politely, then curiously, then eagerly, as I began to feel the fineness of his character,—his calm courtesy, the sweetness of his strength, and his fair blending of the hope and truth of life. Instinctively I bowed before this man, as one bows before the prophets of the world. Some seer he seemed, that came not from the crimson Past or the gray To-come, but from the pulsing Now,—that mocking world which seemed to me at once so light and dark, so splendid and sordid. Four-score years had he wandered in this same world of mine, within the Veil.

He was born with the Missouri Compromise and lay a-dying amid the echoes of Manila and El Caney: stirring times for living, times dark to look back upon, darker to look forward to. The black-faced lad that paused over his mud and marbles seventy years ago saw puzzling vistas as he looked down the world. The slave-ship still groaned across the Atlantic, faint cries burdened the Southern breeze, and the great black father whispered mad tales of cruelty into those young ears. From the low doorway the mother silently watched her boy at play, and at nightfall sought him eagerly lest the shadows bear him away to the land of slaves.

So his young mind worked and winced and shaped curiously a vision of Life; and in the midst of that vision ever stood one dark figure alone,—ever with the hard, thick countenance of that bitter father, and a form that fell in vast and shapeless folds. Thus the temptation of Hate grew and shadowed the growing child,—gliding stealthily into his laughter, fading into his play, and seiz-

ing his dreams by day and night with rough, rude tur-
bulence. So the black boy asked of sky and sun and
flower the never-answered Why? and loved, as he grew,
neither the world nor the world's rough ways.

Strange temptation for a child, you may think; and
yet in this wide land to-day a thousand thousand dark
children brood before this same temptation, and feel its
cold and shuddering arms. For them, perhaps, some one
will some day lift the Veil,—will come tenderly and cheerily into those sad little lives and brush the brooding hate
away, just as Beriah Green strode in upon the life of
Alexander Crummell. And before the bluff, kind-hearted
man the shadow seemed less dark. Beriah Green had a
school in Oneida County, New York, with a score of
mischievous boys. "I'm going to bring a black boy here to
educate," said Beriah Green, as only a crank and an
abolitionist would have dared to say. "Oho!" laughed
the boys. "Ye-es," said his wife; and Alexander came. Once
before, the black boy had sought a school, had travelled,
cold and hungry, four hundred miles up into free New
Hampshire, to Canaan. But the godly farmers hitched
ninety yoke of oxen to the abolition schoolhouse and
dragged it into the middle of the swamp. The black boy
trudged away.

The nineteenth was the first century of human sympathy,—the age when half wonderingly we began to
descry in others that transfigured spark of divinity which
we call Myself; when clodhoppers and peasants, and
tramps and thieves, and millionaires and—sometimes—
Negroes, became throbbing souls whose warm pulsing life
touched us so nearly that we half gasped with surprise,
crying, "Thou too! Hast Thou seen Sorrow and the dull
waters of Hopelessness? Hast Thou known Life?" And
then all helplessly we peered into those Other-worlds, and
wailed, "O World of Worlds, how shall man make you
one?"

So in that little Oneida school there came to those
schoolboys a revelation of thought and longing beneath
one black skin, of which they had not dreamed before.
And to the lonely boy came a new dawn of sympathy and
inspiration. The shadowy, formless thing—the temptation
of Hate, that hovered between him and the world—grew
fainter and less sinister. It did not wholly fade away, but
diffused itself and lingered thick at the edges. Through it

the child now first saw the blue and gold of life,—the sun-swept road that ran 'twixt heaven and earth until in one far-off wan wavering line they met and kissed. A vision of life came to the growing boy,—mystic, wonderful. He raised his head, stretched himself, breathed deep of the fresh new air. Yonder, behind the forests, he heard strange sounds; then glinting through the trees he saw, far, far away, the bronzed hosts of a nation calling,—calling faintly, calling loudly. He heard the hateful clank of their chains; he felt them cringe and grovel, and there rose within him a protest and a prophecy. And he girded himself to walk down the world.

A voice and vision called him to be a priest,—a seer to lead the uncalled out of the house of bondage. He saw the headless host turn toward him like the whirling of mad waters,—he stretched forth his hands eagerly, and then, even as he stretched them, suddenly there swept across the vision the temptation of Despair.

They were not wicked men,—the problem of life is not the problem of the wicked,—they were calm, good men, Bishops of the Apostolic Church of God, and strove toward righteousness. They said slowly, "It is all very natural—it is even commendable; but the General Theological Seminary of the Episcopal Church cannot admit a Negro." And when that thin, half-grotesque figure still haunted their doors, they put their hands kindly, half sorrowfully, on his shoulders, and said, "Now,—of course, we—we know how *you* feel about it; but you see it is impossible,—that is—well—it is premature. Sometime, we trust—sincerely trust—all such distinctions will fade away; but now the world is as it is."

This was the temptation of Despair; and the young man fought it doggedly. Like some grave shadow he flitted by those halls, pleading, arguing, half angrily demanding admittance, until there came the final *No:* until men hustled the disturber away, marked him as foolish, unreasonable, and injudicious, a vain rebel against God's law. And then from that Vision Splendid all the glory faded slowly away, and left an earth gray and stern rolling on beneath a dark despair. Even the kind hands that stretched themselves toward him from out the depths of that dull morning seemed but parts of the purple shadows. He saw them coldly, and asked, "Why should I strive by special grace when the way of the world is closed to me?"

All gently yet, the hands urged him on,—the hands of young John Jay, that daring father's daring son; the hands of the good folk of Boston, that free city. And yet, with a way to the priesthood of the Church open at last before him, the cloud lingered there; and even when in old St. Paul's the venerable Bishop raised his white arms above the Negro deacon—even then the burden had not lifted from that heart, for there had passed a glory from the earth.

And yet the fire through which Alexander Crummell went did not burn in vain. Slowly and more soberly he took up again his plan of life. More critically he studied the situation. Deep down below the slavery and servitude of the Negro people he saw their fatal weaknesses, which long years of mistreatment had emphasized. The dearth of strong moral character, of unbending righteousness, he felt, was their great shortcoming, and here he would begin. He would gather the best of his people into some little Episcopal chapel and there lead, teach, and inspire them, till the leaven spread, till the children grew, till the world hearkened, till—till—and then across his dream gleamed some faint after-glow of that first fair vision of youth—only an after-glow, for there had passed a glory from the earth.

One day—it was in 1842, and the springtide was struggling merrily with the May winds of New England—he stood at last in his own chapel in Providence, a priest of the Church. The days sped by, and the dark young clergyman labored; he wrote his sermons carefully; he intoned his prayers with a soft, earnest voice; he haunted the streets and accosted the wayfarers; he visited the sick, and knelt beside the dying. He worked and toiled, week by week, day by day, month by month. And yet month by month the congregation dwindled, week by week the hollow walls echoed more sharply, day by day the calls came fewer and fewer, and day by day the third temptation sat clearer and still more clearly within the Veil; a temptation, as it were, bland and smiling, with just a shade of mockery in its smooth tones. First it came casually, in the cadence of a voice: "Oh, colored folks? Yes." Or perhaps more definitely: "What do you *expect?*" In voice and gesture lay the doubt—the temptation of Doubt. How he hated it, and stormed at it furiously! "Of course they are capable," he cried; "of course they can learn and strive and achieve—" and "Of course," added the temptation softly, "they do nothing of the sort." Of all the three temptations, this one struck the deepest. Hate? He had outgrown so childish a thing. Despair? He had steeled his right arm against it, and fought it with the vigor of determination. But to doubt the worth of his life-work,—to doubt the destiny and capability of the race his soul loved because it was his; to find listless squalor instead of eager endeavor; to hear his own lips whispering, "They do not care; they cannot know; they are dumb driven cattle,—why cast your pearls before swine?"—this, this seemed more than man could bear; and he closed the door, and sank upon the steps of the chancel, and cast his robe upon the floor and writhed.

The evening sunbeams had set the dust to dancing in the gloomy chapel when he arose. He folded his vestments, put away the hymn-books, and closed the great Bible. He stepped out into the twilight, looked back upon the narrow little pulpit with a weary smile, and locked the door. Then he walked briskly to the Bishop, and told the Bishop what the Bishop already knew. "I have failed," he said simply. And gaining courage by the confession, he added: "What I need is a larger constituency. There are comparatively few Negroes here, and perhaps they are not of the best. I must go where the field is wider, and try again." So the Bishop sent him to Philadelphia, with a letter to Bishop Onderdonk.

Bishop Onderdonk lived at the head of six white steps, —corpulent, red-faced, and the author of several thrilling tracts on Apostolic Succession. It was after dinner, and the Bishop had settled himself for a pleasant season of contemplation, when the bell must needs ring, and there must burst in upon the Bishop a letter and a thin, ungainly Negro. Bishop Onderdonk read the letter hastily and frowned. Fortunately, his mind was already clear on this point; and he cleared his brow and looked at Crummell. Then he said, slowly and impressively: "I will receive you into this diocese on one condition: no Negro priest can sit in my church convention, and no Negro church must ask for representation there."

I sometimes fancy I can see that tableau: the frail black figure, nervously twitching his hat before the massive abdomen of Bishop Onderdonk; his threadbare coat thrown against the dark woodwork of the bookcases,

where Fox's "Lives of the Martyrs" nestled happily beside "The Whole Duty of Man." I seem to see the wide
eyes of the Negro wander past the Bishop's broadcloth to
where the swinging glass doors of the cabinet glow in
the sunlight. A little blue fly is trying to cross the yawn-
ing keyhole. He marches briskly up to it, peers into the
chasm in a surprised sort of way, and rubs his feelers reflectively; then he essays its depths, and, finding it bottom-
less, draws back again. The dark-faced priest finds himself wondering if the fly too has faced its Valley of Humiliation, and if it will plunge into it,—when lo! it spreads its
tiny wings and buzzes merrily across, leaving the watcher
wingless and alone.

Then the full weight of his burden fell upon him. The
rich walls wheeled away, and before him lay the cold
rough moor winding on through life, cut in twain by one
thick granite ridge,—here, the Valley of Humiliation; yonder, the Valley of the Shadow of Death. And I know not
which be darker,—no, not I. But this I know: in yonder
Vale of the Humble stand to-day a million swarthy men,
who willingly would

> ". . . bear the whips and scorns of time,
> The oppressor's wrong, the proud man's contumely,
> The pangs of despised love, the law's delay,
> The insolence of office, and the spurns
> That patient merit of the unworthy takes,"—

all this and more would they bear did they but know that
this were sacrifice and not a meaner thing. So surged the
thought within that lone black breast. The Bishop cleared
his throat suggestively; then, recollecting that there was
really nothing to say, considerately said nothing, only sat
tapping his foot impatiently. But Alexander Crummell
said, slowly and heavily: "I will never enter your diocese
on such terms." And saying this, he turned and passed
into the Valley of the Shadow of Death. You might have
noted only the physical dying, the shattered frame and
hacking cough; but in that soul lay deeper death than
that. He found a chapel in New York,—the church of his
father; he labored for it in poverty and starvation, scorned
by his fellow priests. Half in despair, he wandered across
the sea, a beggar with outstretched hands. Englishmen
clasped them,—Wilberforce and Stanley, Thirwell and

Ingles, and even Froude and Macaulay; Sir Benjamin
Brodie bade him rest awhile at Queen's College in Cambridge, and there he lingered, struggling for health of
body and mind, until he took his degree in '53. Restless
still and unsatisfied, he turned toward Africa, and for
long years, amid the spawn of the slave-smugglers, sought
a new heaven and a new earth.

So the man groped for light; all this was not Life,—it
was the world-wandering of a soul in search of itself, the
striving of one who vainly sought his place in the world,
ever haunted by the shadow of a death that is more than
death,—the passing of a soul that has missed its duty.
Twenty years he wandered,—twenty years and more;
and yet the hard rasping question kept gnawing within
him, "What, in God's name, am I on earth for?" In the narrow New York parish his soul seemed cramped and smoth-
ered. In the fine old air of the English University he
heard the millions wailing over the sea. In the wild fever-
cursed swamps of West Africa he stood helpless and alone.

You will not wonder at his weird pilgrimage,—you who
in the swift whirl of living, amid its cold paradox and
marvellous vision, have fronted life and asked its riddle
face to face. And if you find that riddle hard to read,
remember that yonder black boy finds it just a little
harder; if it is difficult for you to find and face your
duty, it is a shade more difficult for him; if your heart
sickens in the blood and dust of battle, remember that
to him the dust is thicker and the battle fiercer. No wonder the wanderers fall! No wonder we point to thief and
murderer, and haunting prostitute, and the never-ending
throng of unhearsed dead! The Valley of the Shadow of
Death gives few of its pilgrims back to the world.

But Alexander Crummell it gave back. Out of the
temptation of Hate, and burned by the fire of Despair,
triumphant over Doubt, and steeled by Sacrifice against
Humiliation, he turned at last home across the waters,
humble and strong, gentle and determined. He bent to all
the gibes and prejudices, to all hatred and discrimination,
with that rare courtesy which is the armor of pure souls.
He fought among his own, the low, the grasping, and the
wicked, with that unbending righteousness which is the
sword of the just. He never faltered, he seldom complained; he simply worked, inspiring the young, rebuk-
ing the old, helping the weak, guiding the strong.

So he grew, and brought within his wide influence all that was best of those who walk within the Veil. They who live without knew not nor dreamed of that full power within, that mighty inspiration which the dull gauze of caste decreed that most men should not know. And now that he is gone, I sweep the Veil away and cry, Lo! the soul to whose dear memory I bring this little tribute. I can see his face still, dark and heavy-lined beneath his snowy hair; lighting and shading, now with inspiration for the future, now in innocent pain at some human wickedness, now with sorrow at some hard memory from the past. The more I met Alexander Crummell, the more I felt how much that world was losing which knew so little of him. In another age he might have sat among the elders of the land in purple-bordered toga; in another country mothers might have sung him to the cradles.

He did his work,—he did it nobly and well; and yet I sorrow that here he worked alone, with so little human sympathy. His name to-day, in this broad land, means little, and comes to fifty million ears laden with no incense of memory or emulation. And herein lies the tragedy of the age: not that men are poor,—all men know something of poverty; not that men are wicked,—who is good? not that men are ignorant,—what is Truth? Nay, but that men know so little of men.

He sat one morning gazing toward the sea. He smiled and said, "The gate is rusty on the hinges." That night at star-rise a wind came moaning out of the west to blow the gate ajar, and then the soul I loved fled like a flame across the Seas, and in its seat sat Death.

I wonder where he is to-day? I wonder if in that dim world beyond, as he came gliding in, there rose on some wan throne a King,—a dark and pierced Jew, who knows the writhings of the earthly damned, saying, as he laid those heart-wrung talents down, "Well done!" while round about the morning stars sat singing.

XIII

Of the Coming of John

What bring they 'neath the midnight,
    Beside the River-sea?
They bring the human heart wherein
    No nightly calm can be;
That droppeth never with the wind,
    Nor drieth with the dew;
O calm it, God; thy calm is broad
    To cover spirits too.
        The river floweth on.

                          MRS. BROWNING.

Carlisle street runs westward from the centre of Johnstown, across a great black bridge, down a hill and up
again, by little shops and meat-markets, past single-storied
homes, until suddenly it stops against a wide green lawn.
It is a broad, restful place, with two large buildings out-
lined against the west. When at evening the winds come
swelling from the east, and the great pall of the city's
smoke hangs wearily above the valley, then the red west
glows like a dreamland down Carlisle Street, and, at the
tolling of the supper-bell, throws the passing forms of
students in dark silhouette against the sky. Tall and black,
they move slowly by, and seem in the sinister light to
flit before the city like dim warning ghosts. Perhaps they

are; for this is Wells Institute, and these black students
have few dealings with the white city below.

And if you will notice, night after night, there is one
dark form that ever hurries last and late toward the
twinkling lights of Swain Hall,—for Jones is never on
time. A long, straggling fellow he is, brown and hard-
haired, who seems to be growing straight out of his
clothes, and walks with a half-apologetic roll. He used
perpetually to set the quiet dining-room into waves of
merriment, as he stole to his place after the bell had
tapped for prayers; he seemed so perfectly awkward.
And yet one glance at his face made one forgive him
much,—that broad, good-natured smile in which lay no
bit of art or artifice, but seemed just bubbling good-nature
and genuine satisfaction with the world.

He came to us from Altamaha, away down there beneath the gnarled oaks of Southeastern Georgia, where
the sea croons to the sands and the sands listen till they
sink half drowned beneath the waters, rising only here
and there in long, low islands. The white folk of Altamaha voted John a good boy,—fine plough-hand, good in
the rice-fields, handy everywhere, and always good-
natured and respectful. But they shook their heads when
his mother wanted to send him off to school. "It'll spoil
him,—ruin him," they said; and they talked as though
they knew. But full half the black folk followed him
proudly to the station, and carried his queer little trunk
and many bundles. And there they shook and shook
hands, and the girls kissed him shyly and the boys clapped
him on the back. So the train came, and he pinched his
little sister lovingly, and put his great arms about his
mother's neck, and then was away with a puff and a
roar into the great yellow world that flamed and flared
about the doubtful pilgrim. Up the coast they hurried, past
the squares and palmettos of Savannah, through the cotton-
fields and through the weary night, to Millville, and came
with the morning to the noise and bustle of Johnstown.

And they that stood behind, that morning in Altamaha,
and watched the train as it noisily bore playmate and
brother and son away to the world, had thereafter one
ever-recurring word,—"When John comes." Then what
parties were to be, and what speakings in the churches;
what new furniture in the front room,—perhaps even a
new front room; and there would be a new schoolhouse,

with John as teacher; and then perhaps a big wedding; all this and more—when John comes. But the white people shook their heads.

At first he was coming at Christmas-time,—but the vacation proved too short; and then, the next summer,—but times were hard and schooling costly, and so, instead, he worked in Johnstown. And so it drifted to the next summer, and the next,—till playmates scattered, and mother grew gray, and sister went up to the Judge's kitchen to work. And still the legend lingered,—"When John comes."

Up at the Judge's they rather liked this refrain; for they too had a John—a fair-haired, smooth-faced boy, who had played many a long summer's day to its close with his darker namesake. "Yes, sir! John is at Princeton, sir," said the broad-shouldered gray-haired Judge every morning as he marched down to the postoffice. "Showing the Yankees what a Southern gentleman can do," he added; and strode home again with his letters and papers. Up at the great pillared house they lingered long over the Princeton letter,—the Judge and his frail wife, his sister and growing daughters. "It'll make a man of him," said the Judge, "college is the place." And then he asked the shy little waitress, "Well, Jennie, how's your John?" and added reflectively, "Too bad, too bad your mother sent him off,—it will spoil him." And the waitress wondered.

Thus in the far-away Southern village the world lay waiting, half consciously, the coming of two young men, and dreamed in an inarticulate way of new things that would be done and new thoughts that all would think. And yet it was singular that few thought of two Johns,—for the black folk thought of one John, and he was black; and the white folk thought of another John, and he was white. And neither world thought the other world's thought, save with a vague unrest.

Up in Johnstown, at the Institute, we were long puzzled at the case of John Jones. For a long time the clay seemed unfit for any sort of moulding. He was loud and boisterous, always laughing and singing, and never able to work consecutively at anything. He did not know how to study; he had no idea of thoroughness; and with his tardiness, carelessness, and appalling good-humor, we were sore perplexed. One night we sat in faculty-meeting, worried and serious; for Jones was in trouble again. This last escapade was too much, and so we solemnly voted "that Jones, on account of repeated disorder and inattention to work, be suspended for the rest of the term."

It seemed to us that the first time life ever struck Jones as a really serious thing was when the Dean told him he must leave school. He stared at the gray-haired man blankly, with great eyes. "Why,—why," he faltered, "but—I haven't graduated!" Then the Dean slowly and clearly explained, reminding him of the tardiness and the carelessness, of the poor lessons and neglected work, of the noise and disorder, until the fellow hung his head in confusion. Then he said quickly, "But you won't tell mammy and sister,—you won't write mammy, now will you? For if you won't I'll go out into the city and work, and come back next term and show you something." So the Dean promised faithfully, and John shouldered his little trunk, giving neither word nor look to the giggling boys, and walked down Carlisle Street to the great city, with sober eyes and a set and serious face.

Perhaps we imagined it, but someway it seemed to us that the serious look that crept over his boyish face that afternoon never left it again. When he came back to us he went to work with all his rugged strength. It was a hard struggle, for things did not come easily to him,— few crowding memories of early life and teaching came to help him on his new way; but all the world toward which he strove was of his own building, and he builded slow and hard. As the light dawned lingeringly on his new creations, he sat rapt and silent before the vision, or wandered alone over the green campus peering through and beyond the world of men into a world of thought. And the thoughts at times puzzled him sorely; he could not see just why the circle was not square, and carried it out fifty-six decimal places one midnight,—would have gone further, indeed, had not the matron rapped for lights out. He caught terrible colds lying on his back in the meadows of nights, trying to think out the solar system; he had grave doubts as to the ethics of the Fall of Rome, and strongly suspected the Germans of being thieves and rascals, despite his text-books; he pondered long over every new Greek word, and wondered why this meant that and why it couldn't mean something else, and how it must have felt to think all things in Greek. So he thought and puzzled along for himself,—pausing perplexed where others skipped merrily, and walking steadily through the difficulties where the rest stopped and surrendered.

Thus he grew in body and soul, and with him his clothes seemed to grow and arrange themselves; coat sleeves got longer, cuffs appeared, and collars got less soiled. Now and then his boots shone, and a new dignity crept into his walk. And we who saw daily a new thoughtfulness growing in his eyes began to expect something of this plodding boy. Thus he passed out of the preparatory school into college, and we who watched him felt four more years of change, which almost transformed the tall, grave man who bowed to us commencement morning. He had left his queer thought-world and come back to a world of motion and of men. He looked now for the first time sharply about him, and wondered he had seen so little before. He grew slowly to feel almost for the first time the Veil that lay between him and the white world; he first noticed now the oppression that had not seemed oppression before, differences that erstwhile seemed natural, restraints and slights that in his boyhood days had gone unnoticed or been greeted with a laugh. He felt angry now when men did not call him "Mister," he clenched his hands at the "Jim Crow" cars, and chafed at the color-line that hemmed in him and his. A tinge of sarcasm crept into his speech, and a vague bitterness into his life; and he sat long hours wondering and planning a way around these crooked things. Daily he found himself shrinking from the choked and narrow life of his native town. And yet he always planned to go back to Altamaha,—always planned to work there. Still, more and more as the day approached he hesitated with a nameless dread; and even the day after graduation he seized with eagerness the offer of the Dean to send him North with the quartette during the summer vacation, to sing for the Institute. A breath of air before the plunge, he said to himself in half apology.

It was a bright September afternoon, and the streets of New York were brilliant with moving men. They reminded John of the sea, as he sat in the square and watched them, so changelessly changing, so bright and dark, so grave and gay. He scanned their rich and faultless clothes, the way they carried their hands, the shape of their hats; he peered into the hurrying carriages. Then,

leaning back with a sigh, he said, "This is the World." The
notion suddenly seized him to see where the world was
going; since many of the richer and brighter seemed hurrying all one way. So when a tall, light-haired young
man and a little talkative lady came by, he rose half
hesitatingly and followed them. Up the street they went,
past stores and gay shops, across a broad square, until
with a hundred others they entered the high portal of a
great building.

He was pushed toward the ticket-office with the others,
and felt in his pocket for the new five-dollar bill he had
hoarded. There seemed really no time for hesitation, so
he drew it bravely out, passed it to the busy clerk, and
received simply a ticket but no change. When at last he
realized that he had paid five dollars to enter he knew
not what, he stood stockstill amazed. "Be careful," said
a low voice behind him; "you must not lynch the colored
gentleman simply because he's in your way," and a girl
looked up roguishly into the eyes of her fair-haired escort. A shade of annoyance passed over the escort's face.
"You will not understand us in the South," he said half
impatiently, as if continuing an argument. "With all your
professions, one never sees in the North so cordial and
intimate relations between white and black as are everyday
occurrences with us. Why, I remember my closest playfellow in boyhood was a little Negro named after me, and
surely no two,—well!" The man stopped short and flushed
to the roots of his hair, for there directly beside his reserved orchestra chairs sat the negro he had stumbled
over in the hallway. He hesitated and grew pale with
anger, called the usher and gave him his card, with a few
peremptory words, and slowly sat down. The lady deftly
changed the subject.

All this John did not see, for he sat in a half-maze
minding the scene about him; the delicate beauty of the
hall, the faint perfume, the moving myriad of men, the
rich clothing and low hum of talking seemed all a part of
a world so different from his, so strangely more beautiful
than anything he had known, that he sat in dreamland, and
started when, after a hush, rose high and clear the music
of Lohengrin's swan. The infinite beauty of the wail lingered and swept through every muscle of his frame, and
put it all a-tune. He closed his eyes and grasped the elbows of the chair, touching unwittingly the lady's arm.

And the lady drew away. A deep longing swelled in all his heart to rise with that clear music out of the dirt and dust of that low life that held him prisoned and befouled. If he could only live up in the free air where birds sang and setting suns had no touch of blood! Who had called him to be the slave and butt of all? And if he had called, what right had he to call when a world like this lay open before men?

Then the movement changed, and fuller, mightier harmony swelled away. He looked thoughtfully across the hall, and wondered why the beautiful gray-haired woman looked so listless, and what the little man could be whispering about. He would not like to be listless and idle, he thought, for he felt with the music the movement of power within him. If he but had some master-work, some life-service, hard,—aye, bitter hard, but without the cringing and sickening servility, without the cruel hurt that hardened his heart and soul. When at last a soft sorrow crept across the violins, there came to him the vision of a far-off home,—the great eyes of his sister, and the dark drawn face of his mother. And his heart sank below the waters, even as the sea-sand sinks by the shores of Altamaha, only to be lifted aloft again with that last ethereal wail of the swan that quivered and faded away into the sky.

It left John sitting so silent and rapt that he did not for some time notice the usher tapping him lightly on the shoulder and saying politely, "Will you step this way, please, sir?" A little surprised, he arose quickly at the last tap, and, turning to leave his seat, looked full into the face of the fair-haired young man. For the first time the young man recognized his dark boyhood playmate, and John knew that it was the Judge's son. The White John started, lifted his hand, and then froze into his chair; the black John smiled lightly, then grimly, and followed the usher down the aisle. The manager was sorry, very, very sorry,—but he explained that some mistake had been made in selling the gentleman a seat already disposed of; he would refund the money, of course,—and indeed felt the matter keenly, and so forth, and—before he had finished John was gone, walking hurriedly across the square and down the broad streets, and as he passed the park he buttoned his coat and said, "John Jones, you're a natural-born fool." Then he went to his lodgings and wrote a letter, and tore it up; he wrote another, and threw it in the fire. Then he seized a scrap of paper and wrote: "Dear Mother and Sister—I am coming—John"

"Perhaps," said John, as he settled himself on the train, "perhaps I am to blame myself in struggling against my manifest destiny simply because it looks hard and unpleasant. Here is my duty to Altamaha plain before me; perhaps they'll let me help settle the Negro problems there, —perhaps they won't. 'I will go in to the King which is not according to the law; and if I perish, I perish.'" And then he mused and dreamed, and planned a life-work; and the train flew south.

Down in Altamaha, after seven long years, all the world knew John was coming. The homes were scrubbed and scoured,—above all, one; the gardens and yards had an unwonted trimness, and Jennie bought a new gingham. With some finesse and negotiation, all the dark Methodists and Presbyterians were induced to join in a monster welcome at the Baptist Church; and as the day drew near, warm discussions arose on every corner as to the exact extent and nature of John's accomplishments. It was noontide on a gray and cloudy day when he came. The black town flocked to the depot, with a little of the white at the edges,—a happy throng, with "Good-mawnings" and "How dys" and laughing and joking and jostling. Mother sat yonder in the window watching; but sister Jennie stood on the platform, nervously fingering her dress, tall and lithe, with soft brown skin and loving eyes peering from out a tangled wilderness of hair. John rose gloomily as the train stopped, for he was thinking of the "Jim Crow" car; he stepped to the platform, and paused: a little dingy station, a black crowd gaudy and dirty, a half-mile of dilapidated shanties along a straggling ditch of mud. An overwhelming sense of the sordidness and narrowness of it all seized him; he looked in vain for his mother, kissed coldly the tall, strange girl who called him brother, spoke a short, dry word here and there; then, lingering neither for handshaking nor gossip, started silently up the street, raising his hat merely to the last eager old aunty, to her openmouthed astonishment The people were distinctly bewildered. This silent, cold man,—was this John? Where was his smile and hearty hand-grasp? "'Peared kind o' down in the mouf," said the Methodist preacher thoughtfully. "Seemed monstus stuck up," complained a Baptist sister. But the white postmaster from the edge of the crowd expressed the opinion of his folks plainly. "That damn Nigger," said he, as he shouldered the mail and arranged his tobacco, "has gone North and got plum full o' fool notions; but they won't work in Altamaha." And the crowd melted away.

The meeting of welcome at the Baptist Church was a failure. Rain spoiled the barbecue, and thunder turned the milk in the ice-cream. When the speaking came at night, the house was crowded to overflowing. The three preachers had especially prepared themselves, but somehow John's manner seemed to throw a blanket over everything, —he seemed so cold and preoccupied, and had so strange an air of restraint that the Methodist brother could not warm up to his theme and elicited not a single "Amen"; the Presbyterian prayer was but feebly responded to, and even the Baptist preacher, though he wakened faint enthusiasm, got so mixed up in his favorite sentence that he had to close it by stopping fully fifteen minutes sooner than he meant. The people moved uneasily in their seats as John rose to reply. He spoke slowly and methodically. The age, he said, demanded new ideas; we were far different from those men of the seventeenth and eighteenth centuries,—with broader ideas of human brotherhood and destiny. Then he spoke of the rise of charity and popular education, and particularly of the spread of wealth and work. The question was, then, he added reflectively, looking at the low discolored ceiling, what part the Negroes of this land would take in the striving of the new century. He sketched in vague outline the new Industrial School that might rise among these pines, he spoke in detail of the charitable and philanthropic work that might be organized, of money that might be saved for banks and business. Finally he urged unity, and deprecated especially religious and denominational bickering. "To-day," he said, with a smile, "the world cares little whether a man be Baptist or Methodist, or indeed a churchman at all, so long as he is good and true. What difference does it make whether a man be baptized in river or washbowl, or not at all? Let's leave all that littleness, and look higher." Then, thinking of nothing else, he slowly sat down. A painful hush seized that crowded mass. Little had they understood of what he said, for he spoke an unknown tongue, save the last word about baptism; that they knew, and they sat very still while the clock ticked. Then at last a low suppressed snarl came from the Amen corner, and an old bent man arose, walked over the seats, and climbed straight up into the pulpit. He was wrinkled and black, with scant gray and tufted hair; his voice and hands shook as with palsy; but on his face lay the intense rapt look of the religious fanatic. He seized the Bible with his rough, huge hands; twice he raised it inarticulate, and then fairly burst into words, with rude and awful eloquence. He quivered, swayed, and bent; then rose aloft in perfect majesty, till the people moaned and wept, wailed and shouted, and a wild shrieking arose from the corners where all the pent-up feeling of the hour gathered itself and rushed into the air. John never knew clearly what the old man said; he only felt himself held up to scorn and scathing denunciation for trampling on the true religion, and he realized with amazement that all unknowingly he had put rough, rude hands on something this little world held sacred. He arose silently, and passed out into the night. Down toward the sea he went, in the fitful starlight, half conscious of the girl who followed timidly after him. When at last he stood upon the bluff, he turned to his little sister and looked upon her sorrowfully, remembering with sudden pain how little thought he had given her. He put his arm about her and let her passion of tears spend itself on his shoulder.

Long they stood together, peering over the gray un-resting water.

"John," she said, "does it make every one—unhappy when they study and learn lots of things?"

He paused and smiled. "I am afraid it does," he said.

"And, John, are you glad you studied?"

"Yes," came the answer, slowly but positively.

She watched the flickering lights upon the sea, and said thoughtfully, "I wish I was unhappy,—and—and," putting both arms about his neck, "I think I am, a little, John."

It was several days later that John walked up to the Judge's house to ask for the privilege of teaching the Negro school. The Judge himself met him at the front door, stared a little hard at him, and said brusquely, "Go 'round to the kitchen door, John, and wait." Sitting on the kitchen steps, John stared at the corn, thoroughly perplexed. What on earth had come over him? Every step he made offended some one. He had come to save his people, and before he left the depot he had hurt them. He sought to teach them at the church, and had outraged their deepest feelings. He had schooled himself to be respectful to the Judge, and then blundered into his front door. And all the time he had meant right,—and yet, and yet, somehow he found it so hard and strange to fit his old surroundings again, to find his place in the world about him. He could not remember that he used to have any difficulty in the past, when life was glad and gay. The world seemed smooth and easy then. Perhaps,—but his sister came to the kitchen door just then and said the Judge awaited him.

The Judge sat in the dining-room amid his morning's mail, and he did not ask John to sit down. He plunged squarely into the business. "You've come for the school, I suppose. Well, John, I want to speak to you plainly. You know I'm a friend to your people. I've helped you and your family, and would have done more if you hadn't got the notion of going off. Now I like the colored people, and sympathize with all their reasonable aspirations; but you and I both know, John, that in this country the Negro must remain subordinate, and can never expect to be the equal of white men. In their place, your people can be honest and respectful; and God knows, I'll do what I can to help them. But when they want to reverse nature, and rule white men, and marry white women, and sit in my parlor, then, by God! we'll hold them under if we have to lynch every Nigger in the land. Now, John, the question is, are you, with your education and Northern notions, going to accept the situation and teach the darkies to be faithful servants and laborers as your fathers were, —I knew your father, John, he belonged to my brother, and he was a good Nigger. Well—well, are you going to be like him, or are you going to try to put fool ideas of rising and equality into these folks' heads, and make them discontented and unhappy?"

"I am going to accept the situation, Judge Henderson," answered John, with a brevity that did not escape the keen old man. He hesitated a moment, and then said shortly, "Very well,—we'll try you awhile. Good-morning."

It was a full month after the opening of the Negro school that the other John came home, tall, gay, and headstrong. The mother wept, the sisters sang. The whole white town was glad. A proud man was the Judge, and it was a goodly sight to see the two swinging down Main

Street together. And yet all did not go smoothly between them, for the younger man could not and did not veil his contempt for the little town, and plainly had his heart set on New York. Now the one cherished ambition of the Judge was to see his son mayor of Altamaha, representative to the legislature, and—who could say?—governor of Georgia. So the argument often waxed hot between them. "Good heavens, father," the younger man would say after dinner, as he lighted a cigar and stood by the fireplace, "you surely don't expect a young fellow like me to settle down permanently in this—this God-forgotten town with nothing but mud and Negroes?" "*I* did," the Judge would answer laconically; and on this particular day it seemed from the gathering scowl that he was about to add something more emphatic, but neighbors had already begun to drop in to admire his son, and the conversation drifted.

"Heah that John is livenin' things up at the darky school," volunteered the postmaster, after a pause.

"What now?" asked the Judge, sharply.

"Oh, nothin' in particulah,—just his almighty air and uppish ways. B'lieve I did heah somethin' about his givin' talks on the French Revolution, equality, and such like. He's what I call a dangerous Nigger."

"Have you heard him say anything out of the way?"

"Why, no,—but Sally, our girl, told my wife a lot of rot. Then, too, I don't need to heah: a Nigger what won't say 'sir' to a white man, or—"

"Who is this John?" interrupted the son.

"Why, it's little black John, Peggy's son,—your old playfellow."

The young man's face flushed angrily, and then he laughed.

"Oh," said he, "it's the darky that tried to force himself into a seat beside the lady I was escorting—"

But Judge Henderson waited to hear no more. He had been nettled all day, and now at this he rose with a half-smothered oath, took his hat and cane, and walked straight to the schoolhouse.

For John, it had been a long, hard pull to get things started in the rickety old shanty that sheltered his school. The Negroes were rent into factions for and against him, the parents were careless, the children irregular and dirty, and books, pencils, and slates largely missing. Nevertheless, he struggled hopefully on, and seemed to see at last some glimmering of dawn. The attendance was larger and the children were a shade cleaner this week. Even the booby class in reading showed a little comforting progress. So John settled himself with renewed patience this afternoon.

"Now, Mandy," he said cheerfully, "that's better; but you mustn't chop your words up so: 'If—the—man—goes.' Why, your little brother even wouldn't tell a story that way, now would he?"

"Naw, suh, he cain't talk."

"All right; now let's try again: 'If the man—'"

"John!"

The whole school started in surprise, and the teacher half arose, as the red, angry face of the Judge appeared in the open doorway.

"John, this school is closed. You children can go home and get to work. The white people of Altamaha are not spending their money on black folks to have their heads crammed with impudence and lies. Clear out! I'll lock the door myself."

Up at the great pillared house the tall young son wandered aimlessly about after his father's abrupt departure. In the house there was little to interest him; the books were old and stale, the local newspaper flat, and the women had retired with headaches and sewing. He tried a nap, but it was too warm. So he sauntered out into the fields, complaining disconsolately, "Good Lord! how long will this imprisonment last!" He was not a bad fellow,—just a little spoiled and self-indulgent, and as headstrong as his proud father. He seemed a young man pleasant to look upon, as he sat on the great black stump at the edge of the pines idly swinging his legs and smoking. "Why, there isn't even a girl worth getting up a respectable flirtation with," he growled. Just then his eye caught a tall, willowy figure hurrying toward him on the narrow path. He looked with interest at first, and then burst into a laugh as he said, "Well, I declare, if it isn't Jennie, the little brown kitchen-maid! Why, I never noticed before what a trim little body she is. Hello, Jennie! Why, you haven't kissed me since I came home," he said gaily. The young girl stared at him in surprise and confusion,—faltered something inarticulate, and attempted to pass. But a willful mood had seized the young idler, and he caught

at her arm. Frightened, she slipped by; and half mis-
chievously he turned and ran after her through the tall
pines.

Yonder, toward the sea, at the end of the path, came
John slowly, with his head down He had turned wearily
homeward from the schoolhouse; then, thinking to shield
his mother from the blow, started to meet his sister as
she came from work and break the news of his dismissal
to her. "I'll go away," he said slowly; "I'll go away and
find work, and send for them. I cannot live here longer."
And then the fierce, buried anger surged up into his
throat. He waved his arms and hurried wildly up the path.

The great brown sea lay silent. The air scarce breathed.
The dying day bathed the twisted oaks and mighty pines
in black and gold. There came from the wind no warning,
not a whisper from the cloudless sky. There was only a
black man hurrying on with an ache in his heart, seeing
neither sun nor sea, but starting as from a dream at the
frightened cry that woke the pines, to see his dark sister
struggling in the arms of a tall and fair-haired man.

He said not a word, but, seizing a fallen limb, struck
him with all the pent-up hatred of his great black arm;
and the body lay white and still beneath the pines, all
bathed in sunshine and in blood. John looked at it dream-
ily, then walked back to the house briskly, and said in a
soft voice, "Mammy, I'm going away,—I'm going to be
free."

She gazed at him dimly and faltered, "No'th, honey, is
yo' gwine No'th agin?"

He looked out where the North Star glistened pale above
the waters, and said, "Yes, mammy, I'm going—North."

Then, without another word, he went out into the narrow lane, up by the straight pines to the same winding
path and seated himself on the great black stump, looking at the blood where the body had lain. Yonder in the
gray past he had played with that dead boy, romping together under the solemn trees. The night deepened; he
thought of the boys at Johnstown. He wondered how
Brown had turned out, and Carey? And Jones,—Jones?
Why, *he* was Jones, and he wondered what they would all
say when they knew, when they knew, in that great long
dining-room with its hundreds of merry eyes. Then as the
sheen of the starlight stole over him, he thought of the
gilded ceiling of that vast concert hall, and heard stealing

toward him the faint sweet music of the swan. Hark! was it music, or the hurry and shouting of men? Yes, surely! Clear and high the faint sweet melody rose and fluttered like a living thing, so that the very earth trembled as with the tramp of horses and murmur of angry men.

He leaned back and smiled toward the sea, whence rose the strange melody, away from the dark shadows where lay the noise of horses galloping, galloping on. With an effort he roused himself, bent forward, and looked steadily down the pathway, softly humming the "Song of the Bride,"—

"Freudig geführt, ziehet dahin."

Amid the trees in the dim morning twilight he watched their shadows dancing and heard their horses thundering toward him, until at last they came sweeping like a storm, and he saw in front that haggard white-haired man, whose eyes flashed red with fury. Oh, how he pitied him, —pitied him,—and wondered if he had the coiling twisted rope. Then, as the storm burst round him, he rose slowly to his feet and turned his closed eyes toward the Sea.

And the world whistled in his ears.

XIV

Of the Sorrow Songs

> I walk through the churchyard
>   To lay this body down;
> I know moon-rise, I know star-rise;
> I walk in the moonlight, I walk in the starlight;
> I'll lie in the grave and stretch out my arms,
> I'll go to judgment in the evening of the day,
> And my soul and thy soul shall meet that day,
>   When I lay this body down.

NEGRO SONG.

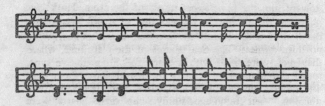

They that walked in darkness sang songs in the olden days—Sorrow Songs—for they were weary at heart. And so before each thought that I have written in this book I have set a phrase, a haunting echo of these weird old songs in which the soul of the black slave spoke to men. Ever since I was a child these songs have stirred me strangely. They came out of the South unknown to me, one by one, and yet at once I knew them as of me and of mine. Then in after years when I came to Nashville I saw the great temple builded of these songs towering over the pale city. To me Jubilee Hall seemed ever made of the songs themselves, and its bricks were red with the blood and dust of toil. Out of them rose for me morning, noon, and night, bursts of wonderful melody, full of the voices of my brothers and sisters, full of the voices of the past.

Little of beauty has America given the world save the rude grandeur God himself stamped on her bosom; the human spirit in this new world has expressed itself in vigor and ingenuity rather than in beauty. And so by fateful chance the Negro folk-song—the rhythmic cry of the slave—stands to-day not simply as the sole American music, but as the most beautiful expression of human experience born this side the seas. It has been neglected, it has been, and is, half despised, and above all it has been persistently mistaken and misunderstood; but notwithstanding, it still remains as the singular spiritual heritage of the nation and the greatest gift of the Negro people.

Away back in the thirties the melody of these slave songs stirred the nation, but the songs were soon half forgotten. Some, like "Near the lake where drooped the willow," passed into current airs and their source was forgotten; others were caricatured on the "minstrel" stage and their memory died away. Then in war-time came the singular Port Royal experiment after the capture of Hilton Head, and perhaps for the first time the North met the Southern slave face to face and heart to heart with no third witness. The Sea Islands of the Carolinas, where they met, were filled with a black folk of primitive type, touched and moulded less by the world about them than any others outside the Black Belt. Their appearance was uncouth, their language funny, but their hearts were human and their singing stirred men with a mighty power. Thomas Wentworth Higginson hastened to tell of these songs, and Miss McKim and others urged upon the world their rare beauty. But the world listened only half credulously until the Fisk Jubilee Singers sang the slave songs so deeply into the world's heart that it can never wholly forget them again.

There was once a blacksmith's son born at Cadiz, New York, who in the changes of time taught school in Ohio and helped defend Cincinnati from Kirby Smith. Then he fought at Chancellorsville and Gettysburg and finally served in the Freedman's Bureau at Nashville. Here he formed a Sunday-school class of black children in 1866, and sang with them and taught them to sing. And then they taught him to sing, and when once the glory of the Jubilee songs passed into the soul of George L. White, he knew his life-work was to let those Negroes sing to the world as they had sung to him. So in 1871 the pilgrimage of the Fisk Jubilee Singers began. North to Cincinnati they rode,—four half-clothed black boys and five girl-women,—led by a man with a cause and a purpose. They stopped at Wilberforce, the oldest of Negro schools, where a black bishop blessed them. Then they went, fighting cold and starvation, shut out of hotels, and cheerfully sneered at, ever northward; and ever the magic of their song kept thrilling hearts, until a burst of applause in the Congregational Council at Oberlin revealed them to the world. They came to New York and Henry Ward Beecher dared to welcome them, even though the metropolitan dailies sneered at his "Nigger Minstrels." So their songs conquered till they sang across the land and across the sea, before Queen and Kaiser, in Scotland and Ireland, Holland and Switzerland. Seven years they sang, and brought back a hundred and fifty thousand dollars to found Fisk University.

Since their day they have been imitated—sometimes

well, by the singers of Hampton and Atlanta, sometimes
ill, by straggling quartettes. Caricature has sought again
to spoil the quaint beauty of the music, and has filled
the air with many debased melodies which vulgar ears
scarce know from the real. But the true Negro folk-song
still lives in the hearts of those who have heard them
truly sung and in the hearts of the Negro people.

What are these songs, and what do they mean? I know
little of music and can say nothing in technical phrase,
but I know something of men, and knowing them, I know
that these songs are the articulate message of the slave
to the world. They tell us in these eager days that life
was joyous to the black slave, careless and happy. I can
easily believe this of some, of many. But not all the
past South, though it rose from the dead, can gainsay
the heart-touching witness of these songs. They are the
music of an unhappy people, of the children of disappointment; they tell of death and suffering and unvoiced
longing toward a truer world, of misty wanderings and
hidden ways.

The songs are indeed the siftings of centuries; the
music is far more ancient than the words, and in it we
can trace here and there signs of development. My grandfather's grandmother was seized by an evil Dutch trader
two centuries ago; and coming to the valleys of the Hudson and Housatonic, black, little, and lithe, she shivered
and shrank in the harsh north winds, looked longingly
at the hills, and often crooned a heathen melody to the
child between her knees, thus:

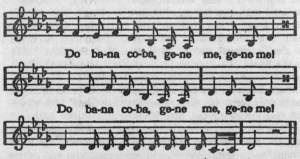

Do ba-na co-ba, ge-ne   me, ge-ne me!

Do ba-na co-ba, ge-ne   me, ge-ne me!

Ben d' nu-li, nu-li, nu-li, nu-li, ben d'   le.

The child sang it to his children and they to their children's children, and so two hundred years it has travelled down to us and we sing it to our children, knowing as little as our fathers what its words may mean, but knowing well the meaning of its music.

This was primitive African music; it may be seen in larger form in the strange chant which heralds "The Coming of John":

"You may bury me in the East,
  You may bury me in the West,
  But I'll hear the trumpet sound in that morning,"

—the voice of exile.

Ten master songs, more or less, one may pluck from this forest of melody—songs of undoubted Negro origin and wide popular currency, and songs peculiarly characteristic of the slave. One of these I have just mentioned. Another whose strains begin this book is "Nobody knows the trouble I've seen." When, struck with a sudden poverty, the United States refused to fulfill its promises of land to the freedmen, a brigadier-general went down to the Sea Islands to carry the news. An old woman on the outskirts of the throng began singing this song; all the mass joined with her, swaying. And the soldier wept.

The third song is the cradle-song of death which all men know,—"Swing low, sweet chariot,"—whose bars begin the life story of "Alexander Crummell." Then there is the song of many waters, "Roll, Jordon, roll," a mighty chorus with minor cadences. There were many songs of the fugitive like that which opens "The Wings of Atalanta," and the more familiar "Been a-listening." The seventh is the song of the End and the Beginning—"My Lord, what a mourning! when the stars begin to fall"; a strain of this is placed before "The Dawn of Freedom." The song of groping—"My way's cloudy"—begins "The Meaning of Progress"; the ninth is the song of this chapter—"Wrestlin' Jacob, the day is a-breaking,"—a pæan of hopeful strife. The last master song is the song of songs—"Steal away,"—sprung from "The Faith of the Fathers."

There are many others of the Negro folk-songs as striking and characteristic as these, as, for instance, the three strains in the third, eighth, and ninth chapters; and others I am sure could easily make a selection on more scientific principles. There are, too, songs that seem to be a step removed from the more primitive types: there is the maze-like medley, "Bright sparkles," one phrase of which heads "The Black Belt"; the Easter carol, "Dust, dust and ashes"; the dirge, "My mother's took her flight and gone home"; and that burst of melody hovering over "The Passing of the First-Born"—"I hope my mother will be there in that beautiful world on high."

These represent a third step in the development of the slave song, of which "You may bury me in the East" is the first, and songs like "March on" (chapter six) and "Steal away" are the second. The first is African music, the second Afro-American, while the third is a blending of Negro music with the music heard in the foster land. The result is still distinctively Negro and the method of blending original, but the elements are both Negro and Caucasian. One might go further and find a fourth step in this development, where the songs of white America have been distinctively influenced by the slave songs or have incorporated whole phrases of Negro melody, as "Swanee River" and "Old Black Joe." Side by side, too, with the growth has gone the debasements and imitations —the Negro "minstrel" songs, many of the "gospel" hymns, and some of the contemporary "coon" songs,—a mass of music in which the novice may easily lose himself and never find the real Negro melodies.

In these songs, as I have said, the slave spoke to the world. Such a message is naturally veiled and half articulate. Words and music have lost each other and new and cant phrases of a dimly understood theology have displaced the older sentiment. Once in a while we catch a strange word of an unknown tongue, as the "Mighty Myo," which figures as a river of death; more often slight words or mere doggerel are joined to music of singular sweetness. Purely secular songs are few in number, partly because many of them were turned into hymns by a change of words, partly because the frolics were seldom heard by the stranger, and the music less often caught. Of nearly all the songs, however, the music is distinctly sorrowful. The ten master songs I have mentioned tell in word and music of trouble and exile, of strife and hiding; they grope toward some unseen power and sigh for rest in the End.

The words that are left to us are not without interest,

and, cleared of evident dross, they conceal much of real poetry and meaning beneath conventional theology and unmeaning rhapsody. Like all primitive folk, the slave stood near to Nature's heart. Life was a "rough and rolling sea" like the brown Atlantic of the Sea Islands; the "Wilderness" was the home of God, and the "lonesome valley" led to the way of life. "Winter'll soon be over," was the picture of life and death to a tropical imagination. The sudden wild thunder-storms of the South awed and impressed the Negroes,—at times the rumbling seemed to them "mournful," at times imperious:

> "My Lord calls me,
>   He calls me by the thunder,
>   The trumpet sounds it in my soul."

The monotonous toil and exposure is painted in many words. One sees the ploughmen in the hot, moist furrow, singing:

> "Dere's no rain to wet you,
>   Dere's no sun to burn you,
>   Oh, push along, believer,
>   I want to go home."

The bowed and bent old man cries, with thrice-repeated wail:

> "O Lord, keep me from sinking down,"

and he rebukes the devil of doubt who can whisper:

> "Jesus is dead and God's gone away."

Yet the soul-hunger is there, the restlessness of the savage, the wail of the wanderer, and the plaint is put in one little phrase:

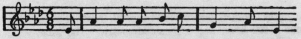

My soul wants something that's new, that's new

Over the inner thoughts of the slaves and their relations one with another the shadow of fear ever hung,
so that we get but glimpses here and there, and also
with them, eloquent omissions and silences. Mother and
child are sung, but seldom father; fugitive and weary wanderer call for pity and affection, but there is little of
wooing and wedding; the rocks and the mountains are
well known, but home is unknown. Strange blending of
love and helplessness signs through the refrain:

> "Yonder 's my ole mudder,
>   Been waggin' at de hill so long;
>   'Bout time she cross over,
>   Git home bime-by."

Elsewhere comes the cry of the "motherless" and the
"Farewell, farewell, my only child."

Love-songs are scarce and fall into two categories—
the frivolous and light, and the sad. Of deep successful
love there is ominous silence, and in one of the oldest
of these songs there is a depth of history and meaning:

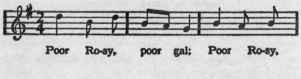

Poor   Ro-sy,   poor gal;   Poor   Ro-sy,

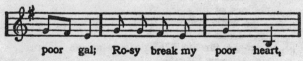

poor   gal;   Ro-sy break my   poor   heart,

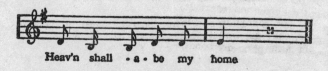

Heav'n shall · a · be   my   home

A black woman said of the song, "It can't be sung without a full heart and a troubled sperrit." The same voice
sings here that sings in the German folk-song:

"Jetz Geh i' an's brunele, trink' aber net."

Of death the Negro showed little fear, but talked of
it familiarly and even fondly as simply a crossing of the
waters, perhaps—who knows?—back to his ancient forests again. Later days transfigured his fatalism, and amid
the dust and dirt the toiler sang:

> "Dust, dust and ashes, fly over my grave,
> But the Lord shall bear my spirit home."

The things evidently borrowed from the surrounding
world undergo characteristic change when they enter the
mouth of the slave. Especially is this true of Bible phrases.
"Weep, O captive daughter of Zion," is quaintly turned
into "Zion, weep-a-low," and the wheels of Ezekiel are
turned every way in the mystic dreaming of the slave,
till he says:

> "There's a little wheel a-turnin' in-a-my heart."

As in olden time, the words of these hymns were improvised by some leading minstrel of the religious band.
The circumstances of the gathering, however, the rhythm
of the songs, and the limitations of allowable thought,
confined the poetry for the most part to single or double
lines, and they seldom were expanded to quatrains or
longer tales, although there are some few examples of
sustained efforts, chiefly paraphrases of the Bible. Three
short series of verses have always attracted me,—the one
that heads this chapter, of one line of which Thomas
Wentworth Higginson has fittingly said, "Never, it seems
to me, since man first lived and suffered was his infinite
longing for peace uttered more plaintively." The second
and third are descriptions of the Last Judgment,—the one
a late improvisation, with some traces of outside influence:

"Oh, the stars in the elements are falling,
  And the moon drips away into blood,
  And the ransomed of the Lord are returning unto God,
  Blessed be the name of the Lord."

And the other earlier and homelier picture from the
low coast lands:

"Michael, haul the boat ashore,
    Then you'll hear the horn they blow,
    Then you'll hear the trumpet sound,
    Trumpet sound the world around,
    Trumpet sound for rich and poor,
    Trumpet sound the Jubilee,
    Trumpet sound for you and me."

Through all the sorrow of the Sorrow Songs there breathes a hope—a faith in the ultimate justice of things. The minor cadences of despair change often to triumph and calm confidence. Sometimes it is faith in life, sometimes a faith in death, sometimes assurance of boundless justice in some fair world beyond. But whichever it is, the meaning is always clear: that sometime, somewhere, men will judge men by their souls and not by their skins. Is such a hope justified? Do the Sorrow Songs sing true?

The silently growing assumption of this age is that the probation of races is past, and that the backward races of to-day are of proven inefficiency and not worth the saving. Such an assumption is the arrogance of peoples irreverent toward Time and ignorant of the deeds of men. A thousand years ago such an assumption, easily possible, would have made it difficult for the Teuton to prove his right to life. Two thousand years ago such dogmatism, readily welcome, would have scouted the idea of blond races ever leading civilization. So woefully unorganized is sociological knowledge that the meaning of progress, the meaning of "swift" and "slow" in human doing, and the limits of human perfectability, are veiled, unanswered sphinxes on the shores of science. Why should Æschylus have sung two thousand years before Shakespeare was born? Why has civilization flourished in Europe, and flickered, flamed, and died in Africa? So long as the world stands meekly dumb before such questions, shall this nation proclaim its ignorance and unhallowed prejudices by denying freedom of opportunity to those who brought the Sorrow Songs to the Seats of the Mighty?

Your country? How came it yours? Before the Pilgrims landed we were here. Here we have brought our three gifts and mingled them with yours: a gift of story and song—soft, stirring melody in an ill-harmonized and unmelodious land; the gift of sweat and brawn to beat back the wilderness, conquer the soil, and lay the foundations of this vast economic empire two hundred years earlier than your weak hands could have done it; the third, a gift of the Spirit. Around us the history of the land has centred for thrice a hundred years; out of the nation's heart we have called all that was best to throttle and subdue all that was worst; fire and blood, prayer and sacrifice, have billowed over this people, and they have found peace only in the altars of the God of Right. Nor has our gift of the Spirit been merely passive. Actively we have woven ourselves with the very warp and woof of this nation,—we fought their battles, shared their sorrow, mingled our blood with theirs, and generation after generation have pleaded with a headstrong, careless people to despise not Justice, Mercy, and Truth, lest the nation be smitten with a curse. Our song, our toil, our cheer, and warning have been given to this nation in blood-brotherhood. Are not these gifts worth the giving? Is not this work and striving? Would America have been America without her Negro people?

Even so is the hope that sang in the songs of my fathers well sung. If somewhere in this whirl and chaos of things there dwells Eternal Good, pitiful yet masterful, then anon in His good time America shall rend the Veil and the prisoned shall go free. Free, free as the sunshine trickling down the morning into these high windows of mine, free as yonder fresh young voices welling up to me from the caverns of brick and mortar below—swelling with song, instinct with life, tremulous treble and darkening bass. My children, my little children, are singing to the sunshine, and thus they sing:

And the traveller girds himself, and sets his face toward the Morning, and goes his way.

THE AFTER-THOUGHT

Hear my cry, O God the Reader; vouchsafe that this my book fall not still-born into the world wilderness. Let there spring, Gentle One, from out its leaves vigor of thought and thoughtful deed to reap the harvest wonderful. Let the ears of a guilty people tingle with truth, and seventy millions sigh for the righteousness which exalteth nations, in this drear day when human brotherhood is mockery and a snare. Thus in Thy good time may infinite reason turn the tangle straight, and these crooked marks on a fragile leaf be not indeed

THE END

The Autobiography of
an Ex-Colored Man

CHAPTER I

I know that in writing the following pages I am divulging the great secret of my life, the secret which for some years I have guarded far more carefully than any of my earthly possessions; and it is a curious study to me to analyse the motives which prompt me to do it. I feel that I am led by the same impulse which forces the un-found-out criminal to take somebody into his confidence, although he knows that the act is likely, even almost certain, to lead to his undoing. I know that I am playing with fire, and I feel the thrill which accompanies that most fascinating pastime; and, back of it all, I think I find a sort of savage and diabolical desire to gather up all the little tragedies of my life, and turn them into a practical joke on society.

And, too, I suffer a vague feeling of unsatisfaction, of regret, of almost remorse, from which I am seeking relief, and of which I shall speak in the last paragraph of this account.

I was born in a little town of Georgia a few years after the close of the Civil War. I shall not mention the name of the town, because there are people still living there who could be connected with this narrative. I have only a faint recollection of the place of my birth. At times I can close my eyes and call up in a dreamlike way things that seem to have happened ages ago in some other world. I can see in this half vision a little house—I am quite sure it was not a large one—I can remember that flowers grew in the front yard, and that around each bed of flowers was a hedge of vari-coloured glass bottles stuck in the ground neck down. I remember that once, while playing round in the sand, I became curious to know whether or not the bottles grew as the flowers did, and I proceeded to dig them up to find out; the investigation brought me a terrific spanking, which indelibly fixed the

incident in my mind. I can remember, too, that behind
the house was a shed under which stood two or three
wooden wash-tubs. These tubs were the earliest aversion
of my life, for regularly on certain evenings I was plunged
into one of them and scrubbed until my skin ached. I
can remember to this day the pain caused by the strong,
rank soap's getting into my eyes.

Back from the house a vegetable garden ran, perhaps
seventy-five or one hundred feet; but to my childish fancy
it was an endless territory. I can still recall the thrill
of joy, excitement, and wonder it gave me to go on an
exploring expedition through it, to find the blackberries,
both ripe and green, that grew along the edge of the
fence.

I remember with what pleasure I used to arrive at, and
stand before, a little enclosure in which stood a patient
cow chewing her cud, how I would occasionally offer her
through the bars a piece of my bread and molasses, and
how I would jerk back my hand in half fright if she
made any motion to accept my offer.

I have a dim recollection of several people who moved
in and about this little house, but I have a distinct mental
image of only two: one, my mother; and the other, a
tall man with a small, dark moustache. I remember that
his shoes or boots were always shiny, and that he wore
a gold chain and a great gold watch with which he was
always willing to let me play. My admiration was almost
equally divided between the watch and chain and the
shoes. He used to come to the house evenings, perhaps
two or three times a week; and it became my appointed
duty whenever he came to bring him a pair of slippers
and to put the shiny shoes in a particular corner; he
often gave me in return for this service a bright coin,
which my mother taught me to promptly drop in a little
tin bank. I remember distinctly the last time this tall man
came to the little house in Georgia; that evening before I
went to bed he took me up in his arms and squeezed
me very tightly; my mother stood behind his chair wiping
tears from her eyes. I remember how I sat upon his knee
and watched him laboriously drill a hole through a ten-
dollar gold piece, and then tie the coin around my neck
with a string. I have worn that gold piece around my
neck the greater part of my life, and still possess it,
but more than once I have wished that some other way

had been found of attaching it to me besides putting a hole through it.

On the day after the coin was put around my neck my mother and I started on what seemed to me an endless journey. I knelt on the seat and watched through the train window the corn- and cotton-fields pass swiftly by until I fell asleep. When I fully awoke, we were being driven through the streets of a large city—Savannah. I sat up and blinked at the bright lights. At Savannah we boarded a steamer which finally landed us in New York. From New York we went to a town in Connecticut, which became the home of my boyhood.

My mother and I lived together in a little cottage which seemed to me to be fitted up almost luxuriously; there were horse-hair covered chairs in the parlour, and a little square piano; there was a stairway with red carpet on it leading to a half second story; there were pictures on the walls, and a few books in a glass-doored case. My mother dressed me very neatly, and I developed that pride which well-dressed boys generally have. She was careful about my associates, and I myself was quite particular. As I look back now I can see that I was a perfect little aristocrat. My mother rarely went to anyone's house, but she did sewing, and there were a great many ladies coming to our cottage. If I was round they would generally call me, and ask me my name and age and tell my mother what a pretty boy I was. Some of them would pat me on the head and kiss me.

My mother was kept very busy with her sewing; sometimes she would have another woman helping her. I think she must have derived a fair income from her work. I know, too, that at least once each month she received a letter; I used to watch for the postman, get the letter, and run to her with it; whether she was busy or not, she would take it and instantly thrust it into her bosom. I never saw her read one of these letters. I knew later that they contained money and what was to her more than money. As busy as she generally was, she found time, however, to teach me my letters and figures and how to spell a number of easy words. Always on Sunday evenings she opened the little square piano and picked out hymns. I can recall now that whenever she played hymns from the book her *tempo* was always decidedly *largo*. Sometimes on other evenings, when she was not sewing, she

would play simple accompaniments to some old Southern
songs which she sang. In these songs she was freer, because she played them by ear. Those evenings on which
she opened the little piano were the happiest hours of my
childhood. Whenever she started toward the instrument,
I used to follow her with all the interest and irrepressible
joy that a pampered pet dog shows when a package is
opened in which he knows there is a sweet bit for him. I
used to stand by her side and often interrupt and annoy
her by chiming in with strange harmonies which I
found on either the high keys of the treble or the low
keys of the bass. I remember that I had a particular
fondness for the black keys. Always on such evenings,
when the music was over, my mother would sit with me
in her arms, often for a very long time. She would hold
me close, softly crooning some old melody without
words, all the while gently stroking her face against my
head; many and many a night I thus fell asleep. I can see
her now, her great dark eyes looking into the fire, to
where? No one knew but her. The memory of that picture
has more than once kept me from straying too far from
the place of purity and safety in which her arms held me.

At a very early age I began to thump on the piano
alone, and it was not long before I was able to pick out a
few tunes. When I was seven years old, I could play by
ear all of the hymns and songs that my mother knew. I
had also learned the names of the notes in both clefs,
but I preferred not to be hampered by notes. About this
time several ladies for whom my mother sewed heard me
play and they persuaded her that I should at once be put
under a teacher; so arrangements were made for me to
study the piano with a lady who was a fairly good musician; at the same time arrangements were made for me
to study my books with this lady's daughter. My music
teacher had no small difficulty at first in pinning me down
to the notes. If she played my lesson over for me, I in-
variably attempted to reproduce the required sounds without the slightest recourse to the written characters. Her
daughter, my other teacher, also had her worries. She
found that, in reading, whenever I came to words that
were difficult or unfamiliar, I was prone to bring my
imagination to the rescue and read from the picture. She
has laughingly told me, since then, that I would sometimes substitute whole sentences and even paragraphs from

what meaning I thought the illustrations conveyed. She said she not only was sometimes amused at the fresh treatment I would give an author's subject, but, when I gave some new and sudden turn to the plot of the story, often grew interested and even excited in listening to hear what kind of a denouement I would bring about. But I am sure this was not due to dullness, for I made rapid progress in both my music and my books.

And so for a couple of years my life was divided between my music and my school-books. Music took up the greater part of my time. I had no playmates, but amused myself with games—some of them my own invention— which could be played alone. I knew a few boys whom I had met at the church which I attended with my mother, but I had formed no close friendships with any of them. Then, when I was nine years old, my mother decided to enter me in the public school, so all at once I found myself thrown among a crowd of boys of all sizes and kinds; some of them seemed to me like savages. I shall never forget the bewilderment, the pain, the heart-sickness, of that first day at school. I seemed to be the only stranger in the place; every other boy seemed to know every other boy. I was fortunate enough, however, to be assigned to a teacher who knew me; my mother made her dresses. She was one of the ladies who used to pat me on the head and kiss me. She had the tact to address a few words directly to me; this gave me a certain sort of standing in the class and put me somewhat at ease.

Within a few days I had made one staunch friend and was on fairly good terms with most of the boys. I was shy of the girls, and remained so; even now a word or look from a pretty woman sets me all a-tremble. This friend I bound to me with hooks of steel in a very simple way. He was a big awkward boy with a face full of freckles and a head full of very red hair. He was perhaps fourteen years of age; that is, four or five years older than any other boy in the class. This seniority was due to the fact that he had spent twice the required amount of time in several of the preceding classes. I had not been at school many hours before I felt that "Red Head"—as I involuntarily called him—and I were to be friends. I do not doubt that this feeling was strengthened by the fact that I had been quick enough to see that a big, strong boy was a friend to be desired at a public school; and, perhaps, in spite of his dullness, "Red Head" had been able to discern that I could be of service to him. At any rate there was a simultaneous mutual attraction.

The teacher had strung the class promiscuously round the walls of the room for a sort of trial heat for places of rank; when the line was straightened out, I found that by skilful manœuvring I had placed myself third and had piloted "Red Head" to the place next to me. The teacher began by giving us to spell the words corresponding to our order in the line. "Spell *first*." "Spell *second*." "Spell *third*." I rattled off: "T-h-i-r-d, third," in a way which said: "Why don't you give us something hard?" As the words went down the line, I could see how lucky I had been to get a good place together with an easy word. As young as I was, I felt impressed with the unfairness of the whole proceeding when I saw the tailenders going down before *twelfth* and *twentieth*, and I felt sorry for those who had to spell such words in order to hold a low position. "Spell *fourth*." "Red Head," with his hands clutched tightly behind his back, began bravely: "F-o-r-t-h." Like a flash a score of hands went up, and the teacher began saying: "No snapping of fingers, no snapping of fingers." This was the first word missed, and it seemed to me that some of the scholars were about to lose their senses; some were dancing up and down on one foot with a hand above their heads, the fingers working furiously, and joy beaming all over their faces; others stood still, their hands raised not so high, their fingers working less rapidly, and their faces expressing not quite so much happiness; there were still others who did not move or raise their hands, but stood with great wrinkles on their foreheads, looking very thoughtful.

The whole thing was new to me, and I did not raise my hand, but slyly whispered the letter "u" to "Red Head" several times. "Second chance," said the teacher. The hands went down and the class became quiet. "Red Head," his face now red, after looking beseechingly at the ceiling, then pitiably at the floor, began very haltingly: "F-u——" Immediately an impulse to raise hands went through the class, but the teacher checked it, and poor "Red Head," though he knew that each letter he added only took him farther out of the way, went doggedly on and finished: "—r-t-h." The handraising was now repeated with more hubbub and excitement than at first. Those who before had not moved a finger were now waving their hands above their heads. "Red Head" felt that he was lost. He looked very big and foolish, and some of the scholars began to snicker. His helpless condition went straight to my heart, and gripped my sympathies. I felt that if he failed, it would in some way be my failure. I raised my hand, and, under cover of the excitement and the teacher's attempts to regain order, I hurriedly shot up into his ear twice, quite distinctly: "F-o-u-r-t-h, f-o-u-r-t-h." The teacher tapped on her desk and said: "Third and last chance." The hands came down, the silence became oppressive. "Red Head" began: "F ——" Since that day I have waited anxiously for many a turn of the wheel of fortune, but never under greater tension than when I watched for the order in which those letters would fall from "Red's" lips "o-u-r-t-h." A sigh of relief and disappointment went up from the class. Afterwards, through all our school-days, "Red Head" shared my wit and quickness and I benefited by his strength and dogged faithfulness.

There were some black and brown boys and girls in the school, and several of them were in my class. One of the boys strongly attracted my attention from the first day I saw him. His face was as black as night, but shone as though it were polished; he had sparkling eyes, and when he opened his mouth, he displayed glistening white teeth. It struck me at once as appropriate to call him "Shiny Face," or "Shiny Eyes," or "Shiny Teeth," and I spoke of him often by one of these names to the other boys. These terms were finally merged into "Shiny," and to that name he answered goodnaturedly during the balance of his public school days.

"Shiny" was considered without question to be the best speller, the best reader, the best penman—in a word, the best scholar, in the class. He was very quick to catch anything, but, nevertheless, studied hard; thus he possessed two powers very rarely combined in one boy. I saw him year after year, on up into the high school, win the majority of the prizes for punctuality, deportment, essay writing, and declamation. Yet it did not take me long to discover that, in spite of his standing as a scholar, he was in some way looked down upon.

The other black boys and girls were still more looked down upon. Some of the boys often spoke of them as "nig-

gers." Sometimes on the way home from school a crowd
would walk behind them repeating:

> "Nigger, nigger, never die,
> Black face and shiny eye."

On one such afternoon one of the black boys turned suddenly on his tormentors and hurled a slate; it struck one
of the white boys in the mouth, cutting a slight gash in
his lip. At sight of the blood the boy who had thrown the
slate ran, and his companions quickly followed. We ran
after them pelting them with stones until they separated
in several directions. I was very much wrought up over
the affair, and went home and told my mother how one
of the "niggers" had struck a boy with a slate. I shall
never forget how she turned on me. "Don't you ever use
that word again," she said, "and don't you ever bother the
coloured children at school. You ought to be ashamed of
yourself." I did hang my head in shame, not because she
had convinced me that I had done wrong, but because I
was hurt by the first sharp word she had ever given me.

My school-days ran along very pleasantly. I stood well
in my studies, not always so well with regard to my be-
haviour. I was never guilty of any serious misconduct, but
my love of fun sometimes got me into trouble. I remember, however, that my sense of humour was so sly that
most of the trouble usually fell on the head of the other
fellow. My ability to play on the piano at school exercises
was looked upon as little short of marvellous in a boy of
my age. I was not chummy with many of my mates, but,
on the whole, was about as popular as it is good for a
boy to be.

One day near the end of my second term at school the
principal came into our room and, after talking to the
teacher, for some reason said: "I wish all of the white
scholars to stand for a moment." I rose with the others.
The teacher looked at me and, calling my name, said:
"You sit down for the present, and rise with the others." I
did not quite understand her, and questioned: "Ma'm?"
She repeated, with a softer tone in her voice: "You sit
down now, and rise with the others." I sat down dazed. I
saw and heard nothing. When the others were asked to
rise, I did not know it. When school was dismissed, I went
out in a kind of stupor. A few of the white boys jeered

me, saying: "Oh, you're a nigger too." I heard some black children say: "We knew he was coloured." "Shiny" said to them: "Come along, don't tease him," and thereby won my undying gratitude.

I hurried on as fast as I could, and had gone some distance before I perceived that "Red Head" was walking by my side. After a while he said to me: "Le' me carry your books." I gave him my strap without being able to answer. When we got to my gate, he said as he handed me my books: "Say, you know my big red agate? I can't shoot with it any more. I'm going to bring it to school for you tomorrow." I took my books and ran into the house. As I passed through the hallway, I saw that my mother was busy with one of her customers; I rushed up into my own little room, shut the door, and went quickly to where my looking-glass hung on the wall. For an instant I was afraid to look, but when I did, I looked long and earnestly. I had often heard people say to my mother: "What a pretty boy you have!" I was accustomed to hear remarks about my beauty; but now, for the first time, I became conscious of it and recognized it. I noticed the ivory whiteness of my skin, the beauty of my mouth, the size and liquid darkness of my eyes, and how the long, black lashes that fringed and shaded them produced an effect that was strangely fascinating even to me. I noticed the softness and glossiness of my dark hair that fell in waves over my temples, making my forehead appear whiter than it really was. How long I stood there gazing at my image I do not know. When I came out and reached the head of the stairs, I heard the lady who had been with my mother going out. I ran downstairs and rushed to where my mother was sitting, with a piece of work in her hands. I buried my head in her lap and blurted out: "Mother, mother, tell me, am I a nigger?" I could not see her face, but I knew the piece of work dropped to the floor and I felt her hands on my head. I looked up into her face and repeated: "Tell me, mother, am I a nigger?" There were tears in her eyes and I could see that she was suffering for me. And then it was that I looked at her critically for the first time. I had thought of her in a childish way only as the most beautiful woman in the world; now I looked at her searching for defects. I could see that her skin was almost brown, that her hair was not so soft as mine, and that she did differ in some way from the other ladies who came to the house; yet, even so, I could see that she was very beautiful, more beautiful than any of them. She must have felt that I was examining her, for she hid her face in my hair and said with difficulty: "No, my darling, you are not a nigger." She went on: "You are as good as anybody; if anyone calls you a nigger, don't notice them." But the more she talked, the less was I reassured, and I stopped her by asking: "Well, mother, am I white? Are you white?" She answered tremblingly: "No, I am not white, but you—your father is one of the greatest men in the country—the best blood of the South is in you——" This suddenly opened up in my heart a fresh chasm of misgiving and fear, and I almost fiercely demanded: "Who is my father? Where is he?" She stroked my hair and said: "I'll tell you about him some day." I sobbed: "I want to know now." She answered: "No, not now."

Perhaps it had to be done, but I have never forgiven the woman who did it so cruelly. It may be that she never knew that she gave me a sword-thrust that day in school which was years in healing.

CHAPTER II

Since I have grown older I have often gone back and tried to analyse the change that came into my life after that fateful day in school. There did come a radical change, and, young as I was, I felt fully conscious of it, though I did not fully comprehend it. Like my first spanking, it is one of the few incidents in my life that I can remember clearly. In the life of everyone there is a limited number of unhappy experiences which are not written upon the memory, but stamped there with a die; and in long years after, they can be called up in detail, and every emotion that was stirred by them can be lived through anew; these are the tragedies of life. We may grow to include some of them among the trivial incidents of childhood—a broken toy, a promise made to us which was not kept, a harsh, heart-piercing word—but these, too, as well as the bitter experiences and disappointments of mature years, are the tragedies of life.

And so I have often lived through that hour, that day, that week, in which was wrought the miracle of my transition from one world into another; for I did indeed pass into another world. From that time I looked out through other eyes, my thoughts were coloured, my words dictated, my actions limited by one dominating, all-pervading idea which constantly increased in force and weight until I finally realized in it a great, tangible fact.

And this is the dwarfing, warping, distorting influence which operates upon each and every coloured man in the United States. He is forced to take his outlook on all things, not from the view-point of a citizen, or a man, or even a human being, but from the view-point of a *coloured* man. It is wonderful to me that the race has progressed so broadly as it has, since most of its thought and all of its activity must run through the narrow neck of this one funnel.

And it is this, too, which makes the coloured people of this country, in reality, a mystery to the whites. It is a difficult thing for a white man to learn what a coloured man really thinks; because, generally, with the latter an additional and different light must be brought to bear on what he thinks; and his thoughts are often influenced by considerations so delicate and subtle that it would be impossible for him to confess or explain them to one of the opposite race. This gives to every coloured man, in proportion to his intellectuality, a sort of dual personality; there is one phase of him which is disclosed only in the freemasonry of his own race. I have often watched with interest and sometimes with amazement even ignorant coloured men under cover of broad grins and minstrel antics maintain this dualism in the presence of white men.

I believe it to be a fact that the coloured people of this country know and understand the white people better than the white people know and understand them.

I now think that this change which came into my life was at first more subjective than objective. I do not think my friends at school changed so much toward me as I did toward them. I grew reserved, I might say suspicious. I grew constantly more and more afraid of laying myself open to some injury to my feelings or my pride. I frequently saw or fancied some slight where, I am sure, none was intended. On the other hand, my friends and teachers were, if anything different, more considerate of me; but I can remember that it was against this very attitude in particular that my sensitiveness revolted. "Red" was the only one who did not so wound me; up to this day I recall with a swelling heart his clumsy efforts to make me understand that nothing could change his love for me.

I am sure that at this time the majority of my white school-mates did not understand or appreciate any differences between me and themselves; but there were a few who had evidently received instructions at home on the matter, and more than once they displayed their knowledge in word and action. As the years passed, I noticed that the most innocent and ignorant among the others grew in wisdom.

I myself would not have so clearly understood this difference had it not been for the presence of the other coloured children at school; I had learned what their status was, and now I learned that theirs was mine. I had had no particular like or dislike for these black and brown boys and girls; in fact, with the exception of "Shiny," they had occupied very little of my thought; but I do know that when the blow fell, I had a very strong aversion to being classed with them. So I became something of a solitary. "Red" and I remained inseparable, and there was between "Shiny" and me a sort of sympathetic bond, but my intercourse with the others was never entirely free from a feeling of constraint. I must add, however, that this feeling was confined almost entirely to my intercourse with boys and girls of about my own age; I did not experience it with my seniors. And when I grew to manhood, I found myself freer with elderly white people than with those near my own age.

I was now about eleven years old, but these emotions and impressions which I have just described could not have been stronger or more distinct at an older age. There were two immediate results of my forced loneliness: I began to find company in books, and greater pleasure in music. I made the former discovery through a big, gilt-bound, illustrated copy of the Bible, which used to lie in splendid neglect on the centre table in our little parlour. On top of the Bible lay a photograph album. I had often looked at the pictures in the album, and one day, after taking the larger book down and opening it on the floor, I was overjoyed to find that it contained what seemed to be an inexhaustible supply of pictures. I looked at these pictures many times; in fact, so often that I knew the story of each one without having to read the subject, and then, somehow, I picked up the thread of history on which are strung the trials and tribulations of the Hebrew children; this I followed with feverish interest and excitement. For a long time King David, with Samson a close second, stood at the head of my list of heroes; he was not displaced until I came to know Robert the Bruce. I read a good portion of the Old Testament, all that part treating of wars and rumours of wars, and then started in on the New. I became interested in the life of Christ, but became impatient and disappointed when I found that, notwithstanding the great power he possessed, he did not make use of it when, in my judgment, he most needed to do so. And so my first general impression of the Bible was what my later impression has been of a number of modern books, that the authors put their best work in the first part, and grew either exhausted or careless toward the end.

After reading the Bible, or those parts which held my attention, I began to explore the glass-doored bookcase which I have already mentioned. I found there *Pilgrim's Progess*, Peter Parley's *History of the United States*, Grimm's *Household Stories, Tales of a Grandfather*, a bound volume of an old English publication (I think it was called *The Mirror*), a little volume called *Familiar Science*, and somebody's *Natural Theology*, which last, of course, I could not read, but which, nevertheless, I tackled, with the result of gaining a permanent dislike for all kinds of theology. There were several other books of no particular name or merit, such as agents sell to people who know nothing of buying books. How my mother came by this little library which, considering all things, was so well suited to me I never sought to know. But she was far from being an ignorant woman and had herself, very likely, read the majority of these books, though I do not remember ever seeing her with a book in her hand, with the exception of the Episcopal Prayer-book. At any rate she encouraged in me the habit of reading, and when I had about exhausted those books in the little library which interested me, she began to buy books for me. She also regularly gave me money to buy a weekly paper which was then very popular for boys.

At this time I went in for music with an earnestness worthy of maturer years; a change of teachers was largely responsible for this. I began now to take lessons of the organist of the church which I attend with my mother; he was a good teacher and quite a thorough musician. He was so skilful in his instruction and filled me with such enthusiasm that my progress—these are his words—was marvellous. I remember that when I was barely twelve years old I appeared on a program with a number of adults at an entertainment given for some charitable purpose, and carried off the honours. I did more, I brought upon myself through the local newspapers the handicapping title of "infant prodigy."

I can believe that I did astonish my audience, for I never played the piano like a child; that is, in the "one-two-three" style with accelerated motion. Neither did I depend upon mere brilliancy of technique, a trick by which children often surprise their listeners; but I always tried to interpret a piece of music; I always played with feeling. Very early I acquired that knack of using the pedals, which makes the piano a sympathetic, singing instrument, quite a different thing from the source of hard or blurred sounds it so generally is. I think this was due not entirely to natural artistic temperament, but largely to the fact that I did not begin to learn the piano by counting out exercises, but by trying to reproduce the quaint songs which my mother used to sing, with all their pathetic turns and cadences.

Even at a tender age, in playing I helped to express what I felt by some of the mannerisms which I afterwards observed in great performers; I had not copied them. I have often heard people speak of the mannerisms of musicians as affectations adopted for mere effect; in some cases they may be so; but a true artist can no more play upon the piano or violin without putting his whole body in accord with the emotions he is striving to express than a swallow can fly without being graceful. Often when playing I could not keep the tears which formed in my eyes from rolling down my cheeks. Sometimes at the end or even in the midst of a composition, as big a boy as I was, I would jump from the piano, and throw myself sobbing into my mother's arms. She, by her caresses and

often her tears, only encouraged these fits of sentimental
hysteria Of course, to counteract this tendency to temper-
amental excesses I should have been out playing ball or
in swimming with other boys of my age; but my mother
didn't know that. There was only once when she was really
firm with me, making me do what she considered was
best; I did not want to return to school after the unpleasant episode which I have related, and she was inflexible.

I began my third term, and the days ran along as I
have already indicated. I had been promoted twice, and
had managed each time to pull "Red" along with me. I
think the teachers came to consider me the only hope of
his ever getting through school. and I believe they secret-
ly conspired with me to bring about the desired end. At
any rate, I know it became easier in each succeeding
examination for me not only to assist "Red," but absolutely to do his work. It is strange how in some things honest
people can be dishonest without the slightest compunction.
I knew boys at school who were too honourable to tell a
fib even when one would have been just the right thing,
but could not resist the temptation to assist or receive
assistance in an examination I have long considered it
the highest proof of honesty in a man to hand his street-
car fare to the conductor who had overlooked it.

One afternoon after school, during my third term, I
rushed home in a great hurry to get my dinner and go to
my music teacher's. I was never reluctant about going
there, but on this particular afternoon I was impetuous.
The reason of this was I had been asked to play the accompaniment for a young lady who was to play a violin
solo at a concert given by the young people of the church,
and on this afternoon we were to have our first rehearsal.
At that time playing accompaniments was the only thing
in music I did not enjoy. later this feeling grew into
positive dislike. I have never been a really good accom-
panist because my ideas of interpretation were always too
strongly individual. I constantly forced my *accelerandos*
and *rubatos* upon the soloist, often throwing the duet
entirely out of gear.

Perhaps the reader has already guessed why I was so
willing and anxious to play the accompaniment to this
violin solo; if not—the violinist was a girl of seventeen
or eighteen whom I had first heard play a short time be-

fore on a Sunday afternoon at a special service of some
kind, and who had moved me to a degree which now I
can hardly think of as possible. At present I do not
think it was due to her wonderful playing, though I
judge she must have been a very fair performer, but
there was just the proper setting to produce the effect
upon a boy such as I was; the half-dim church, the air of
devotion on the part of the listeners, the heaving tremor
of the organ under the clear wail of the violin, and she,
her eyes almost closing, the escaping strands of her dark
hair wildly framing her pale face, and her slender body
swaying to the tones she called forth, all combined to fire
my imagination and my heart with a passion, though
boyish, yet strong and, somehow, lasting. I have tried to
describe the scene; if I have succeeded, it is only half
success, for words can only partially express what I
wish to convey. Always in recalling that Sunday afternoon I am subconscious of a faint but distinct fragrance
which, like some old memory-awakening perfume, rises
and suffuses my whole imagination, inducing a state of
reverie so airy as just to evade the powers of expression.

She was my first love, and I loved her as only a boy
loves. I dreamed of her, I built air castles for her, she
was the incarnation of each beautiful heroine I knew;
when I played the piano, it was to her, not even music
furnished an adequate outlet for my passion; I bought a
new notebook and, to sing her praises, made my first and
last attempts at poetry. I remember one day at school,
after we had given in our note-books to have some exercises corrected, the teacher called me to her desk and
said: "I couldn't correct your exercises because I found
nothing in your book but a rhapsody on somebody's
brown eyes." I had passed in the wrong note-book. I
don't think I have felt greater embarrassment in my whole
life than I did at that moment. I was ashamed not only
that my teacher should see this nakedness of my heart,
but that she should find out that I had any knowledge of
such affairs. It did not then occur to me to be ashamed
of the kind of poetry I had written.

Of course, the reader must know that all of this adora-
tion was in secret; next to my great love for this young
lady was the dread that in some way she would find it
out. I did not know what some men never find out, that
the woman who cannot discern when she is loved has

never lived. It makes me laugh to think how successful I was in concealing it all; within a short time after our duet all of the friends of my dear one were referring to me as her "little sweetheart," or her "little beau," and she laughingly encouraged it. This did not entirely satisfy me; I wanted to be taken seriously. I had definitely made up my mind that I should never love another woman, and that if she deceived me I should do something desperate —the great difficulty was to think of something sufficiently desperate—and the heartless jade, how she led me on!

So I hurried home that afternoon, humming snatches of the violin part of the duet, my heart beating with pleasurable excitement over the fact that I was going to be near her, to have her attention placed directly upon me; that I was going to be of service to her, and in a way in which I could show myself to advantage—this last consideration has much to do with cheerful service.—The anticipation produced in me a sensation somewhat between bliss and fear. I rushed through the gate, took the three steps to the house at one bound, threw open the door, and was about to hang my cap on its accustomed peg of the hall rack when I noticed that that particular peg was occupied by a black derby hat. I stopped suddenly and gazed at this hat as though I had never seen an object of its description. I was still looking at it in open-eyed wonder when my mother, coming out of the parlour into the hallway, called me and said there was someone inside who wanted to see me. Feeling that I was being made a party to some kind of mystery, I went in with her, and there I saw a man standing leaning with one elbow on the mantel, his back partly turned toward the door. As I entered, he turned and I saw a tall, handsome, well-dressed gentleman of perhaps thirty-five; he advanced a step toward me with a smile on his face. I stopped and looked at him with the same feelings with which I had looked at the derby hat, except that they were greatly magnified. I looked at him from head to foot, but he was an absolute blank to me until my eyes rested on his slender, elegant polished shoes; then it seemed that indistinct and partly obliterated films of memory began, at first slowly, then rapidly, to unroll, forming a vague panorama of my childhood days in Georgia.

My mother broke the spell by calling me by name and saying: "This is your father."

"Father, father," that was the word which had been to me a source of doubt and perplexity ever since the interview with my mother on the subject. How often I had wondered about my father, who he was, what he was like, whether alive or dead, and, above all, why she would not tell me about him. More than once I had been on the point of recalling to her the promise she had made me, but I instinctively felt that she was happier for not telling me and that I was happier for not being told; yet I had not the slightest idea what the real truth was. And here he stood before me, just the kind of looking father I had wishfully pictured him to be; but I made no advance toward him; I stood there feeling embarrassed and foolish, not knowing what to say or do. I am not sure but that he felt pretty much the same. My mother stood at my side with one hand on my shoulder, almost pushing me forward, but I did not move. I can well remember the look of disappointment, even pain, on her face; and I can now understand that she could expect nothing else but that at the name "father" I should throw myself into his arms. But I could not rise to this dramatic, or, better, melodramatic, climax. Somehow I could not arouse any considerable feeling of need for a father. He broke the awkward tableau by saying: "Well, boy, aren't you glad to see me?" He evidently meant the words kindly enough, but I don't know what he could have said that would have had a worse effect; however, my good breeding came to my rescue, and I answered: "Yes, sir," and went to him and offered him my hand. He took my hand into one of his, and, with the other, stroked my head, saying that I had grown into a fine youngster. He asked me how old I was; which, of course, he must have done merely to say something more, or perhaps he did so as a test of my intelligence. I replied: "Twelve, sir." He then made the trite observation about the flight of time, and we lapsed into another awkward pause.

My mother was all in smiles; I believe that was one of the happiest moments of her life. Either to put me more at ease or to show me off, she asked me to play something for my father. There is only one thing in the world that can make music, at all times and under all circumstances, up to its general standard; that is a hand-organ, or one of its variations. I went to the piano and played something in a listless, half-hearted way. I simply was not in the mood. I was wondering, while playing, when my mother would dismiss me and let me go; but my father was so enthusiastic in his praise that he touched my vanity—which was great—and more than that; he displayed that sincere appreciation which always arouses an artist to his best effort, and, too, in an unexplainable manner, makes him feel like shedding tears. I showed my gratitude by playing for him a Chopin waltz with all the feeling that was in me. When I had finished, my mother's eyes were glistening with tears; my father stepped across the room, seized me in his arms, and squeezed me to his breast. I am certain that for that moment he was proud to be my father. He sat and held me standing between his knees while he talked to my mother. I, in the mean time, examined him with more curiosity, perhaps, than politeness. I interrupted the conversation by asking: "Mother, is he going to stay with us now?" I found it impossible to frame the word "father"; it was too new to me; so I asked the question through my mother. Without waiting for her to speak, my father answered: "I've got to go back to New York this afternoon, but I'm coming to see you again." I turned abruptly and went over to my mother, and almost in a whisper reminded her that I had an appointment which I should not miss; to my pleasant surprise she said that she would give me something to eat at once so that I might go. She went out of the room and I began to gather from off the piano the music I needed. When I had finished, my father, who had been watching me, asked: "Are you going?" I replied: "Yes, sir, I've got to go to practise for a concert." He spoke some words of advice to me about being a good boy and taking care of my mother when I grew up, and added that he was going to send me something nice from New York. My mother called, and I said good-bye to him and went out. I saw him only once after that.

I quickly swallowed down what my mother had put on the table for me, seized my cap and music, and hurried off to my teacher's house. On the way I could think of nothing but this new father, where he came from, where he had been, why he was here, and why he would not stay. In my mind I ran over the whole list of fathers I had become acquainted with in my reading, but I could not classify him. The thought did not cross my mind that he was different from me, and even if it had, the mystery would not thereby have been explained; for, not-withstanding my changed relations with most of my school-mates, I had only a faint knowledge of prejudice and no idea at all how it ramified and affected our entire social organism. I felt, however, that there was something about the whole affair which had to be hid.

When I arrived, I found that she of the brown eyes had been rehearsing with my teacher and was on the point of leaving. My teacher, with some expressions of surprise, asked why I was late, and I stammered out the first deliberate lie of which I have any recollection. I told him that when I reached home from school, I found my mother quite sick, and that I had stayed with her awhile before coming. Then unnecessarily and gratuitously—to give my words force of conviction, I suppose—I added: "I don't think she'll be with us very long." In speaking these words I must have been comical; for I noticed that my teacher, instead of showing signs of anxiety or sorrow, half hid a smile. But how little did I know that in that lie I was speaking a prophecy!

She of the brown eyes unpacked her violin, and we went through the duet several times. I was soon lost to all other thoughts in the delights of music and love. I saw delights of love without reservation; for at no time of life is love so pure, so delicious, so poetic, so romantic, as it is in boyhood. A great deal has been said about the heart of a girl when she stands "where the brook and river meet," but what she feels is negative; more interesting is the heart of a boy when just at the budding dawn of manhood he stands looking wide-eyed into the long vistas opening before him; when he first becomes conscious of the awakening and quickening of strange desires and unknown powers; when what he sees and feels is still shadowy and mystical enough to be intangible, and, so, more beautiful, when his imagination is unsullied, and his faith new and whole—then it is that love wears a halo. The man who has not loved before he was fourteen has missed a foretaste of Elysium.

When I reached home, it was quite dark and I found my mother without a light, sitting rocking in a chair, as she so often used to do in my childhood days, looking into the fire and singing softly to herself. I nestled close to her, and, with her arms round me, she haltingly told me who my father was—a great man, a fine gentleman—he loved me and loved her very much; he was going to make a great man of me. All she said was so limited by reserve and so coloured by her feelings that it was but half truth; and so I did not yet fully understand.

CHAPTER III

Perhaps I ought not pass on in this narrative without mentioning that the duet was a great success, so great that we were obliged to respond with two encores. It seemed to me that life could hold no greater joy than it contained when I took her hand and we stepped down to the front of the stage bowing to our enthusiastic audience. When we reached the little dressing-room, where the other performers were applauding as wildly as the audience, she impulsively threw both her arms round me and kissed me, while I struggled to get away.

One day a couple of weeks after my father had been to see us, a wagon drove up to our cottage loaded with a big box. I was about to tell the men on the wagon that they had made a mistake, when my mother, acting darkly wise, told them to bring their load in; she had them unpack the box, and quickly there was evolved from the boards, paper, and other packing-material a beautiful, brand-new, upright piano. Then she informed me that it was a present to me from my father. I at once sat down and ran my fingers over the keys; the full, mellow tone of the instrument was ravishing. I thought, almost remorsefully, of how I had left my father; but, even so, there momentarily crossed my mind a feeling of disappointment that the piano was not a grand. The new instrument greatly increased the pleasure of my hours of study and practice at home.

Shortly after this I was made a member of the boys' choir, it being found that I possessed a clear, strong soprano voice. I enjoyed the singing very much. About a year later I began the study of the pipe organ and the theory of music; and before I finished the grammar-school, I had written out several simple preludes for

organ which won the admiration of my teacher, and which
he did me the honour to play at services.

The older I grew, the more thought I gave to the question of my mother's and my position, and what was our
exact relation to the world in general. My idea of the
whole matter was rather hazy. My study of United States
history had been confined to those periods which were
designated in my book as "Discovery," "Colonial," "Rev-
olutionary," and "Constitutional." I now began to study
about the Civil War, but the story was told in such a con-
densed and skipping style that I gained from it very little
real information. It is a marvel how children ever learn
any history out of books of that sort. And, too, I began
now to read the newspapers; I often saw articles which
aroused my curiosity, but did not enlighten me. But one
day I drew from the circulating library a book that
cleared the whole mystery, a book that I read with the
same feverish intensity with which I had read the old
Bible stories, a book that gave me my first perspective
of the life I was entering; that book was *Uncle Tom's
Cabin*.

This work of Harriet Beecher Stowe has been the object of much unfavourable criticism. It has been assailed,
not only as fiction of the most imaginative sort, but as
being a direct misrepresentation. Several successful attempts have lately been made to displace the book from
Northern school libraries. Its critics would brush it aside
with the remark that there never was a Negro as good as
Uncle Tom, nor a slave-holder as bad as Legree. For my
part, I was never an admirer of Uncle Tom, nor of his
type of goodness; but I believe that there were lots of old
Negroes as foolishly good as he; the proof of which is
that they knowingly stayed and worked the plantations
that furnished sinews for the army which was fighting
to keep them enslaved. But in these later years several
cases have come to my personal knowledge in which old
Negroes have died and left what was a considerable fortune to the descendants of their former masters. I do not
think it takes any great stretch of the imagination to
believe there was a fairly large class of slave-holders
typified in Legree. And we must also remember that the
author depicted a number of worthless if not vicious
Negroes, and a slave-holder who was as much of a Christian and a gentleman as it was possible for one in his

position to be; that she pictured the happy, singing, shuffling "darky" as well as the mother wailing for her child sold "down river."

I do not think it is claiming too much to say that *Uncle Tom's Cabin* was a fair and truthful panorama of slavery; however that may be, it opened my eyes as to who and what I was and what my country considered me; in fact, it gave me my bearing. But there was no shock; I took the whole revelation in a kind of stoical way. One of the greatest benefits I derived from reading the book was that I could afterwards talk frankly with my mother on all the questions which had been vaguely troubling my mind. As a result, she was entirely freed from reserve, and often herself brought up the subject, talking of things directly touching her life and mine and of things which had come down to her through the "old folks." What she told me interested and even fascinated me, and, what may seem strange, kindled in me a strong desire to see the South. She spoke to me quite frankly about herself, my father, and myself: she, the sewing girl of my father's mother; he, an impetuous young man home from college; I, the child of this unsanctioned love. She told me even the principal reason for our coming north. My father was about to be married to a young lady of another great Southern family. She did not neglect to add that another reason for our being in Connecticut was that he intended to give me an education and make a man of me. In none of her talks did she ever utter one word of complaint against my father. She always endeavoured to impress upon me how good he had been and still was, and that he was all to us that custom and the law would allow. She loved him; more, she worshipped him, and she died firmly believing that he loved her more than any other woman in the world. Perhaps she was right. Who knows?

All of these newly-awakened ideas and thoughts took the form of a definite aspiration on the day I graduated from the grammar-school. And what a day that was! The girls in white dresses, with fresh ribbons in their hair; the boys in new suits and creaky shoes; the great crowd of parents and friends; the flowers, the prizes and congratulations, made the day seem to me one of the greatest importance. I was on the program, and played a piano solo which was received by the audience with that amount of applause which I had come to look upon as being only the just due of my talent.

But the real enthusiasm was aroused by "Shiny." He was the principal speaker of the day, and well did he measure up to the honour. He made a striking picture, that thin little black boy standing on the platform, dressed in clothes that did not fit him any too well, his eyes burning with excitement, his shrill, musical voice vibrating in tones of appealing defiance, and his black face alight with such great intelligence and earnestness as to be positively handsome. What were his thoughts when he stepped forward and looked into that crowd of faces, all white with the exception of a score or so that were lost to view? I do not know, but I fancy he felt his loneliness. I think there must have rushed over him a feeling akin to that of a gladiator tossed into the arena and bade to fight for his life. I think that solitary little black figure standing there felt that for the particular time and place he bore the weight and responsibility of his race; that for him to fail meant general defeat; but he won, and nobly. His oration was Wendell Phillips's "Toussaint L'Ouverture," a speech which may now be classed as rhetorical—even, perhaps, bombastic; but as the words fell from "Shiny's" lips their effect was magical. How so young an orator could stir so great enthusiasm was to be wondered at. When, in the famous peroration, his voice, trembling with suppressed emotion, rose higher and higher and then rested on the name "Toussaint L'Ouverture," it was like touching an electric button which loosed the pent-up feelings of his listeners. They actually rose to him.

I have since known of coloured men who have been chosen as class orators in our leading universities, of others who have played on the varsity football and baseball teams, of coloured speakers who have addressed great white audiences. In each of these instances I believe the men were stirred by the same emotions which actuated "Shiny" on the day of his graduation; and, too, in each case where the efforts have reached any high standard of excellence they have been followed by the same phenomenon of enthusiasm. I think the explanation of the latter lies in what is a basic, though often dormant, principle of the Anglo-Saxon heart, love of fair play. "Shiny," it is true, was what is so common in his race, a natural orator; but

I doubt that any white boy of equal talent could have wrought the same effect. The sight of that boy gallantly waging with puny, black arms so unequal a battle touched the deep springs in the hearts of his audience, and they were swept by a wave of sympathy and admiration.

But the effect upon me of "Shiny's" speech was double; I not only shared the enthusiasm of his audience, but he imparted to me some of his own enthusiasm. I felt leap within me pride that I was coloured; and I began to form wild dreams of bringing glory and honour to the Negro race. For days I could talk of nothing else with my mother except my ambitions to be a great man, a great coloured man, to reflect credit on the race and gain fame for myself. It was not until years after that I formulated a definite and feasible plan for realizing my dreams.

I entered the high school with my class, and still continued my study of the piano, the pipe organ, and the theory of music. I had to drop out of the boys' choir on account of a changing voice; this I regretted very much. As I grew older, my love for reading grew stronger. I read with studious interest everything I could find relating to coloured men who had gained prominence My heroes had been King David, then Robert the Bruce; now Frederick Douglass was enshrined in the place of honour. When I learned that Alexandre Dumas was a coloured man, I re-read *Monte Cristo* and *The Three Guardsmen* with magnified pleasure. I lived between my music and books, on the whole a rather unwholesome life for a boy to lead. I dwelt in a world of imagination, of dreams and air castles—the kind of atmosphere that sometimes nourishes a genius, more often men unfitted for the practical struggles of life. I never played a game of ball, never went fishing or learned to swim; in fact, the only outdoor exercise in which I took any interest was skating. Nevertheless, though slender, I grew well formed and in perfect health. After I entered the high school, I began to notice the change in my mother's health, which I suppose had been going on for some years. She began to complain a little and to cough a great deal; she tried several remedies, and finally went to see a doctor; but though she was failing in health, she kept her spirits up. She still did a great deal of sewing, and in the busy seasons hired two women to help her. The purpose she had formed of having me go through college without financial worries kept her at work when she was not fit for it. I was so fortunate as to be able to organize a class of eight or ten beginners on the piano, and so start a separate little fund of my own. As the time for my graduation from the high school grew nearer, the plans for my college career became the chief subject of our talks. I sent for catalogues of all the prominent schools in the East and eagerly gathered all the information I could concerning them from different sources. My mother told me that my father wanted me to go to Harvard or Yale; she herself had a half desire for me to go to Atlanta University, and even had me write for a catalogue of that school. There were two reasons, however, that inclined her to my father's choice; the first, that at Harvard or Yale I should be near her; the second, that my father had promised to pay for a part of my college education.

Both "Shiny" and "Red" came to my house quite often of evenings, and we used to talk over our plans and prospects for the future. Sometimes I would play for them, and they seemed to enjoy the music very much. My mother often prepared sundry Southern dishes for them, which I am not sure but that they enjoyed more. "Shiny" had an uncle in Amherst, Mass., and he expected to live with him and work his way through Amherst College. "Red" declared that he had enough of school and that after he got his high school diploma, he would get a position in a bank. It was his ambition to become a banker and he felt sure of getting the opportunity through certain members of his family.

My mother barely had strength to attend the closing exercises of the high school when I graduated, and after that day she was seldom out of bed. She could no longer direct her work, and under the expense of medicines, doctors, and someone to look after her our college fund began to diminish rapidly. Many of her customers and some of the neighbours were very kind, and frequently brought her nourishment of one kind or another. My mother realized what I did not, that she was mortally ill, and she had me write a long letter to my father. For some time past she had heard from him only at irregular intervals; we never received an answer. In those last days I often sat at her bed-side and read to her until she fell asleep. Sometimes I would leave the parlour door open and play on the piano, just loud enough for the music to reach her. This she always enjoyed.

One night, near the end of July, after I had been watching beside her for some hours, I went into the parlour and, throwing myself into the big arm-chair, dozed off into a fitful sleep. I was suddenly aroused by one of the neighbours, who had come in to sit with her that night. She said: "Come to your mother at once." I hurried upstairs. and at the bedroom door met the woman who was acting as nurse. I noted with a dissolving heart the strange look of awe on her face. From my first glance at my mother I discerned the light of death upon her countenance. I fell upon my knees beside the bed and, burying my face in the sheets, sobbed convulsively. She died with the fingers of her left hand entwined in my hair.

I will not rake over this, one of the two sacred sorrows of my life; nor could I describe the feeling of unutterable loneliness that fell upon me. After the funeral I went to the house of my music teacher; he had kindly offered me the hospitality of his home for so long as I might need it. A few days later I moved my trunk, piano, my music, and most of my books to his home; the rest of my books I divided between "Shiny" and "Red." Some of the household effects I gave to "Shiny's" mother and to two or three of the neighbours who had been kind to us during my mother's illness; the others I sold. After settling up my little estate I found that, besides a good supply of clothes, a piano, some books and trinkets, I had about two hundred dollars in cash.

The question of what I was to do now confronted me. My teacher suggested a concert tour; but both of us realized that I was too old to be exploited as an infant prodigy and too young and inexperienced to go before the public as a finished artist. He, however, insisted that the people of the town would generously patronize a benefit concert; so he took up the matter and made arrangements for such an entertainment. A more than sufficient number of people with musical and elocutionary talent volunteered their services to make a program. Among these was my brown-eyed violinist. But our relations were not the same as they were when we had played our first duet together. A year or so after that time she had dealt me a crushing blow by getting married. I was partially avenged, however, by the fact that, though she was growing more beautiful, she was losing her ability to play the violin.

I was down on the program for one number. My selection might have appeared at that particular time as a bit of affectation, but I considered it deeply appropriate; I played Beethoven's "Sonata Pathétique." When I sat down at the piano and glanced into the faces of the several hundreds of people who were there solely on account of love or sympathy for me, emotions swelled in my heart which enabled me to play the "Pathétique" as I could never again play it. When the last tone died away, the few who began to applaud were hushed by the silence of the others; and for once I played without receiving an encore.

The benefit yielded me a little more than two hundred dollars, thus raising my cash capital to about four hundred dollars. I still held to my determination of going to college; so it was now a question of trying to squeeze through a year at Harvard or going to Atlanta, where the money I had would pay my actual expenses for at least two years. The peculiar fascination which the South held over my imagination and my limited capital decided me in favour of Atlanta University; so about the last of September I bade farewell to the friends and scenes of my boyhood and boarded a train for the south.

CHAPTER IV

The farther I got below Washington, the more disappointed I became in the appearance of the country. I peered through the car windows, looking in vain for the luxuriant semi-tropical scenery which I had pictured in my mind. I did not find the grass so green, nor the woods so beautiful, nor the flowers so plentiful, as they were in Connecticut. Instead, the red earth partly covered by tough, scrawny grass, the muddy, straggling roads, the cottages of unpainted pine boards, and the clay-daubed huts imparted a "burnt up" impression. Occasionally we ran through a little white and green village that was like an oasis in a desert.

When I reached Atlanta, my steadily increasing disappointment was not lessened. I found it a big, dull, red town. This dull red colour of that part of the South I was then seeing had much, I think, to do with the extreme depression of my spirits—no public squares, no fountains, dingy streetcars, and, with the exception of three or four principal thoroughfares, unpaved streets. It was raining when I arrived and some of these unpaved streets were absolutely impassable. Wheels sank to the hubs in red mire, and I actually stood for an hour and watched four or five men work to save a mule, which had stepped into a deep sink, from drowning, or, rather, suffocating in the mud. The Atlanta of today is a new city.

On the train I had talked with one of the Pullman-car porters, a bright young fellow who was himself a student, and told him that I was going to Atlanta to attend school. I had also asked him to tell me where I might stop for a day or two until the University opened. He said I might go with him to the place where he stopped during his "layovers" in Atlanta. I gladly accepted his offer and went with him along one of those muddy streets until we came to a rather rickety-looking frame-house, which we entered. The proprietor of the house was a big, fat, greasy-looking brown-skin man. When I asked him if he could give me accommodation, he wanted to know how long I would stay. I told him perhaps two days, not more than three. In reply he said: "Oh, dat's all right den," at the same time leading the way up a pair of creaky stairs. I followed him and the porter to a room, the door of which the proprietor opened while continuing, it seemed, his remark, "Oh, dat's all right den," by adding "You kin sleep in dat cot in de corner der. Fifty cents, please." The porter interrupted by saying: "You needn't collect from him now, he's got a trunk." This seemed to satisfy the man, and he went down, leaving me and my porter friend in the room. I glanced round the apartment and saw that it contained a double bed and two cots, two wash-stands, three chairs, and a time-worn bureau, with a looking-glass that would have made Adonis appear hideous. I looked at the cot in which I was to sleep and suspected, not without good reasons, that I should not be the first to use the sheets and pillow-case since they had last come from the wash. When I thought of the clean, tidy, comfortable surroundings in which I

had been reared, a wave of homesickness swept over me
that made me feel faint. Had it not been for the presence
of my companion, and that I knew this much of his history—that he was not yet quite twenty, just three years older than myself, and that he had been fighting his own way
in the world, earning his own living and providing for his
own education since he was fourteen—I should not have
been able to stop the tears that were welling up in my eyes.

I asked him why it was that the proprietor of the house
seemed unwilling to accommodate me for more than a
couple of days. He informed me that the man ran a lodging-house especially for Pullman porters, and, as their
stays in town were not longer than one or two nights, it
would interfere with his arrangements to have anyone stay
longer. He went on to say: "You see this room is fixed up
to accommodate four men at a time. Well, by keeping a sort
of table of trips, in and out, of the men, and working them
like checkers, he can accommodate fifteen or sixteen in
each week and generally avoid having an empty bed. You
happen to catch a bed that would have been empty for a
couple of nights." I asked him where he was going to sleep.
He answered: "I sleep in that other cot tonight; tomorrow
night I go out." He went on to tell me that the man who
kept the house did not serve meals, and that if I was hungry, we would go out and get something to eat.

We went into the street, and in passing the railroad
station I hired a wagon to take my trunk to my lodging-
place. We passed along until, finally, we turned into a
street that stretched away, up and down hill, for a mile or
two; and here I caught my first sight of coloured people
in large numbers. I had seen little squads around the railroad stations on my way south, but here I saw a street
crowded with them. They filled the shops and thronged
the sidewalks and lined the curb. I asked my companion
if all the coloured people in Atlanta lived in this street. He
said they did not and assured me that the ones I saw were
of the lower class. I felt relieved, in spite of the size of the
lower class. The unkempt appearance, the shambling,
slouching gait and loud talk and laughter of these people
aroused in me a feeling of almost repulsion. Only one
thing about them awoke a feeling of interest; that was
their dialect. I had read some Negro dialect and had heard
snatches of it on my journey down from Washington; but
here I heard it in all of its fullness and freedom. I was

particularly struck by the way in which it was punctuated by such exclamatory phrases as "Lawd a mussy!" "G'wan, man!" "Bless ma soul!" "Look heah, chile!" These people talked and laughed without restraint. In fact, they talked straight from their lungs and laughed from the pits of their stomachs. And this hearty laughter was often justified by the droll humour of some remark. I paused long enough to hear one man say to another: "W'at's de mattah wid you an' yo' fr'en' Sam?" and the other came back like a flash: "Ma fr'en'? He ma fr'en'? Man! I'd go to his funeral jes' de same as I'd go to a minstrel show." I have since learned that this ability to laugh heartily is, in part, the salvation of the American Negro; it does much to keep him from going the way of the Indian.

The business places of the street along which we were passing consisted chiefly of low bars, cheap dry-goods and notion stores, barber shops, and fish and bread restaurants. We, at length, turned down a pair of stairs that led to a basement and I found myself in an eating-house somewhat better than those I had seen in passing; but that did not mean much for its excellence. The place was smoky, the tables were covered with oilcloth, the floor with sawdust, and from the kitchen came a rancid odour of fish fried over several times, which almost nauseated me. I asked my companion if this was the place where we were to eat. He informed me that it was the best place in town where a coloured man could get a meal. I then wanted to know why somebody didn't open a place where respectable coloured people who had money could be accommodated. He answered: "It wouldn't pay; all the respectable coloured people eat at home, and the few who travel generally have friends in the towns to which they go, who entertain them." He added: "Of course, you could go in any place in the city; they wouldn't know you from white."

I sat down with the porter at one of the tables, but was not hungry enough to eat with any relish what was put before me. The food was not badly cooked; but the iron knives and forks needed to be scrubbed, the plates and dishes and glasses needed to be washed and well dried. I minced over what I took on my plate while my companion ate. When we finished, we paid the waiter twenty cents each and went out. We walked round until the lights of the city were lit. Then the porter said that he must get to bed and have some rest, as he had not had six hours' sleep since he left Jersey City. I went back to our lodging-house with him.

When I awoke in the morning, there were, besides my new-found friend, two other men in the room, asleep in the double bed. I got up and dressed myself very quietly, so as not to awake anyone. I then drew from under the pillow my precious roll of greenbacks, took out a ten-dollar bill, and, very softly unlocking my trunk, put the remainder, about three hundred dollars, in the inside pocket of a coat near the bottom, glad of the opportunity to put it unobserved in a place of safety. When I had carefully locked my trunk, I tiptoed toward the door with the intention of going out to look for a decent restaurant where I might get something fit to eat. As I was easing the door open, my porter friend said with a yawn: "Hello! You're going out?" I answered him: "Yes." "Oh!" he yawned again, "I guess I've had enough sleep; wait a minute, I'll go with you." For the instant his friendship bored and embarrassed me. I had visions of another meal in the greasy restaurant of the day before. He must have divined my thoughts, for he went on to say: "I know a woman across town who takes a few boarders; I think we can go over there and get a good breakfast." With a feeling of mingled fears and doubts regarding what the breakfast might be, I waited until he had dressed himself.

When I saw the neat appearance of the cottage we entered, my fears vanished, and when I saw the woman who kept it, my doubts followed the same course. Scrupulously clean, in a spotless white apron and coloured head-hand-kerchief, her round face beaming with motherly kindness, she was picturesquely beautiful. She impressed me as one broad expanse of happiness and good nature. In a few minutes she was addressing me as "chile" and "honey." She made me feel as though I should like to lay my head on her capacious bosom and go to sleep.

And the breakfast, simple as it was, I could not have had at any restaurant in Atlanta at any price. There was fried chicken, as it is fried only in the South, hominy boiled to the consistency where it could be eaten with a fork, and biscuits so light and flaky that a fellow with any appetite at all would have no difficulty in disposing of eight or ten. When I had finished, I felt that I had experienced the realization of, at least, one of my dreams of Southern life.

During the meal we found out from our hostess, who had two boys in school, that Atlanta University opened on that very day. I had somehow mixed my dates. My friend the porter suggested that I go out to the University at once and offered to walk over and show me the way. We had to walk because, although the University was not more than twenty minutes' distance from the centre of the city, there were no street-cars running in that direction. My first sight of the school-grounds made me feel that I was not far from home; here the red hills had been terraced and covered with green grass; clean gravel walks, well shaded, led up to the buildings; indeed, it was a bit of New England transplanted. At the gate my companion said he would bid me good-bye, because it was likely that he would not see me again before his car went out. He told me that he would make two more trips to Atlanta and that he would come out and see me; that after his second trip he would leave the Pullman service for the winter and return to school in Nashville. We shook hands, I thanked him for all his kindness, and we said good-bye.

I walked up to a group of students and made some inquiries. They directed me to the president's office in the main building. The president gave me a cordial welcome; it was more than cordial; he talked to me, not as the official head of a college, but as though he were adopting me into what was his large family, personally to look after my general welfare as well as my education. He seemed especially pleased with the fact that I had come to them all the way from the North. He told me that I could have come to the school as soon as I had reached the city and that I had better move my trunk out at once. I gladly promised him that I would do so. He then called a boy and directed him to take me to the matron, and to show me round afterwards. I found the matron even more motherly than the president was fatherly. She had me register, which was in effect to sign a pledge to abstain from the use of intoxicating beverages, tobacco, and profane language while I was a student in the school. This act caused me no sacrifice, as, up to that time, I was free from all three habits. The boy who was with me then showed me about the grounds. I was especially interested in the industrial building.

The sounding of a bell, he told me, was the signal for the students to gather in the general assembly hall, and he asked me if I would go. Of course I would. There were between three and four hundred students and perhaps all of the teachers gathered in the room. I noticed that several of the latter were coloured. The president gave a talk addressed principally to new-comers; but I scarcely heard what he said, I was so much occupied in looking at those around me. They were of all types and colours, the more intelligent types predominating. The colours ranged from jet black to pure white, with light hair and eyes. Among the girls especially there were many so fair that it was difficult to believe that they had Negro blood in them. And, too, I could not help noticing that many of the girls, particularly those of the delicate brown shades, with black eyes and wavy dark hair, were decidedly pretty. Among the boys many of the blackest were fine specimens of young manhood, tall, straight, and muscular, with magnificent heads; these were the kind of boys who developed into the patriarchal "uncles" of the old slave regime.

When I left the University, it was with the determination to get my trunk and move out to the school before night. I walked back across the city with a light step and a light heart. I felt perfectly satisfied with life for the first time since my mother's death. In passing the railroad station I hired a wagon and rode with the driver as far as my stopping-place. I settled with my landlord and went upstairs to put away several articles I had left out. As soon as I opened my trunk, a dart of suspicion shot through my heart; the arrangement of things did not look familiar. I began to dig down excitedly to the bottom till I reached the coat in which I had concealed my treasure. My money was gone! Every single bill of it. I knew it was useless to do so, but I searched through every other coat, every pair of trousers, every vest, and even each pair of socks. When I had finished my fruitless search, I sat down dazed and heart-sick. I called the landlord up and informed him of my loss; he comforted me by saying that I ought to have better sense than to keep money in a trunk and that he was not responsible for his lodgers' personal effects. His cooling words brought me enough to my senses to cause me to look and see if anything else was missing. Several small articles were gone, among them a black and grey necktie of odd design upon which my heart was set; almost as much as the loss of my money I felt the loss of my tie.

After thinking for a while as best I could, I wisely decided to go at once back to the University and lay my troubles before the president. I rushed breathlessly back to the school. As I neared the grounds, the thought came across me, would not my story sound fishy? Would it not place me in the position of an impostor or beggar? What right had I to worry these busy peeople with the results of my carelessness? If the money could not be recovered, and I doubted that it could, what good would it do to tell them about? The shame and embarrassment which the whole situation gave me caused me to stop at the gate. I paused, undecided, for a moment; then turned and slowly retraced my steps, and so changed the whole course of my life.

If the reader has never been in a strange city without money or friends, it is useless to try to describe what my feelings were; he could not understand. If he has been, it is equally useless, for he understands more than words could convey. When I reached my lodgings, I found in the room one of the porters who had slept there the night before. When he heard what misfortune had befallen me, he offered many words of sympathy and advice. He asked me how much money I had left. I told him that I had ten or twelve dollars in my pocket. He said: "That won't last you very long here, and you will hardly be able to find anything to do in Atlanta. I'll tell you what you do, go down to Jacksonville and you won't have any trouble to get a job in one of the big hotels there, or in St. Augustine." I thanked him, but intimated my doubts of being able to get to Jacksonville on the money I had. He reassured me by saying: "Oh, that's all right. You express your trunk on through, and I'll take you down in my closet." I thanked him again, not knowing then what it was to travel in a Pullman porter's closet. He put me under a deeper debt of gratitude by lending me fifteen dollars, which he said I could pay back after I had secured work. His generosity brought tears to my eyes, and I concluded that, after all, there were some kind hearts in the world.

I now forgot my troubles in the hurry and excitement of getting my trunk off in time to catch the train, which went out at seven o'clock. I even forgot that I hadn't eaten anything since morning. We got a wagon—the porter went with me—and took my trunk to the express office. My new friend then told me to come to the station at about a quarter of seven and walk straight to the car where I should see him standing, and not to lose my nerve. I found my role not so difficult to play as I thought it would be, because the train did not leave from the central station, but from a smaller one, where there were no gates and guards to pass. I followed directions, and the porter took me on his car and locked me in his closet. In a few minutes the train pulled out for Jacksonville.

I may live to be a hundred years old, but I shall never forget the agonies I suffered that night. I spent twelve hours doubled up in the porter's basket for soiled linen, not being able to straighten up on account of the shelves for clean linen just over my head. The air was hot and suffocating and the smell of damp towels and used linen was sickening. At each lurch of the car over the none too smooth track I was bumped and bruised against the narrow walls of my narrow compartment. I became acutely conscious of the fact that I had not eaten for hours. Then nausea took possession of me, and at one time I had grave doubts about reaching my destination alive. If I had the trip to make again, I should prefer to walk.

CHAPTER V

The next morning I got out of the car at Jacksonville with a stiff and aching body. I determined to ask no more porters, not even my benefactor, about stopping-places; so I found myself on the street not knowing where to go. I walked along listlessly until I met a coloured man who had the appearance of a preacher. I asked him if he could direct me to a respectable boarding-house for coloured people. He said that if I walked along with him in the

direction he was going, he would show me such a place:
I turned and walked at his side. He proved to be a minister,
and asked me a great many direct questions about myself.
I answered as many as I saw fit to answer; the others
I evaded or ignored. At length we stopped in front of a
frame-house, and my guide informed me that it was the
place. A woman was standing in the doorway, and he
called to her saying that he had brought her a new boarder.
I thanked him for his trouble, and after he had urged
upon me to attend his church while I was in the city,
he went on his way.

I went in and found the house neat and not uncomfortable. The parlour was furnished with cane-bottomed
chairs, each of which was adorned with a white crocheted
tidy. The mantel over the fireplace had a white crocheted
cover; a marble-topped centre table held a lamp, a photograph album and several trinkets, each of which was set
upon a white crocheted mat. There was a cottage organ
in a corner of the room, and I noted that the lamp-racks
upon it were covered with white crocheted mats. There
was a matting on the floor, but a white crocheted carpet
would not have been out of keeping. I made arrangements with the landlady for my board and lodging; the
amount was, I think, three dollars and a half a week.
She was a rather fine-looking, stout, brown-skin woman
of about forty years of age. Her husband was a light-
coloured Cuban, a man about one half her size, and one
whose age could not be guessed from his appearance. He
was small in size, but a handsome black moustache and
typical Spanish eyes redeemed him from insignificance.

I was in time for breakfast, and at the table I had
the opportunity to see my fellow boarders. There were
eight or ten of them. Two, as I afterwards learned, were
coloured Americans. All of them were cigar-makers and
worked in one of the large factories—cigar-making is one
trade in which the colour line is not drawn. The conversation was carried on entirely in Spanish, and my ignorance of the language subjected me more to alarm than
embarrassment. I had never heard such uproarious conversation; everybody talked at once, loud exclamations,
rolling *"carambas,"* menacing gesticulations with knives,
forks, and spoons. I looked every moment for the clash
of blows. One man was emphasizing his remarks by
flourishing a cup in his hand, seemingly forgetful of the

fact that it was nearly full of hot coffee. He ended by
emptying it over what was, relatively, the only quiet man
at the table excepting myself, bringing from him a volley
of language which made the others appear dumb by comparison. I soon learned that in all of this clatter of voices
and table utensils they were discussing purely ordinary
affairs and arguing about mere trifles, and that not the
least ill feeling was aroused. It was not long before I
enjoyed the spirited chatter and *badinage* at the table as
much as I did my meals—and the meals were not bad.

I spent the afternoon in looking round the town. The
streets were sandy, but were well shaded by fine oak-
trees and far preferable to the clay roads of Atlanta.
One or two public squares with green grass and trees
gave the city a touch of freshness. That night after supper
I spoke to my landlady and her husband about my intentions. They told me that the big winter hotels would
not open within two months. It can easily be imagined
what effect this news had on me. I spoke to them frankly
about my financial condition and related the main fact
of my misfortune in Atlanta. I modestly mentioned my
ability to teach music and asked if there was any likelihood of my being able to get some scholars. My landlady
suggested that I speak to the preacher who had shown
me her house; she felt sure that through his influence
I should be able to get up a class in piano. She added,
however, that the coloured people were poor, and that the
general price for music lessons was only twenty-five cents.
I noticed that the thought of my teaching white pupils
did not even remotely enter her mind. None of this information made my prospects look much brighter.

The husband, who up to this time had allowed the
woman to do most of the talking, gave me the first bit of
tangible hope; he said that he could get me a job as a
"stripper" in the factory where he worked, and that if
I succeeded in getting some music pupils, I could teach
a couple of them every night, and so make a living until
something better turned up. He went on to say that it
would not be a bad thing for me to stay at the factory
and learn my trade as a cigar-maker, and impressed on
me that, for a young man knocking about the country,
a trade was a handy thing to have. I determined to accept
his offer and thanked him heartily. In fact, I became enthusiastic, not only because I saw a way out of my fi-

nancial troubles, but also because I was eager and curious over the new experience I was about to enter. I wanted to know all about the cigar-making business. This narrowed the conversation down to the husband and myself, so the wife went in and left us talking.

He was what is called a *regalía* workman, and earned from thirty-five to forty dollars a week. He generally worked a sixty-dollar job; that is, he made cigars for which he was paid at the rate of sixty dollars per thousand. It was impossible for him to make a thousand in a week because he had to work very carefully and slowly. Each cigar was made entirely by hand. Each piece of filler and each wrapper had to be selected with care. He was able to make a bundle of one hundred cigars in a day, not one of which could be told from the others by any difference in size or shape, or even by any appreciable difference in weight. This was the acme of artistic skill in cigar-making. Workmen of this class were rare, never more than three or four in one factory, and it was never necessary for them to remain out of work. There were men who made two, three, and four hundred cigars of the cheaper grades in a day; they had to be very fast in order to make decent week's wages. Cigar-making was a rather independent trade; the men went to work when they pleased and knocked off when they felt like doing so. As a class the workmen were careless and improvident; some very rapid makers would not work more than three or four days out of the week, and there were others who never showed up at the factory on Mondays. "Strippers" were the boys who pulled the long stems from the tobacco leaves. After they had served at that work for a certain time they were given tables as apprentices.

All of this was interesting to me; and we drifted along in conversation until my companion struck the subject nearest his heart, the independence of Cuba. He was an exile from the island, and a prominent member of the Jacksonville Junta. Every week sums of money were collected from juntas all over the country. This money went to buy arms and ammunition for the insurgents. As the man sat there nervously smoking his long, "green" cigar, and telling me of the Gómezes, both the white one and the black one, of Macéo and Bandera, he grew positively eloquent. He also showed that he was a man of considerable education and reading. He spoke English excellently, and frequently surprised me by using words one would hardly expect from a foreigner. The first one of this class of words he employed almost shocked me, and I never forgot it; 'twas "ramify." We sat on the piazza until after ten o'clock. When we arose to go in to bed, it was with the understanding that I should start in the factory on the next day.

I began work the next morning seated at a barrel with another boy, who showed me how to strip the stems from the leaves, to smooth out each half leaf, and to put the "rights" together in one pile, and the "lefts" together in another pile on the edge of the barrel. My fingers, strong and sensitive from their long training, were well adapted to this kind of work, and within two weeks I was accounted the fastest "stripper" in the factory. At first the heavy odour of the tobacco almost sickened me, but when I became accustomed to it, I liked the smell. I was now earning four dollars a week, and was soon able to pick up a couple more by teaching a few scholars at night, whom I had secured through the good offices of the preacher I had met on my first morning in Jacksonville.

At the end of about three months, through my skill as a "stripper" and the influence of my landlord, I was advanced to a table and began to learn my trade; in fact, more than my trade; for I learned not only to make cigars, but also to smoke, to swear, and to speak Spanish. I discovered that I had a talent for languages as well as for music. The rapidity and ease with which I acquired Spanish astonished my associates. In a short time I was able not only to understand most of what was said at the table during meals, but to join in the conversation. I bought a method for learning the Spanish language, and with the aid of my landlord as a teacher, by constant practice with my fellow workmen, and by regularly reading the Cuban newspapers and finally some books of standard Spanish literature which were at the house, I was able in less than a year to speak like a native. In fact, it was my pride that I spoke better Spanish than many of the Cuban workmen at the factory.

After I had been in the factory a little over a year, I was repaid for all the effort I had put forth to learn Spanish by being selected as "reader." The "reader" is quite an institution in all cigar factories which employ Spanish-speaking workmen. He sits in the centre of the large room in which the cigar-makers work and reads to them for a certain number of hours each day all the important news from the papers and whatever else he may consider would be interesting. He often selects an exciting novel and reads it in daily instalments. He must, of course, have a good voice, but he must also have a reputation among the men for intelligence, for being well posted and having in his head a stock of varied information. He is generally the final authority on all arguments which arise, and in a cigar factory these arguments are many and frequent, ranging from the respective and relative merits of rival baseball clubs to the duration of the sun's light and energy—cigar-making is a trade in which talk does not interfere with work. My position as "reader" not only released me from the rather monotonous work of rolling cigars, and gave me something more in accord with my tastes, but also added considerably to my income. I was now earning about twenty-five dollars a week, and was able to give up my peripatetic method of giving music lessons. I hired a piano and taught only those who could arrange to take their lessons where I lived. I finally gave up teaching entirely, as what I made scarcely paid for my time and trouble. I kept the piano, however, in order to keep up my own studies, and occasionally I played at some church concert or other charitable entertainment.

Through my music teaching and my not absolutely irregular attendance at church I became acquainted with the best class of coloured people in Jacksonville. This was really my entrance into the race. It was my initiation into what I have termed the freemasonry of the race. I had formulated a theory of what it was to be coloured; now I was getting the practice. The novelty of my position caused me to observe and consider things which, I think, entirely escaped the young men I associated with; or, at least, were so commonplace to them as not to attract their attention. And of many of the impressions which came to me then I have realized the full import only within the past few years, since I have had a broader knowledge of men and history, and a fuller comprehension of the tremendous struggle which is going on between the races in the South.

It is a struggle; for though the black man fights passively, he nevertheless fights; and his passive resistance is more effective at present than active resistance could possibly be. He bears the fury of the storm as does the willow-tree.

It is a struggle; for though the white man of the South may be too proud to admit it, he is, nevertheless, using in the contest his best energies; he is devoting to it the greater part of his thought and much of his endeavour. The South today stands panting and almost breathless from its exertions.

And how the scene of the struggle has shifted! The battle was first waged over the right of the Negro to be classed as a human being with a soul; later, as to whether he had sufficient intellect to master even the rudiments of learning; and today it is being fought out over his social recognition.

I said somewhere in the early part of this narrative that because the coloured man looked at everything through the prism of his relationship to society as a *coloured* man, and because most of his mental efforts ran through the narrow channel bounded by his rights and his wrongs, it was to be wondered at that he has progressed so broadly as he has. The same thing may be said of the white man of the South; most of his mental efforts run through one narrow channel; his life as a man and a citizen, many of his financial activities, and all of his political activities are impassably limited by the ever present "Negro question." I am sure it would be safe to wager that no group of Southern white men could get together and talk for sixty minutes without bringing up the "race question." If a Northern white man happened to be in the group, the time could be safely cut to thirty minutes. In this respect I consider the conditions of the whites more to be deplored than that of the blacks. Here, a truly great people, a people that produced a majority of the great historic Americans from Washington to Lincoln, now forced to use up its energies in a conflict as lamentable as it is violent.

I shall give the observations I made in Jacksonville as seen through the light of after years; and they apply generally to every Southern community. The coloured people may be said to be roughly divided into three classes, not so much in respect to themselves as in respect to their relations with the whites. There are those constituting what might be called the desperate class—the men who work in the lumber and turpentine camps, the ex-convicts, the bar-room loafers are all in this class. These men conform to the requirements of civilization much as a trained lion with low muttered growls goes through his stunts under the crack of the trainer's whip. They cherish a sullen hatred for all white men, and they value life as cheap. I have heard more than one of them say: "I'll go to hell for the first white man that bothers me." Many who have expressed that sentiment have kept their word, and it is that fact which gives such prominence to this class; for in numbers it is only a small proportion of the coloured people, but it often dominates public opinion concerning the whole race. Happily, this class represents the black people of the South far below their normal physical and moral condition, but in its increase lies the possibility of grave dangers. I am sure there is no more urgent work before the white South, not only for its present happiness, but for its future safety, than the decreasing of this class of blacks. And it is not at all a hopeless class; for these men are but the creatures of conditions, as much so as the slum and criminal elements of all the great cities of the world are creatures of conditions. Decreasing their number by shooting and burning them off will not be successful; for these men are truly desperate, and thoughts of death, however terrible, have little effect in deterring them from acts the result of hatred or degeneracy. This class of blacks hate everything covered by a white skin, and in return they are loathed by the whites. The whites regard them just about as a man would a vicious mule, a thing to be worked, driven, and beaten, and killed for kicking.

The second class, as regards the relation between blacks and whites, comprises the servants, the washerwomen, the waiters, the cooks, the coachmen, and all who are connected with the whites by domestic service. These may be generally characterized as simple, kind-hearted, and faithful; not over-fine in their moral deductions, but intensely religious, and relatively—such matters can be judged only relatively—about as honest and wholesome in their lives as any other grade of society. Any white person is "good" who treats them kindly, and they love him for that kindness. In return, the white people with whom they have to do regard them with indulgent affection. They come into close daily contact with the whites, and may be called the connecting link between whites and blacks; in fact, it is through them that the whites know the rest of their col-

oured neighbours. Between this class of the blacks and
the whites there is little or no friction.

The third class is composed of the independent workmen and tradesmen, and of the well-to-do and educated
coloured people; and, strange to say, for a directly opposite reason they are as far removed from the whites as
the members of the first class I mentioned. These people
live in a little world of their own; in fact, I concluded
that if a coloured man wanted to separate himself from
his white neighbours, he had but to acquire some money,
education, and culture, and to live in accordance. For example, the proudest and fairest lady in the South could
with propriety—and it is what she would most likely do
—go to the cabin of Aunt Mary, her cook, if Aunt Mary
was sick, and minister to her comfort with her own hands;
but if Mary's daughter, Eliza, a girl who used to run
round my lady's kitchen, but who has received an education and married a prosperous young coloured man, were
at death's door, my lady would no more think of crossing
the threshold of Eliza's cottage than she would of going
into a bar-room for a drink.

I was walking down the street one day with a young
man who was born in Jacksonville, but had been away to
prepare himself for a professional life. We passed a young
white man, and my companion said to me: "You see that
young man? We grew up together; we have played, hunted,
and fished together; we have even eaten and slept together; and now since I have come back home, he barely
speaks to me." The fact that the whites of the South despise and ill-treat the desperate class of blacks is not only
explainable according to the ancient laws of human nature, but it is not nearly so serious or important as the
fact that as the progressive coloured people advance, they
constantly widen the gulf between themselves and their
white neighbours. I think that the white people somehow
feel that coloured people who have education and money,
who wear good clothes and live in comfortable houses,
are "putting on airs," that they do these things for the
sole purpose of "spiting the white folks," or are, at best,
going through a sort of monkey-like imitation. Of course,
such feelings can only cause irritation or breed disgust.
It seems that the whites have not yet been able to realize
and understand that these people in striving to better their
physical and social surroundings in accordance with their

financial and intellectual progress are simply obeying an
impulse which is common to human nature the world over.
I am in grave doubt as to whether the greater part of the
friction in the South is caused by the whites having a
natural antipathy to Negroes as a race, or an acquired
antipathy to Negroes in certain relations to themselves.
However that may be, there is to my mind no more pathetic side of this many-sided question than the isolated
position into which are forced the very coloured people
who most need and who could best appreciate sympathetic
co-operation; and their position grows tragic when the effort is made to couple them, whether or no, with the Negroes of the first class I mentioned.

This latter class of coloured people are well-disposed
towards the whites, and always willing to meet them more
than half-way. They, however, feel keenly any injustice or
gross discrimination, and generally show their resentment.
The effort is sometimes made to convey the impression
that the better class of coloured people fight against riding
in "Jim Crow" cars because they want to ride with white
people or object to being with humbler members of their
own race. The truth is they object to the humiliation of
being forced to ride in a *particular* car, aside from the
fact that that car is distinctly inferior, and that they are
required to pay full first-class fare. To say that the whites
are forced to ride in the superior car is less than a joke.
And, too, odd as it may sound, refined coloured people
get no more pleasure out of riding with offensive Negroes
than anybody else would get.

I can realize more fully than I could years ago that the
position of the advanced element of the coloured race is
often very trying. They are the ones among the blacks who
carry the entire weight of the race question; it worries the
others very little, and I believe the only thing which at
times sustains them is that they know that they are in the
right. On the other hand, this class of coloured people
get a good deal of pleasure out of life; their existence is
far from being one long groan about their condition. Out
of a chaos of ignorance and poverty they have evolved a
social life of which they need not be ashamed. In cities
where the professional and well-to-do class is large they
have formed society—society as discriminating as the
actual conditions will allow it to be; I should say, perhaps,
society possessing discriminating tendencies which become

rules as fast as actual conditions allow. This statement will, I know, sound preposterous, even ridiculous, to some persons; but as this class of coloured people is the least known of the race it is not surprising. These social circles are connected throughout the country, and a person in good standing in one city is readily accepted in another. One who is on the outside will often find it a difficult matter to get in. I know personally of one case in which money to the extent of thirty or forty thousand dollars and a fine house, not backed up by a good reputation, after several years of repeated effort, failed to gain entry for the possessor. These people have their dances and dinners and card parties, their musicals, and their literary societies. The women attend social affairs dressed in good taste, and the men in dress suits which they own; and the reader will make a mistake to confound these entertainments with the "Bellman's Balls" and "Whitewashers' Picnics" and "Lime-kiln Clubs" with which the humorous press of the country illustrates "Cullud Sassiety."

Jacksonville, when I was there, was a small town, and the number of educated and well-to-do coloured people was small; so this society phase of life did not equal what I have since seen in Boston, Washington, Richmond, and Nashville; and it is upon what I have more recently seen in these cities that I have made the observations just above. However, there were many comfortable and pleasant homes in Jacksonville to which I was often invited. I belonged to the literary society—at which we generally discussed the race question—and attended all of the church festivals and other charitable entertainments. In this way I passed three years which were not at all the least enjoyable of my life. In fact, my joy took such an exuberant turn that I fell in love with a young school-teacher and began to have dreams of matrimonial bliss; but another turn in the course of my life brought these dreams to an end.

I do not wish to mislead my readers into thinking that I led a life in Jacksonville which would make copy for the hero of a Sunday-school library book. I was a hail fellow well met with all of the workmen at the factory, most of whom knew little and cared less about social distinctions. From their example I learned to be careless about money, and for that reason I constantly postponed and finally abandoned returning to Atlanta University. It seemed impossible for me to save as much as two hundred dollars. Several of the men at the factory were my intimate friends, and I frequently joined them in their pleasures. During the summer months we went almost every Monday on an excursion to a seaside resort called Pablo Beach. These excursions were always crowded. There was a dancing-pavilion, a great deal of drinking, and generally a fight or two to add to the excitement. I also contracted the cigar-maker's habit of riding round in a hack on Sunday after-noons I sometimes went with my cigar-maker friends to public balls that were given at a large hall on one of the main streets. I learned to take a drink occasionally and paid for quite a number that my friends took; but strong liquors never appealed to my appetite. I drank them only when the company I was in required it, and suffered for it afterwards On the whole, though I was a bit wild, I can't remember that I ever did anything disgraceful, or, as the usual standard for young men goes, anything to forfeit my claim to respectability.

At one of the first public balls I attended I saw the Pull-man-car porter who had so kindly assisted me in getting to Jacksonville I went immediately to one of my factory friends and borrowed fifteen dollars with which to repay the loan my benefactor had made me After I had given him the money and was thanking him, I noticed that he wore what was, at least, an exact duplicate of my lamented black and grey tie It was somewhat worn, but distinct enough for me to trace the same odd design which had first attracted my eye This was enough to arouse my strongest suspicions, but whether it was sufficient for the law to take cognizance of I did not consider My astonish-ment and the ironical humour of the situation drove every-thing else out of my mind

These balls were attended by a great variety of people. They were generally given by the waiters of some one of the big hotels, and were often patronized by a number of hotel guests who came to "see the sights.' The crowd was always noisy, but good-natured; there was much quadrille-dancing, and a strong-lunged man called figures in a voice which did not confine itself to the limits of the hall It is not worth the while for me to describe in detail how these people acted; they conducted themselves in about the same manner as I have seen other people at similar balls conduct themselves. When one has seen something of the world and human nature, one must conclude, after all, that between people in like stations of life there is very little difference the world over.

However, it was at one of these balls that I first saw the cake-walk. There was a contest for a gold watch, to be awarded to the hotel head-waiter receiving the greatest number of votes. There was some dancing while the votes were being counted. Then the floor was cleared for the cake-walk. A half-dozen guests from some of the hotels took seats on the stage to act as judges, and twelve or fourteen couples began to walk for a sure enough, highly decorated cake, which was in plain evidence. The spectators crowded about the space reserved for the contestants and watched them with interest and excitement. The couples did not walk round in a circle, but in a square, with the men on the inside. The fine points to be considered were the bearing of the men, the precision with which they turned the corners, the grace of the women, and the ease with which they swung round the pivots. The men walked with stately and soldierly step, and the women with considerable grace. The judges arrived at their decision by a process of elimination. The music and the walk continued for some minutes; then both were stopped while the judges conferred; when the walk began again, several couples were left out. In this way the contest was finally narrowed down to three or four couples. Then the excitement became intense; there was much partisan cheering as one couple or another would execute a turn in extra elegant style. When the cake was finally awarded, the spectators were about evenly divided between those who cheered the winners and those who muttered about the unfairness of the judges. This was the cake-walk in its original form, and it is what the coloured performers on the theatrical stage developed into the prancing movements now known all over the world, and which some Parisian critics pronounced the acme of poetic motion.

There are a great many coloured people who are ashamed of the cake-walk, but I think they ought to be proud of it. It is my opinion that the coloured people of this country have done four things which refute the oft-advanced theory that they are an absolutely inferior race, which demonstrate that they have originality and artistic conception, and, what is more, the power of creating that which can influence and appeal universally. The first two of these are the Uncle Remus stories, collected by Joel Chandler Harris, and the Jubilee songs, to which the Fisk singers made the public and the skilled musicians of both America and Europe listen. The other two are rag-time music and the cake-walk. No one who has travelled can question the world-conquering influence of rag-time, and I do not think it would be an exaggeration to say that in Europe the United States is popularly known better by rag-time than by anything else it has produced in a generation. In Paris they call it American music. The newspapers have already told how the practice of intricate cake-walk steps has taken up the time of European royalty and nobility. These are lower forms of art, but they give evidence of a power that will some day be applied to the higher forms. In this measure, at least, and aside from the number of prominent individuals the coloured people of the United States have produced, the race has been a world influence; and all of the Indians between Alaska and Patagonia haven't done as much.

Just when I was beginning to look upon Jacksonville as my permanent home and was beginning to plan about marrying the young school-teacher, raising a family, and working in a cigar factory the rest of my life, for some reason, which I do not now remember, the factory at which I worked was indefinitely shut down. Some of the men got work in other factories in town; some decided to go to Key West and Tampa, others made up their minds to go to New York for work. All at once a desire like a fever seized me to see the North again and I cast my lot with those bound for New York.

CHAPTER VI

We steamed up into New York harbour late one afternoon in spring. The last efforts of the sun were being put forth in turning the waters of the bay to glistening gold; the green islands on either side, in spite of their warlike mountings, looked calm and peaceful; the buildings of the town shone out in a reflected light which gave the city an air of enchantment; and, truly, it is an enchanted spot.

New York City is the most fatally fascinating thing in America. She sits like a great witch at the gate of the country, showing her alluring white face and hiding her crooked hands and feet under the folds of her wide garments—constantly enticing thousands from far within, and tempting those who come from across the seas to go no farther. And all these become the victims of her caprice. Some she at once crushes beneath her cruel feet; others she condemns to a fate like that of galley-slaves; a few she favours and fondles, riding them high on the bubbles of fortune; then with a sudden breath she blows the bubbles out and laughs mockingly as she watches them fall.

Twice I had passed through it, but this was really my first visit to New York; and as I walked about that evening, I began to feel the dread power of the city; the crowds, the lights, the excitement, the gaiety, and all its subtler stimulating influences began to take effect upon me. My blood ran quicker and I felt that I was just beginning to live. To some natures this stimulant of life in a great city becomes a thing as binding and necessary as opium is to one addicted to the habit. It becomes their breath of life; they cannot exist outside of it; rather than be deprived of it they are content to suffer hunger, want, pain, and misery; they would not exchange even a ragged and wretched condition among the great crowd for any degree of comfort away from it.

As soon as we landed, four of us went directly to a lodging-house in Twenty-seventh Street, just west of Sixth Avenue. The house was run by a short, stout mulatto man, who was exceedingly talkative and inquisitive. In fifteen minutes he not only knew the history of the past life of each one of us, but had a clearer idea of what we intended to do in the future than we ourselves. He sought this information so much with an air of being very particular as to whom he admitted into his house that we tremblingly answered every question that he asked. When we had become located, we went out and got supper, then walked round until about ten o'clock. At that hour we met a couple of young fellows who lived in New York and were known to one of the members of our party. It was suggested we go to a certain place which was known by the proprietor's name. We turned into one of the cross streets and mounted the stoop of a house in about the middle of a block between Sixth and Seventh Avenues. One of the young men whom we had met rang a ball, and a man on the inside cracked the door a couple of inches; then opened it and let us in. We found ourselves in the hallway of what had once been a residence. The front parlour had been converted into a bar, and a half-dozen or so well-dressed men were in the room. We went in and after a general introduction had several rounds of beer. In the back parlour a crowd was sitting and standing round the walls of the room watching an exciting and noisy game of pool. I walked back and joined this crowd to watch the game, and principally to get away from the drinking party. The game was really interesting, the players being quite expert, and the excitement was heightened by the bets which were being made on the result. At times the antics and remarks of both players and spectators were amusing. When, at a critical point, a player missed a shot, he was deluged, by those financially interested in his making it, with a flood of epithets synonymous with "chump"; while from the others he would be jeered by such remarks as "Nigger, dat cue ain't no hoe-handle." I noticed that among this class of coloured men the word "nigger" was freely used in about the same sense as the word "fellow," and sometimes as a term of almost endearment; but I soon learned that its use was positively and absolutely prohibited to white men.

I stood watching this pool game until I was called by my friends, who were still in the bar-room, to go upstairs. On the second floor there were two large rooms. From the hall I looked into the one on the front. There was a large, round table in the centre, at which five or six men were seated playing poker. The air and conduct here were greatly in contrast to what I had just seen in the pool-room; these men were evidently the aristocrats of the place; they were well, perhaps a bit flashily, dressed and spoke in low modulated voices, frequently using the word "gentlemen"; in fact, they seemed to be practising a sort of Chesterfieldian politeness towards each other. I was watching these men with a great deal of interest and some degree of admiration when I was again called by the members of our party, and I followed them on to the back room. There was a door-keeper at this room, and we were admitted only after inspection. When we got inside, I saw a crowd of men of all ages and kinds grouped about

an old billiard-table, regarding some of whom, in sup-
posing them to be white, I made no mistake. At first I did
not know what these men were doing; they were using
terms that were strange to me. I could hear only a confusion of voices exclaiming: "Shoot the two!" "Shoot the
four!" "Fate me! Fate me!" "I've got you fated!" "Twenty-
five cents he don't turn!" This was the ancient and terribly fascinating game of dice, popularly known as
"craps." I myself had played pool in Jacksonville—it is a
favourite game among cigar-makers—and I had seen
others play cards; but here was something new. I edged
my way in to the table and stood between one of my new-
found New York friends and a tall, slender, black fellow,
who was making side bets while the dice were at the
other end of the table. My companion explained to me the
principles of the game; and they are so simple that they
hardly need to be explained twice. The dice came round
the table until they reached the man on the other side of
the tall, black fellow. He lost, and the latter said: "Gimme
the bones." He threw a dollar on the table and said:
"Shoot the dollar." His style of play was so strenuous that
he had to be allowed plenty of room. He shook the dice
high above his head, and each time he threw them on the
table, he emitted a grunt such as men give when they are
putting forth physical exertion with a rhythmic regular-
ity. He frequently whirled completely round on his heels,
throwing the dice the entire length of the table, and
talking to them as though they were trained animals. He appealed to them in short singsong phrases. "Come, dice,"
he would say. "Little Phœbe," "Little Joe," " 'Way down
yonder in the cornfield." Whether these mystic incantations
were efficacious or not I could not say, but, at any rate,
his luck was great, and he had what gamblers term
"nerve." "Shoot the dollar!" "Shoot the two!" "Shoot the
four!" "Shoot the eight!" came from his lips as quickly as
the dice turned to his advantage. My companion asked me
if I had ever played. I told him no. He said that I ought
to try my luck: that everybody won at first. The tall man
at my side was waving his arms in the air, exclaiming:
"Shoot the sixteen!" "Shoot the sixteen!" "Fate me!"
Whether it was my companion's suggestion or some latent
dare-devil strain in my blood which suddenly sprang into
activity I do not know; but with a thrill of excitement which
went through my whole body I threw a twenty-dollar bill

on the table and said in a trembling voice: "I fate you."

I could feel that I had gained the attention and respect of everybody in the room, every eye was fixed on me, and the widespread question, "Who is he?" went round. This was gratifying to a certain sense of vanity of which I have never been able to rid myself, and I felt that it was worth the money even if I lost. The tall man, with a whirl on his heels and a double grunt, threw the dice; four was the number which turned up. This is considered as a hard "point" to make. He redoubled his contortions and his grunts and his pleadings to the dice; but on his third or fourth throw the fateful seven turned up, and I had won. My companion and all my friends shouted to me to follow up my luck. The fever was on me. I seized the dice. My hands were so hot that the bits of bone felt like pieces of ice. I shouted as loudly as I could: "Shoot it all!" but the blood was tingling so about my ears that I could not hear my own voice. I was soon "fated." I threw the dice—seven—I had won. "Shoot it all!" I cried again. There was a pause; the stake was more than one man cared to or could cover. I was finally "fated" by several men taking each a part of it. I then threw the dice again. Seven. I had won. "Shoot it all!" I shouted excitedly. After a short delay I was "fated." Again I rolled the dice. Eleven. Again I won. My friends now surrounded me and, much against my inclination, forced me to take down all of the money except five dollars. I tried my luck once more, and threw some small "point" which I failed to make, and the dice passed on to the next man.

In less than three minutes I had won more than two hundred dollars, a sum which afterwards cost me dearly. I was the hero of the moment and was soon surrounded by a group of men who expressed admiration for my "nerve" and predicted for me a brilliant future as a gambler. Although at the time I had no thought of becoming a gambler, I felt proud of my success. I felt a bit ashamed, too, that I had allowed my friends to persuade me to take down my money so soon. Another set of men also got round me and begged me for twenty-five or fifty cents to put them back into the game. I gave each of them something. I saw that several of them had on linen dusters, and as I looked about, I noticed that there were perhaps a dozen men in the room similarly clad. I

asked the fellow who had been my prompter at the dice
table why they dressed in such a manner. He told me that
men who had lost all the money and jewellery they possessed, frequently, in an effort to recoup their losses, would
gamble away all their outer clothing and even their shoes;
and that the proprietor kept on hand a supply of linen
dusters for all who were so unfortunate. My informant
went on to say that sometimes a fellow would become almost completely dressed and then, by a turn of the dice,
would be thrown back into a state of semi-nakedness. Some
of them were virtually prisoners and unable to get into
the streets for days at a time. They ate at the lunch
counter, where their credit was good so long as they
were fair gamblers and did not attempt to jump their
debts, and they slept round in chairs. They importuned
friends and winners to put them back in the game, and
kept at it until fortune again smiled on them. I laughed
heartily at this, not thinking the day was coming which
would find me in the same ludicrous predicament.

On passing downstairs I was told that the third and top
floor of the house was occupied by the proprietor. When
we passed through the bar, I treated everybody in the
room—and that was no small number, for eight or ten
had followed us down. Then our party went out. It was
now about half past twelve, but my nerves were at such
a tension that I could not endure the mere thought of
going to bed. I asked if there was no other place to which
we could go; our guides said yes, and suggested that we
go to the "Club." We went to Sixth Avenue, walked two
blocks, and turned to the west into another street. We
stopped in front of a house with three stories and a basement. In the basement was a Chinese chop-suey restaurant.
There was a red lantern at the iron gate to the areaway,
inside of which the Chinaman's name was printed. We
went up the steps of the stoop, rang the bell, and were
admitted without any delay. From the outside the house
bore a rather gloomy aspect, the windows being absolutely
dark, but within, it was a veritable house of mirth. When
we had passed through a small vestibule and reached the
hallway, we heard mingled sounds of music and laughter,
the clink of glasses, and the pop of bottles. We went into
the main room and I was little prepared for what I saw.
The brilliancy of the place, the display of diamond rings,
scarf-pins, ear-rings, and breast-pins, the big rolls of money

that were brought into evidence when drinks were paid for, and the air of gaiety that pervaded the place, all completely dazzled and dazed me. I felt positively giddy, and it was several minutes before I was able to make any clear and definite observations.

We at length secured places at a table in a corner of the room and, as soon as we could attract the attention of one of the busy waiters, ordered a round of drinks. When I had somewhat collected my senses, I realized that in a large back room into which the main room opened, there was a young fellow singing a song, accompanied on the piano by a short, thickset, dark man. After each verse he did some dance steps, which brought forth great applause and a shower of small coins at his feet. After the singer had responded to a rousing encore, the stout man at the piano began to run his fingers up and down the keyboard. This he did in a manner which indicated that he was master of a good deal of technique. Then he began to play; and such playing! I stopped talking to listen. It was music of a kind I had never heard before. It was music that demanded physical response, patting of the feet, drumming of the fingers, or nodding of the head in time with the beat. The barbaric harmonies, the audacious resolutions, often consisting of an abrupt jump from one key to another, the intricate rhythms in which the accents fell in the most unexpected places, but in which the beat was never lost, produced a most curious effect. And, too, the player——the dexterity of his left hand in making rapid octave runs and jumps was little short of marvellous; and with his right hand he frequently swept half the keyboard with clean-cut chromatics which he fitted in so nicely as never to fail to arouse in his listeners a sort of pleasant surprise at the accomplishment of the feat.

This was rag-time music, then a novelty in New York, and just growing to be a rage, which has not yet subsided. It was originated in the questionable resorts about Memphis and St. Louis by Negro piano-players who knew no more of the theory of music than they did of the theory of the universe, but were guided by natural musical instinct and talent. It made its way to Chicago, where it was popular some time before it reached New York. These players often improvised crude and, at times, vulgar words to fit the melodies. This was the beginning of the

rag-time song. Several of these improvisations were taken
down by white men, the words slightly altered, and published under the names of the arrangers. They sprang
into immediate popularity and earned small fortunes, of
which the Negro originators got only a few dollars. But
I have learned that since that time a number of coloured
men, of not only musical talent, but training, are writing
out their own melodies and words and reaping the reward
of their work. I have learned also that they have a
large number of white imitators and adulterators.

American musicians, instead of investigating rag-time,
attempt to ignore it, or dismiss it with a contemptuous
word. But that has always been the course of scholasticism
in every branch of art. Whatever new thing the *people*
like is pooh-poohed; whatever is *popular* is spoken of as
not worth the while. The fact is, nothing great or enduring,
especially in music, has ever sprung full-fledged and unprecedented from the brain of any master; the best that
he gives to the world he gathers from the hearts of the
people, and runs it through the alembic of his genius. In
spite of the bans which musicians and music teachers
have placed upon it, the people still demand and enjoy
rag-time. One thing cannot be denied; it is music which
possesses at least one strong element of greatness: it appeals universally; not only the American, but the English,
the French, and even the German people find delight in
it. In fact, there is not a corner of the civilized world
in which it is not known, and this proves it originality; for
if it were an imitation, the people of Europe, anyhow,
would not have found it a novelty. Anyone who doubts
that there is a peculiar heel-tickling, smile-provoking, joy-
awakening charm in rag-time needs only to hear a skilful
performer play the genuine article to be convinced. I believe that it has its place as well as the music which draws
from us sighs and tears.

I became so interested in both the music and the player
that I left the table where I was sitting, and made my
way through the hall into the back room, where I could
see as well as hear. I talked to the piano-player between
the musical numbers and found out that he was just a
natural musician, never having taken a lesson in his life.
Not only could he play almost anything he heard, but
he could accompany singers in songs he had never heard.
He had, by ear alone, composed some pieces, several of

which he played over for me; each of them was properly proportioned and balanced. I began to wonder what this man with such a lavish natural endowment would have done had he been trained. Perhaps he wouldn't have done anything at all; he might have become, at best, a mediocre imitator of the great masters in what they have already done to a finish, or one of the modern innovators who strive after originality by seeing how cleverly they can dodge about through the rules of harmony and at the same time avoid melody. It is certain that he would not have been so delightful as he was in rag-time.

I sat by, watching and listening to this man until I was dragged away by my friends. The place was now almost deserted; only a few stragglers hung on, and they were all the worse for drink. My friends were well up in this class. We passed into the street; the lamps were pale against the sky; day was just breaking. We went home and got into bed. I fell into a fitful sort of sleep, with rag-time music ringing continually in my ears.

CHAPTER VII

I shall take advantage of this pause in my narrative to describe more closely the "Club" spoken of in the later part of the preceding chapter—to describe it as I afterwards came to know it, as an habitué. I shall do this not only because of the direct influence it had on my life, but also because it was at that time the most famous place of its kind in New York, and was well known to both white and coloured people of certain classes.

I have already stated that in the basement of the house there was a Chinese restaurant. The Chinaman who kept it did an exceptionally good business; for chop-suey was a favourite dish among the frequenters of the place. It is a food that, somehow, has the power of absorbing alcoholic liquors that have been taken into the stomach. I have heard men claim that they could sober up on chop-suey. Perhaps that accounted, in some degree, for its popularity. On the main floor there were two large rooms: a parlour about thirty feet in length, and a large, square back

room into which the parlour opened. The floor of the
parlour was carpeted; small tables and chairs were arranged about the room; the windows were draped with
lace curtains, and the walls were literally covered with
photographs or lithographs of every coloured man in
America who had ever "done anything." There were pictures of Frederick Douglass and of Peter Jackson, of all
the lesser lights of the prize-fighting ring, of all the famous jockeys and the stage celebrities, down to the newest
song and dance team. The most of these photographs were
autographed and, in a sense, made a really valuable
collection. In the back room there was a piano, and tables
were placed round the wall. The floor was bare and the
centre was left vacant for singers, dancers, and others
who entertained the patrons. In a closet in this room
which jutted out into the hall the proprietor kept his
buffet. There was no open bar, because the place had no
liquor licence. In this back room the tables were sometimes pushed aside, and the floor given over to general
dancing. The front room on the next floor was a sort of
private party room; a back room on the same floor contained no furniture and was devoted to the use of new
and ambitious performers. In this room song and dance
teams practised their steps, acrobatic teams practised
their tumbles, and many other kinds of "acts" rehearsed
their "turns." The other rooms of the house were used as
sleeping-apartments.

No gambling was allowed, and the conduct of the place
was surprisingly orderly. It was, in short, a centre of
coloured Bohemians and sports. Here the great prize-
fighters were wont to come, the famous jockeys, the noted
minstrels, whose names and faces were familiar on every
bill-board in the country; and these drew a multitude of
those who love to dwell in the shadow of greatness.
There were then no organizations giving performances of
such order as are now given by several colored companies; that was because no manager could imagine that
audiences would pay to see Negro performers in any
other role than that of Mississippi River roustabouts; but
there was lots of talent and ambition. I often heard the
younger and brighter men discussing the time when they
would compel the public to recognize that they could do
something more than grin and cut pigeon-wings.

Sometimes one or two of the visiting stage-professionals,

after being sufficiently urged, would go into the back room and take the places of the regular amateur entertainers, but they were very sparing with these favours, and the patrons regarded them as special treats. There was one man, a minstrel, who, whenever he responded to a request to "do something," never essayed anything below a reading from Shakespeare. How well he read I do not know, but he greatly impressed me; and I can say that at least he had a voice which strangely stirred those who heard it. Here was a man who made people laugh at the size of his mouth, while he carried in his heart a burning ambition to be a tragedian; and so after all he did play a part in a tragedy.

These notables of the ring, the turf, and the stage, drew to the place crowds of admirers, both white and coloured. Whenever one of them came in, there were awe-inspired whispers from those who knew him by sight, in which they enlightened those round them as to his identity, and hinted darkly at their great intimacy with the noted one. Those who were on terms of approach immediately showed their privilege over others less fortunate by gathering round their divinity. I was, at first, among those who dwelt in darkness. Most of these celebrities I had never heard of. This made me an object of pity among many of my new associates. I soon learned, however, to fake a knowledge for the benefit of those who were greener than I; and, finally, I became personally acquainted with the majority of the famous personages who came to the "Club."

A great deal of money was spent here, so many of the patrons were men who earned large sums. I remember one night a dapper little brown-skin fellow was pointed out to me and I was told that he was the most popular jockey of the day, and that he earned $12,000 a year. This latter statement I couldn't doubt, for with my own eyes I saw him spending at about thirty times that rate. For his friends and those who were introduced to him he bought nothing but wine—in sporting circles, "wine" means champagne—and paid for it at five dollars a quart. He sent a quart to every table in the place with his compliments; and on the table at which he and his party were seated there were more than a dozen bottles. It was the custom at the "Club" for the waiter not to remove the bottles when champagne was being drunk until the party had finished. There were reasons for this;

it advertised the brand of wine, it advertised that the party was drinking wine, and advertised how much they had bought. This jockey had won a great race that day, and he was rewarding his admirers for the homage they paid him, all of which he accepted with a fine air of condescension.

Besides the people I have just been describing, there was at the place almost every night one or two parties of white people, men and women, who were out sight-seeing, or slumming. They generally came in cabs; some of them would stay only for a few minutes, while others sometimes stayed until morning. There was also another set of white people who came frequently; it was made up of variety performers and others who delineated "darky characters"; they came to get their imitations first-hand from the Negro entertainers they saw there.

There was still another set of white patrons, composed of women; these were not occasional visitors, but five or six of them were regular habituées. When I first saw them, I was not sure that they were white. In the first place, among the many coloured women who came to the "Club" there were several just as fair; and, secondly, I always saw these women in company with coloured men. They were all good-looking and well dressed, and seemed to be women of some education. One of these in particular attracted my attention; she was an exceedingly beautiful woman of perhaps thirty-five; she had glistening copper-coloured hair, very white skin, and eyes very much like Du Maurier's conception of Trilby's "twin grey stars." When I came to know her, I found that she was a woman of considerable culture; she had travelled in Europe, spoke French, and played the piano well. She was always dressed elegantly, but in absolute good taste. She always came to the "Club" in a cab, and was soon joined by a well-set-up, very black young fellow. He was always faultlessly dressed; one of the most exclusive tailors in New York made his clothes, and he wore a number of diamonds in about as good taste as they could be worn in by a man. I learned that she paid for his clothes and his diamonds. I learned, too, that he was not the only one of his kind. More that I learned would be better suited to a book on social phenomena than to a narrative of my life.

This woman was known at the "Club" as the rich

widow. She went by a very aristocratic-sounding name, which corresponded to her appearance. I shall never forget how hard it was for me to get over my feelings of surprise, perhaps more than surprise, at seeing her with her black companion; somehow I never exactly enjoyed the sight. I have devoted so much time to this pair, the "widow" and her companion, because it was through them that another decided turn was brought about in my life.

CHAPTER VIII

On the day following our night at the "Club" we slept until late in the afternoon; so late that beginning search for work was entirely out of the question. This did not cause me much worry, for I had more than three hundred dollars, and New York had impressed me as a place where there was lots of money and not much difficulty in getting it. It is needless to inform my readers that I did not long hold this opinion. We got out of the house about dark, went to a restaurant on Sixth Avenue and ate something, then walked round for a couple of hours. I finally suggested that we visit the same places we had been in the night before. Following my suggestion, we started first to the gambling-house. The man on the door let us in without any question; I accredited this to my success of the night before. We went straight to the "crap" room, and I at once made my way to a table, where I was rather flattered by the murmur of recognition which went round. I played in up and down luck for three or four hours; then, worn with nervous excitement, quit, having lost about fifty dollars. But I was so strongly possessed with the thought that I would make up my losses the next time I played that I left the place with a light heart.

When we got into the street our party was divided against itself; two were for going home at once and getting to bed. They gave as a reason that we were to get up early and look for jobs. I think the real reason was that they had each lost several dollars in the game. I lived to learn that in the world of sport all men win alike, but lose differently; and so gamblers are rated, not by the way in which they win, but by the way in which they lose. Some men lose with a careless smile, recognizing that losing is a part of the game; others curse their luck and rail at fortune; and others, still, lose sadly; after each such experience they are swept by a wave of reform; they resolve to stop gambling and be good. When in this frame of mind it would take very little persuasion to lead them into a prayer-meeting. Those in the first class are looked upon with admiration; those in the second class are merely commonplace; while those in the third are regarded with contempt. I believe these distinctions hold good in all the ventures of life. After some minutes one of my friends and I succeeded in convincing the other two that a while at the "Club" would put us all in better spirits; and they consented to go, on our promise not to stay longer than an hour. We found the place crowded, and the same sort of thing going on which we had seen the night before. I took a seat at once by the side of the piano-player, and was soon lost to everything except the novel charm of the music. I watched the performer with the idea of catching the trick, and during one of his intermissions I took his place at the piano and made an attempt to imitate him, but even my quick ear and ready fingers were unequal to the task on first trial.

We did not stay at the "Club" very long, but went home to bed in order to be up early the next day. We had no difficulty in finding work, and my third morning in New York found me at a table rolling cigars. I worked steadily for some weeks, at the same time spending my earnings between the "crap" game and the "Club." Making cigars became more and more irksome to me; perhaps my more congenial work as a "reader" had unfitted me for work at the table. And, too, the late hours I was keeping made such a sedentary occupation almost beyond the powers of will and endurance. I often found it hard to keep my eyes open and sometimes had to get up and move round to keep from falling asleep. I began to miss whole days from the factory, days on which I was compelled to stay at home and sleep.

My luck at the gambling-table was varied; sometimes I was fifty to a hundred dollars ahead, and at other times I had to borrow money from my fellow workmen to settle my room rent and pay for my meals. Each night after leaving the dice game I went to the "Club" to hear the music and watch the gaiety. If I had won, this was in accord with my mood; if I had lost, it made me forget. I at last realized that making cigars for a living and gambling for a living could not both be carried on at the same time, and I resolved to give up the cigar-making. This resolution led me into a life which held me bound more than a year. During that period my regular time for going to bed was somewhere between four and six o'clock in the mornings. I got up late in the afternoons, walked about a little, then went to the gambling-house or the "Club." My New York was limited to ten blocks; the boundaries were Sixth Avenue from Twenty-third to Thirty-third Streets, with the cross streets one block to the west. Central park was a distant forest, and the lower part of the city a foreign land. I look back upon the life I then led with a shudder when I think what would have been had I not escaped it. But had I not escaped it, I should have been no more unfortunate than are many young coloured men who come to New York. During that dark period I became acquainted with a score of bright, intelligent young fellows who had come up to the great city with high hopes and ambitions and who had fallen under the spell of this under life, a spell they could not throw off. There was one popularly known as "the doctor"; he had had two years in the Harvard Medical School, but here he was, living this gaslight life, his will and moral sense so enervated and deadened that it was impossible for him to break away. I do not doubt that the same thing is going on now, but I have sympathy rather than censure for these victims, for I know how easy it is to slip into a slough from which it takes a herculean effort to leap.

I regret that I cannot contrast my views of life among coloured people of New York; but the truth is, during my entire stay in this city I did not become acquainted with a single respectable family. I knew that there were several coloured men worth a hundred or so thousand dollars each, and some families who proudly dated their free ancestry back a half-dozen generations. I also learned that in Brooklyn there lived quite a large colony in comfortable homes which they owned; but at no point did my life come in contact with theirs.

In my gambling-experiences I passed through all the states and conditions that a gambler is heir to. Some days

found me able to peel ten- and twenty-dollar bills from
a roll, and others found me clad in a linen duster and
carpet slippers. I finally caught up another method of earning money, and so did not have to depend entirely upon
the caprices of fortune at the gaming-table. Through continually listening to the music at the "Club," and through
my own previous training, my natural talent and perseverance, I developed into a remarkable player of rag-time;
indeed, I had the name at that time of being the best-rag-
time-player in New York. I brought all my knowledge of
classic music to bear and, in so doing, achieved some
novelties which pleased and even astonished my listeners.
It was I who first made rag-time transcriptions of familiar
classic selections. I used to play Mendelssohn's "Wedding
March" in a manner that never failed to arouse enthusiasm
among the patrons of the "Club." Very few nights passed
during which I was not asked to play it. It was no secret
that the great increase in slumming visitors was due to my
playing. By mastering rag-time I gained several things: first
of all, I gained the title of professor. I was known as
"the professor" as long as I remained in that world. Then,
too, I gained the means of earning a rather fair livelihood.
This work took up much of my time and kept me almost
entirely away from the gambling-table. Through it I also
gained a friend who was the means by which I escaped
from this lower world. And, finally, I secured a wedge
which has opened to me more doors and made me a welcome guest than my playing of Beethoven and Chopin
could ever have done.

The greater part of the money I now began to earn came
through the friend to whom I alluded in the foregoing paragraph. Among the other white "slummers" there came into
the "Club" one night a clean-cut, slender, but athletic-look-
ing man, who would have been taken for a youth had it not
been for the tinge of grey about his temples. He was clean-
shaven and had regular features, and all of his movements
bore the indefinable but unmistakable stamp of culture.
He spoke to no one, but sat languidly puffing cigarettes and
sipping a glass of beer. He was the centre of a great deal of
attention; all of the old-timers were wondering who he
was. When I had finished playing, he called a waiter and
by him sent me a five-dollar bill. For about a month after
that he was at the "Club" one or two nights each week, and
each time after I had played, he gave me five dollars. One

night he sent for me to come to his table; he asked me several questions about myself; then told me that he had an engagement which he wanted me to fill. He gave me a card containing his address and asked me to be there on a certain night.

I was on hand promptly and found that he was giving a dinner in his own apartments to a party of ladies and gentlemen and that I was expected to furnish the musical entertainment. When the grave, dignified man at the door let me in, the place struck me as being almost dark, my eyes had been so accustomed to the garish light of the "Club." He took my coat and hat, bade me take a seat, and went to tell his master that I had come. When my eyes were adjusted to the soft light, I saw that I was in the midst of elegance and luxury in a degree such as I had never seen; but not the elegance which makes one ill at ease. As I sank into a great chair, the subdued tone, the delicately sensuous harmony of my surroundings, drew from me a deep sigh of relief and comfort. How long the man was gone I do not know, but I was startled by a voice saying: "Come this way, if you please, sir," and I saw him standing by my chair. I had been asleep; and I awoke very much confused and a little ashamed, because I did not know how many times he may have called me. I followed him through into the dining-room, where the butler was putting the finishing touches to a table which already looked like a big jewel. The doorman turned me over to the butler, and I passed with the butler on back to where several waiters were busy polishing and assorting table utensils. Without being asked where I was hungry or not, I was placed at a table and given something to eat. Before I had finished eating, I heard the laughter and talk of the guests who were arriving. Soon afterwards I was called in to begin my work.

I passed in to where the company was gathered and went directly to the piano. According to a suggestion from the host, I began with classic music. During the first number there was absolute quiet and appreciative attention, and when I had finished, I was given a round of generous applause. After that the talk and the laughter began to grow until the music was only an accompaniment to the chatter. This, however, did not disconcert me as it once would have done, for I had become accustomed to playing in the midst of uproarious noise. As the guests began to pay less attention to me, I was enabled to pay more to them. There were about a dozen of them. The men ranged in appearance from a girlish-looking youth to a big grizzled man whom everybody addressed as "Judge." None of the women appeared to be under thirty, but each of them struck me as being handsome. I was not long in finding out that they were all decidedly blasé. Several of the women smoked cigarettes, and with a careless grace which showed they were used to the habit. Occasionally a "Damn it!" escaped from the lips of some one of them, but in such a charming way as to rob it of all vulgarity. The most notable thing which I observed was that the reserve of the host increased in direct proportion with the hilarity of his guests. I thought that there was something going wrong which displeased him. I afterwards learned that it was his habitual manner on such occasions. He seemed to take cynical delight in watching and studying others indulging in excess. His guests were evidently accustomed to his rather non-participating attitude, for it did not seem in any degree to dampen their spirits.

When dinner was served, the paino was moved and the door left open, so that the company might hear the music while eating. At a word from the host I struck up one of my liveliest rag-time pieces. The effect was surprising, perhaps even to the host; the rag-time music came very near spoiling the party so far as eating the dinner was concerned. As soon as I began, the conversation suddenly stopped. It was a pleasure to me to watch the expression of astonishment and delight that grew on the faces of everybody. These were people—and they represented a large class—who were ever expecting to find happiness in novelty, each day restlessly exploring and exhausting every resource of this great city that might possibly furnish a new sensation or awaken a fresh emotion, and who were always grateful to anyone who aided them in their quest. Several of the women left the table and gathered about the piano. They watched my fingers and asked what kind of music it was that I was playing, where I had learned it, and a host of other questions. It was only by being repeatedly called back to the table that they were induced to finish their dinner. When the guests arose, I struck up my rag-time transcription of Mendelssohn's "Wedding March," playing it with a terrific chromatic octave runs in the base. This raised everybody's spirits to the highest point of gaiety, and the whole company involuntarily and unconsciously did an impromptu cake-walk. From that time on until the time of leaving they kept me so busy that my arms ached. I obtained a little respite when the girlish-looking youth and one or two of the ladies sang several songs, but after each of these it was "back to rag-time."

In leaving, the guests were enthusiastic in telling the host that he had furnished them the most unusual entertainment they had ever enjoyed. When they had gone, my millionaire friend—for he was reported to be a millionaire—said to me with a smile: "Well, I have given them something they've never had before." After I had put on my coat and was ready to leave, he made me take a glass of wine; he then gave me a cigar and twenty dollars in bills. He told me that he would give me lots of work, his only stipulation being that I should not play any engagements such as I had just filled for him, except by his instructions. I readily accepted the proposition, for I was sure that I could not be the loser by such a contract.

I afterwards played for him at many dinners and parties of one kind or another. Occasionally he "loaned" me to some of his friends. And, too, I often played for him alone at his apartments. At such times he was quite a puzzle to me until I became accustomed to his manners. He would sometimes sit for three or four hours hearing me play, his eyes almost closed, making scarcely a motion except to light a fresh cigarette, and never commenting one way or another on the music. At first I sometimes thought he had fallen asleep and would pause in playing. The stopping of the music always aroused him enough to tell me to play this or that; and I soon learned that my task was not to be considered finished until he got up from his chair and said: "That will do." The man's powers of endurance in listening often exceed mine in performing—yet I am not sure that he was always listening. At times I became so oppressed with fatigue and sleepiness that it took almost superhuman effort to keep my fingers going; in fact, I believe I sometimes did so while dozing. During such moments this man sitting there so mysteriously silent, almost hid in a cloud of heavy-scented smoke, filled me with a sort of unearthly terror. He seemed to be some grim, mute, but relentless tyrant, possessing over me a supernatural power which he used to drive me on mercilessly to exhaustion. But these feelings came very rarely; besides, he paid me so liberally I could forget much. There at length grew between us a familiar and warm relationship, and I am sure he had a decided personal liking for me. On my part, I looked upon him at that time as about all a man could wish to be.

The "Club" still remained my headquarters, and when I was not playing for my good patron, I was generally to be found there. However, I no longer depended on playing at the "Club" to earn my living; I rather took rank with the visiting celebrities and, occasionally, after being sufficiently urged, would favour my old and new admirers with a number or two. I say, without any egotistic pride, that among my admirers were several of the best-looking women who frequented the place, and who made no secret of the fact that they admired me as much as they did my playing. Among these was the "widow"; indeed, her attentions became so marked that one of my friends warned me to beware of her black companion, who was generally known as a "bad man." He said there was much more reason to be careful because the pair had lately quarrelled and had not been together at the "Club" for some nights. This warning greatly impressed me and I resolved to stop the affair before it should go any further; but the woman was so beautiful that my native gallantry and delicacy would not allow me to repulse her; my finer feelings entirely overcame my judgment. The warning also opened my eyes sufficiently to see that though my artistic temperament and skill made me interesting and attractive to the woman, she was, after all, using me only to excite the jealousy of her companion and revenge herself upon him. It was this surly, black despot who held sway over her deepest emotions.

One night, shortly afterwards, I went into the "Club" and saw the "widow" sitting at a table in company with another woman. She at once beckoned for me to come to her. I went, knowing that I was committing worse than folly. She ordered a quart of campagne and insisted that I sit down and drink with her. I took a chair on the opposite side of the table and began to sip a glass of the wine. Suddenly I noticed by an expression on the "widow's" face that something had occurred. I instinctively glanced round and saw that her companion had just entered. His ugly look completely frightened me. My back was turned to him, but by watching the "widow's" eyes I judged that he was

pacing back and forth across the room. My feelings were
far from being comfortable; I expected every moment to
feel a blow on my head. She, too, was very nervous; she
was trying hard to appear unconcerned, but could not
succeed in hiding her real feelings. I decided that it was
best to get out of such a predicament even at the expense of
appearing cowardly, and I made a motion to rise. Just as
I partly turned in my chair, I saw the black fellow approaching; he walked directly to our table and leaned
over. The "widow" evidently feared he was going to strike
her, and she threw back her head. Instead of striking her
he whipped out a revolver and fired; the first shot went
straight into her throat. There were other shots fired, but
how many I do not know; for the first knowledge I had
of my surroundings and actions was that I was rushing
through the chop-suey restaurant into the street. Just which
streets I followed when I got outside I do not know, but I
think I must have gone towards Eighth Avenue, then down
towards Twenty-third Street and across towards Fifth
Avenue. I travelled, not by sight, but instinctively. I felt
like one fleeing in a horrible nightmare.

How long and far I walked I cannot tell; but on Fifth
Avenue, under a light, I passed a cab containing a solitary
occupant, who called to me, and I recognized the voice
and face of my millionaire friend. He stopped the cab and
asked: "What on earth are you doing strolling in this part
of the town?" For answer I got into the cab and related to
him all that had happened. He reassured me by saying that
no charge of any kind could be brought against me; then
added: "But of course you don't want to be mixed up in
such an affair." He directed the driver to turn round and go
into the park, and then went on to say: "I decided last
night that I'd go to Europe tomorrow. I think I'll take you
along instead of Walter." Walter was his valet. It was settled that I should go to his apartments for the rest of the
night and sail with him in the morning.

We drove round through the park, exchanging only an
occasional word. The cool air somewhat calmed my nerves
and I lay back and closed my eyes; but still I could see
that beautiful white throat with the ugly wound. The jet
of blood pulsing from it had placed an indelible red stain
on my memory.

CHAPTER IX

I did not feel at ease until the ship was well out of New York harbour; and, notwithstanding the repeated reassurances of my millionaire friend and my own knowledge of the facts in the case, I somehow could not rid myself of the sentiment that I was, in a great degree, responsible for the "widow's" tragic end. We had brought most of the morning papers aboard with us, but my great fear of seeing my name in connexion with the killing would not permit me to read the accounts, although, in one of the papers, I did look at the picture of the victim, which did not in the least resemble her. This morbid state of mind, together with sea-sickness, kept me miserable for three or four days. At the end of that time my spirits began to revive, and I took an interest in the ship, my fellow passengers, and the voyage in general. On the second or third day out we passed several spouting whales, but I could not arouse myself to make the effort to go to the other side of the ship to see them. A little later we ran in close proximity to a large iceberg. I was curious enough to get up and look at it, and I was fully repaid for my pains. The sun was shining full upon it, and it glistened like a mammoth diamond, cut with a million facets. As we passed, it constantly changed its shape; at each different angle of vision it assumed new and astonishing forms of beauty. I watched it through a pair of glasses, seeking to verify my early conception of an iceberg—in the geographies of my grammar-school days the pictures of icebergs always included a stranded polar bear, standing desolately upon one of the snowy crags. I looked for the bear, but if he was there, he refused to put himself on exhibition.

It was not, however, until the morning that we entered the harbour of Le Havre that I was able to shake off my gloom. Then the strange sights, the chatter in an unfamiliar tongue, and the excitement of landing and passing the customs officials caused me to forget completely the events of a few days before. Indeed, I grew so lighthearted that when I caught my first sight of the train which was to take us to Paris, I enjoyed a hearty laugh. The toy-looking engine, the stuffy little compartment cars, with tiny, old-fashioned wheels, struck me as being extremely funny. But before we reached Paris my respect for our train rose considerably. I found that the "tiny" engine made remarkably fast time, and that the old-fashioned wheels ran very smoothly. I even began to appreciate the "stuffy" cars for their privacy. As I watched the passing scenery from the car window, it seemed too beautiful to be real. The bright-coloured houses against the green background impressed me as the work of some idealistic painter. Before we arrived in Paris, there was awakened in my heart a love for France which continued to grow stronger, a love which today makes that country for me the one above all others to be desired.

We rolled into the station Saint Lazare about four o'clock in the afternoon and drove immediately to the Hôtel Continental. My benefactor, humouring my curiosity and enthusiasm, which seemed to please him very much, suggested that we take a short walk before dinner. We stepped out of the hotel and turned to the right into the rue de Rivoli. When the vista of the Place de la Concorde and the Champs Élysées suddenly burst on me, I could hardly credit my own eyes. I shall attempt no such supererogatory task as a description of Paris. I wish only to give briefly the impressions which that wonderful city made upon me. It impressed me as the perfect and perfectly beautiful city; and even after I had been there for some time, and seen not only its avenues and palaces, but its most squalid alleys and hovels, this impression was not weakened. Paris became for me a charmed spot, and whenever I have returned there, I have fallen under the spell, a spell which compels admiration for all of its manners and customs and justification of even its follies and sins.

We walked a short distance up the Champs Élysées and sat for a while in chairs along the sidewalk, watching the passing crowds on foot and in carriages. It was with reluctance that I went back to the hotel for dinner. After dinner we went to one of the summer theatres, and after the performance my friend took me to a large café on one of the Grands Boulevards. Here it was that I had my first glimpse of the French life of popular literature, so different from real French life. There were several hundred people, men and women, in the place drinking, smoking, talking, and listening to the music. My millionaire friend and I took seats at a table, where we sat smoking and watching the crowd. It was not long before we were joined by two or three good-looking, well-dressed young women. My friend talked to them in French and bought drinks for the whole party. I tried to recall my high-school French, but the effort availed me little. I could stammer out a few phrases, but, very naturally, could not understand a word that was said to me. We stayed at the café a couple of hours, then went back to the hotel. The next day we spent several hours in the shops and at the tailor's. I had no clothes except what I had been able to gather together at my benefactor's apartments the night before we sailed. He bought me the same kind of clothes which he himself wore, and that was the best; and he treated me in every way as he dressed me, as an equal, not as a servant. In fact, I don't think anyone could have guessed that such a relationship existed. My duties were light and few, and he was a man full of life and vigour, who rather enjoyed doing things for himself. He kept me supplied with money far beyond what ordinary wages would have amounted to. For the first two weeks we were together almost constantly, seeing the sights, sights old to him, but from which he seemed to get new pleasure in showing them to me. During the day we took in the places of interest, and at night the theatres and cafés. This sort of life appealed to me as ideal, and I asked him one day how long he intended to stay in Paris. He answered: "Oh, until I get tired of it." I could not understand how that could ever happen. As it was, including several short trips to the Mediterranean, to Spain, to Brussels, and to Ostend, we did remain there fourteen or fifteen months. We stayed at the Hôtel Continental about two months of this time. Then my millionaire took apartments, hired a piano, and lived almost the same life he lived in New York. He entertained a great deal, some of the parties being a good deal more blasé than the New York ones. I played for the guests at all of them with an effect which to relate would be but a tiresome repetition to the reader. I played not only for the guests, but continued, as I used to do in New York, to play often for the host when he was alone. This man of the world, who grew weary of everything and was always searching for something new, appeared never to grow tired of my music; he seemed to take it as a drug. He fell into a habit which caused me no little annoyance; sometimes he would come in during the early hours of the morning and, finding me in bed asleep, would wake me up and ask me to play something. This, so far as I can remember, was my only hardship during my whole stay with him in Europe.

After the first few weeks spent in sight-seeing I had a great deal of time left to myself; my friend was often I did not know where. When not with him, I spent the day nosing about all the curious nooks and corners of Paris; of this I never grew tired. At night I usually went to some theatre, but always ended up at the big café on the Grands Boulevards. I wish the reader to know that it was not alone the gaiety which drew me there; aside from that I had a laudable purpose. I had purchased an English-French conversational dictionary, and I went there every night to take a language lesson. I used to get three or four of the young women who frequented the place at a table and buy beer and cigarettes for them. In return I received my lesson. I got more than my money's worth, for they actually compelled me to speak the language. This, together with reading the papers every day, enabled me within a few months to express myself fairly well, and, before I left Paris, to have more than an ordinary command of French. Of course, every person who goes to Paris could not dare to learn French in this manner, but I can think of no easier or quicker way of doing it. The acquiring of another foreign language awoke me to the fact that with a little effort I could secure an added accomplishment as fine and as valuable as music; so I determined to make myself as much of a linguist as possible. I bought a Spanish newspaper every day in order to freshen my memory of that language, and, for French, devised what was, so far as I knew, an original system of study. I compiled a list which I termed "Three hundred necessary words." These I thoroughly committed to memory, also the conjugation of the verbs which were included in the list. I studied these words over and over, much as children of a couple of generations ago studied the alphabet. I also practised a set of phrases like the following: "How!" "What did you say?" "What does the

word——mean?" "I understand all you say except——."
"Please repeat." "What do you call——?" "How do you
say——?" These I called my working sentences. In an
astonishingly short time I reached the point where the
language taught itself—where I learned to speak merely
by speaking. This point is the place which students taught
foreign languages in our schools and colleges find great
difficulty in reaching. I think the main trouble is that
they learn too much of a language at a time. A French
child with a vocabulary of two hundred words can express more spoken ideas than a student of French can
with a knowledge of two thousand. A small vocabulary,
the smaller the better, which embraces the common, every-
day-used ideas, thoroughly mastered, is the key to a
language. When that much is acquired the vocabulary can
be increased simply by talking. And it is easy. Who cannot
commit three hundred words to memory? Later I tried my
method, if I may so term it, with German, and found
that it worked in the same way.

I spent a good many evenings at the Opéra. The music
there made me strangely reminiscent of my life in Connecticut; it was an atmosphere in which I caught a fresh
breath of my boyhood days and early youth. Generally,
in the morning after I had attended a performance, I
would sit at the piano and for a couple of hours play
the music which I used to play in my mother's little
parlour.

One night I went to hear *Faust*. I got into my seat just
as the lights went down for the first act. At the end of the
act I noticed that my neighbour on the left was a young
girl. I cannot describe her either as to feature, or colour
of her hair, or of her eyes; she was so young, so fair, so
ethereal, that I felt to stare at her would be a violation;
yet I was distinctly conscious of her beauty. During the
intermission she spoke English in a low voice to a gentleman and a lady who sat in the seats to her left, addressing them as father and mother. I held my program
as though studying it, but listened to catch every sound
of her voice. Her observations on the performance and
the audience were so fresh and naïve as to be almost
amusing. I gathered that she was just out of school, and
that this was her first trip to Paris. I occasionally stole
a glance at her, and each time I did so my heart leaped
into my throat. Once I glanced beyond to the gentleman

who sat next to her. My glance immediately turned into a stare. Yes, there he was, unmistakably, my father! looking hardly a day older than when I had seen him some ten years before. What a strange coincidence! What should I say to him? What would he say to me? Before I had recovered from my first surprise, there came another shock in the realization that the beautiful, tender girl at my side was my sister. Then all the springs of affection in my heart, stopped since my mother's death, burst out in fresh and terrible torrents, and I could have fallen at her feet and worshipped her. They were singing the second act, but I did not hear the music. Slowly the desolate loneliness of my position became clear to me. I knew that I could not speak, but I would have given a part of my life to touch her hand with mine and call her "sister." I sat through the opera until I could stand it no longer. I felt that I was suffocating. Valentine's love seemed like mockery, and I felt an almost uncontrollable impulse to rise up and scream to the audience: "Here, here in your very midst, is a tragedy, a real tragedy!" This impulse grew so strong that I became afraid of myself, and in the darkness of one of the scenes I stumbled out of the theatre. I walked aimlessly about for an hour or so, my feelings divided between a desire to weep and a desire to curse. I finally took a cab and went from café to café, and for one of the very few times in my life drank myself into a stupor.

It was unwelcome news for me when my benefactor— I could not think of him as employer—informed me that he was at last tired of Paris. This news gave me, I think, a passing doubt as to his sanity. I had enjoyed life in Paris, and, taking all things into consideration, enjoyed it wholesomely. One thing which greatly contributed to my enjoyment was the fact that I was an American. Americans are immensely popular in Paris; and this is not due solely to the fact that they spend lots of money there, for they spend just as much or more in London, and in the latter city they are merely tolerated because they do spend The Londoner seems to think that Americans are people whose only claim to be classed as civilized is that they have money, and the regrettable thing about that is that the money is not English. But the French are more logical and freer from prejudices than the British; so the difference of attitude is easily explained. Only once in Paris did I have cause to blush for my American citizenship. I had become quite friendly with a young man from Luxemburg whom I had met at the big café. He was a stolid, slow-witted fellow, but, as we say, with a heart of gold. He and I grew attached to each other and were together frequently. He was a great admirer of the United States and never grew tired of talking to me about the country and asking for information. It was his intention to try his fortune there some day. One night he asked me in a tone of voice which indicated that he expected an authoritative denial of an ugly rumour: "Did they really burn a man alive in the United States?" I never knew what I stammered out to him as an answer. I should have felt relieved if I could even have said to him: "Well, only one."

When we arrived in London, my sadness at leaving Paris was turned into despair. After my long stay in the French capital, huge, ponderous, massive London seemed to me as ugly a thing as man could contrive to make. I thought of Paris as a beauty-spot on the face of the earth, and of London as a big freckle. But soon London's massiveness, I might say its very ugliness, began to impress me. I began to experience that sense of grandeur which one feels when he looks at a great mountain or a mighty river. Beside London Paris becomes a toy, a pretty plaything. And I must own that before I left the world's metropolis I discovered much there that was beautiful. The beauty in and about London is entirely different from that in and about Paris; and I could not but admit that the beauty of the French city seemed hand-made, artificial, as though set up for the photographer's camera, everything nicely adjusted so as not to spoil the picture; while that of the English city was rugged, natural, and fresh.

How these two cities typify the two peoples who built them! Even the sound of their names expresses a certain racial difference. Paris is the concrete expression of the gaiety, regard for symmetry, love of art, and, I might well add, of the morality of the French people. London stands for the conservatism, the solidarity, the utilitarianism, and, I might well add, the hypocrisy of the Anglo-Saxon. It may sound odd to speak of the morality of the French, if not of the hypocrisy of the English; but this seeming paradox impresses me as a deep truth.

I saw many things in Paris which were immoral according to English standards, but the absence of hypocrisy, the absence of the spirit to do the thing if it might only be done in secret, robbed these very immoralities of the the damning influence of the same evils in London. I have walked along the terrace cafés of Paris and seen hundreds of men and women sipping their wine and beer, without observing a sign of drunkenness. As they drank, they chatted and laughed and watched the passing crowds; the drinking seemed to be a secondary thing. This I have witnessed, not only in the cafés along the Grands Boulevards, but in the out-of-the-way places patronized by the working-classes. In London I have seen in the "pubs" men and women crowded in stuffy little compartments, drinking seemingly only for the pleasure of swallowing as much as they could hold. I have seen there women from eighteen to eighty, some in tatters, and some clutching babes in their arms, drinking the heavy English ales and whiskies served to them by women. In the whole scene, not one ray of brightness, not one flash of gaiety, only maudlin joviality or grim despair. And I have thought, if some men and women will drink—and it is certain that some will—is it not better that they do so under the open sky, in the fresh air, than huddled together in some close, smoky room? There is a sort of frankness about the evils of Paris which robs them of much of the seductiveness ot things forbidden, and with that frankness goes a certain cleanliness of thought belonging to things not hidden. London will do whatever Paris does, provided exterior morals are not shocked. As a result, Paris has the appearance only of being the more immoral city. The difference may be summed up in this: Paris practises its sins as lightly as it does its religion, while London practises both very seriously.

I should not neglect to mention what impressed me most forcibly during my stay in London. It was not St. Paul's nor the British Museum nor Westminster Abbey. It was nothing more or less than the simple phrase "Thank you," or sometimes more elaborated, "Thank you very kindly, sir." I was continually surprised by the varied uses to which it was put; and, strange to say, its use as an expression of politeness seemed more limited than any other. One night I was in a cheap music-hall and accidentally bumped into a waiter who was carrying a tray-

load of beer, almost bringing him to several shillings'
worth of grief. To my amazement he righted himself
and said: "Thank ye, sir," and left me wondering whether
he meant that he thanked me for not completely spilling
his beer, or that he would thank me for keeping out
of his way.

I also found cause to wonder upon what ground the
English accuse Americans of corrupting the language by
introducing slang words. I think I heard more and more
different kinds of slang during my few weeks' stay in
London than in my whole "tenderloin" life in New York.
But I suppose the English feel that the language is theirs,
and that they may do with it as they please without at the
same time allowing that privilege to others.

My millionaire was not so long in growing tired of
London as of Paris. After a stay of six or eight weeks we
went across into Holland. Amsterdam was a great surprise to me. I had always thought of Venice as the city
of canals; it had never entered my mind that I should
find similar conditions in a Dutch town. I don't suppose
the comparison goes far beyond the fact that there are
canals in both cities—I have never seen Venice—but Amsterdam struck me as being extremely picturesque. From
Holland we went to Germany, where we spent five or
six months, most of the time in Berlin. I found Berlin
more to my taste than London, and occasionally I had
to admit that in some things it was superior to Paris.

In Berlin I especially enjoyed the orchestral concerts,
and I attended a large number of them. I formed the
acquaintance of a good many musicians, several of whom
spoke of my playing in high terms. It was in Berlin that
my inspiration was renewed. One night my millionaire
entertained a party of men composed of artists, musicians,
writers, and, for aught I know, a count or two. They
drank and smoked a great deal, talked art and music,
and discussed, it seemed to me, everything that ever entered man's mind. I could only follow the general drift
of what they were saying. When they discussed music, it
was more interesting to me; for then some fellow would
run excitedly to the piano and give a demonstration of
his opinions, and another would follow quickly, doing the
same. In this way, I learned that, regardless of what his
specialty might be, every man in the party was a musician.
I was at the same time impressed with the falsity of the

general idea that Frenchmen are excitable and emotional, and that Germans are calm and phlegmatic. Frenchmen are merely gay and never overwhelmed by their emotions. When they talk loud and fast, it is merely talk, while Germans get worked up and red in the face when sustaining an opinion, and in heated discussions are likely to allow their emotions to sweep them off their feet.

My millionaire planned, in the midst of the discussion on music, to have me play the "new American music" and astonish everybody present. The result was that I was more astonished than anyone else. I went to the piano and played the most intricate rag-time piece I knew. Before there was time for anybody to express an opinion on what I had done, a big bespectacled, bushy-headed man rushed over, and, shoving me out of the chair, exclaimed: "Get up! Get up!" He seated himself at the piano, and, taking the theme of my rag-time, played it through first in straight chords; then varied and developed it through every known musical form. I sat amazed. I had been turning classic music into rag-time, a comparatively easy task; and this man had taken rag-time and made it classic. The thought came across me like a flash—It can be done, why can't I do it? From that moment my mind was made up. I clearly saw the way of carrying out the ambition I had formed when a boy.

I now lost interest in our trip. I thought: "Here I am a man, no longer a boy, and what am I doing but wasting my time and abusing my talent? What use am I making of my gifts? What future have I before me following my present course?" These thoughts made me feel remorseful and put me in a fever to get to work, to begin to do something. Of course I know now that I was not wasting time; that there was nothing I could have done at that age which would have benefited me more than going to Europe as I did. The desire to begin work grew stronger each day. I could think of nothing else. I made up my mind to go back into the very heart of the South, to live among the people, and drink in my inspiration firsthand. I gloated over the immense amount of material I had to work with, not only modern rag-time, but also the old slave songs—material which no one had yet touched.

The more decided and anxious I became to return to the United States, the more I dreaded the ordeal of breaking with my millionaire. Between this peculiar man and

me there had grown a very strong bond of affection,
backed up by a debt which each owed to the other. He
had taken me from a terrible life in New York and, by
giving me the opportunity of travelling and of coming in
contact with the people with whom he associated, had
made me a polished man of the world. On the other hand,
I was his chief means of disposing of the thing which
seemed to sum up all in life that he dreaded—time. As
I remember him now, I can see that time was what he
was always endeavouring to escape, to bridge over, to
blot out; and it is not strange that some years later he did
escape it for ever, by leaping into eternity.

For some weeks I waited for just the right moment in
which to tell my patron of my decision. Those weeks
were a trying time to me. I felt that I was playing the
part of a traitor to my best friend. At length, one day he
said to me: "Well, get ready for a long trip; we are going
to Egypt, and then to Japan." The temptation was for an
instant almost overwhelming, but I summoned determination enough to say: "I don't think I want to go."
"What!" he exclaimed, "you want to go back to your
dear Paris? You still think that the only spot on earth?
Wait until you see Cairo and Tokio, you may change
your mind." "No," I stammered, "it is not because I
want to go back to Paris. I want to go back to the
United States." He wished to know my reason, and I
told him, as best I could, my dreams, my ambition, and my
decision. While I was talking, he watched me with a
curious, almost cynical, smile growing on his lips. When
I had finished he put his hand on my shoulder—this was
the first physical expression of tender regard he had ever
shown me—and looking at me in a big-brotherly way,
said: "My boy, you are by blood, by appearance, by
education, and by tastes a white man. Now, why do you
want to throw your life away amidst the poverty and
ignorance, in the hopeless struggle, of the black people
of the United States? Then look at the terrible handicap
you are placing on yourself by going home and working
as a Negro composer; you can never be able to get the
hearing for your work which it might deserve. I doubt
that even a white musician of recognized ability could
succeed there by working on the theory that American
music should be based on Negro themes. Music is a universal art; anybody's music belongs to everybody; you

can't limit it to race or country. Now, if you want to become a composer, why not stay right here in Europe? I will put you under the best teachers on the Continent. Then if you want to write music on Negro themes, why, go ahead and do it."

We talked for some time on music and the race question. On the latter subject I had never before heard him express any opinion. Between him and me no suggestion of racial differences had ever come up. I found that he was a man entirely free from prejudice, but he recognized that prejudice was a big stubborn entity which had to be taken into account. He went on to say: "This idea you have of making a Negro out of yourself is nothing more than a sentiment; and you do not realize the fearful import of what you intend to do. What kind of a Negro would you make now, especially in the South? If you had remained there, or perhaps even in your club in New York, you might have succeeded very well; but now you would be miserable. I can imagine no more dissatisfied human being than an educated, cultured, and refined coloured man in the United States. I have given more study to the race question in the United States than you may suppose, and I sympathize with the Negroes there; but what's the use? I can't right their wrongs, and neither can you; they must do that themselves. They are unfortunate in having wrongs to right, and you would be foolish to take their wrongs unnecessarily on your shoulders. Perhaps some day, through study and observation, you will come to see that evil is a force, and, like the physical and chemical forces, we cannot annihilate it; we may only change its form. We light upon one evil and hit it with all the might of our civilization, but only succeed in scattering it into a dozen other forms. We hit slavery through a great civil war. Did we destroy it? No, we only changed it into hatred between sections of the country: in the South, into political corruption and chicanery, the degradation of the blacks through peonage, unjust laws, unfair and cruel treatment; and the degradation of the whites by their resorting to these practices, the paralysation of the public conscience, and the ever overhanging dread of what the future may bring. Modern civilization hit ignorance of the masses through the means of popular education. What has it done but turn ignorance into anarchy, socialism, strikes, hatred between poor and

rich, and universal discontent? In like manner, modern
philanthropy hit at suffering and disease through asylums
and hospitals; it prolongs the sufferers' lives, it is true, but
is, at the same time, sending down strains of insanity and
weakness into future generations. My philosophy of life
is this: make yourself as happy as possible, and try to
make those happy whose lives come in touch with yours;
but to attempt to right the wrongs and ease the suffering
of the world in general is a waste of effort. You had
just as well try to bale the Atlantic by pouring the water
into the Pacific."

This tremendous flow of serious talk from a man I was
accustomed to see either gay or taciturn so surprised and
overwhelmed me that I could not frame a reply. He left
me thinking over what he had said. Whatever was the
soundness of his logic or the moral tone of his philosophy,
his argument greatly impressed me. I could see, in spite
of the absolute selfishness upon which it was based, that
there was reason and common sense in it. I began to
analyse my own motives, and found that they, too, were
very largely mixed with selfishness. Was it more a desire
to help those I considered my people, or more a desire to
distinguish myself, which was leading me back to the
United States? That is a question I have never definitely
answered.

For several weeks longer I was in a troubled state of
mind. Added to the fact that I was loath to leave my
good friend was the weight of the question he had aroused
in my mind, whether I was not making a fatal mistake. I
suffered more than one sleepless night during that time.
Finally, I settled the question on purely selfish grounds,
in accordance with my millionaire's philosophy. I argued
that music offered me a better future than anything else
I had any knowledge of, and, in opposition to my friend's
opinion, that I should have greater chances of attracting
attention as a coloured composer than as a white one.
But I must own that I also felt stirred by an unselfish
desire to voice all the joys and sorrows, the hopes and
ambitions, of the American Negro, in classic musical form.

When my mind was fully made up, I told my friend.
He asked me when I intended to start. I replied that I
would do so at once. He then asked me how much money
I had. I told him that I had saved several hundred dollars
out of sums he had given me. He gave me a cheque for

five hundred dollars, told me to write to him in care of his Paris bankers if I ever needed his help, wished me good luck, and bade me good-bye. All this he did almost coldly; and I often wondered whether he was in a hurry to get rid of what he considered a fool, or whether he was striving to hide deeper feelings.

And so I separated from the man who was, all in all, the best friend I ever had, except my mother, the man who exerted the greatest influence ever brought into my life, except that exerted by my mother. My affection for him was so strong, my recollections of him are so distinct, he was such a peculiar and striking character, that I could easily fill several chapters with reminiscences of him; but for fear of tiring the reader I shall go on with my narration.

I decided to go to Liverpool and take ship for Boston. I still had an uneasy feeling about returning to New York; and in a few days I found myself aboard ship headed for home.

CHAPTER X

Among the first of my fellow-passengers of whom I took any particular notice was a tall, broad-shouldered, almost gigantic, coloured man. His dark-brown face was clean-shaven; he was well-dressed and bore a decidedly distinguished air. In fact, if he was not handsome, he at least compelled admiration for his fine physical proportions. He attracted general attention as he strode the deck in a sort of majestic loneliness. I became curious to know who he was and determined to strike up an acquaintance with him at the first opportune moment. The chance came a day or two later. He was sitting in the smoking-room, with a cigar, which had gone out, in his mouth, reading a novel. I sat down beside him and, offering him a fresh cigar, said: "You don't mind my telling you something unpleasant, do you?" He looked at me with a smile, accepted the proffered cigar, and replied in a voice which comported perfectly with his size and appearance: "I think my curiosity overcomes any objections I might have." "Well," I said, "have you noticed

that the man who sat at your right in the saloon during the
first meal has not sat there since?" He frowned slightly
without answering my question. "Well," I continued, "he
asked the steward to remove him; and not only that, he
attempted to persuade a number of the passengers to protest
against your presence in the dining-saloon." The big man
at my side took a long draw from his cigar, threw his head
back, and slowly blew a great cloud of smoke toward
the ceiling. Then turning to me he said: "Do you know, I
don't object to anyone's having prejudices so long as those
prejudices don't interfere with my personal liberty. Now,
the man you are speaking of had a perfect right to
change his seat if I in any way interfered with his appetite or his digestion. I should have no reason to complain if he removed to the farthest corner of the saloon, or
even if he got off the ship; but when his prejudice attempts to move *me* one foot, one inch, out of the place
where I am comfortably located, then I object." On the
word "object" he brought his great fist down on the table in
front of us with such a crash that everyone in the room
turned to look. We both covered up the slight embarrassment with a laugh and strolled out on the deck.

We walked the deck for an hour or more, discussing different phases of the Negro question. In referring to the race
I used the personal pronoun "we"; my companion made no
comment about it, nor evinced any surprise, except to
raise his eyebrows slightly the first time he caught the significance of the word. He was the broadest-minded coloured
man I have ever talked with on the Negro question. He even
went so far as to sympathize with and offer excuses for
some white Southern points of view. I asked him what were
his main reasons for being so hopeful. He replied: "In spite
of all that is written, said, and done, this great, big, incon-
trovertible fact stands out—the Negro is progressing, and
that disproves all the arguments in the world that he is in-
capable of progress. I was born in slavery, and at emancipation was set adrift a ragged, penniless bit of humanity. I
have seen the Negro in every grade, and I know what I am
talking about. Our detractors point to the increase of crime
as evidence against us; certainly we have progressed in
crime as in other things; what less could be expected? And
yet, in this respect, we are far from the point which has been
reached by the more highly civilized white race. As we
continue to progress, crime among us will gradually lose

much of its brutal, vulgar, I might say healthy, aspect, and become more delicate, refined, and subtle. Then it will be less shocking and noticeable, although more dangerous to society." Then dropping his tone of irony, he continued with some show of eloquence: "But, above all, when I am discouraged and disheartened, I have this to fall back on: if there is a principle of right in the world, which finally prevails, and I believe that there is; if there is a merciful but justice-loving God in heaven, and I believe that there is, we shall win; for we have right on our side, while those who oppose us can defend themselves by nothing in the moral law, nor even by anything in the enlightened thought of the present age."

For several days, together with other topics, we discussed the race problem, not only of the United States, but as it affected native Africans and Jews. Finally, before we reached Boston, our conversation had grown familiar and personal. I had told him something of my past and much about my intentions for the future. I learned that he was a physician, a graduate of Howard University, Washington, and had done post-graduate work in Philadelphia; and this was his second trip abroad to attend professional courses. He had practised for some years in the city of Washington, and though he did not say so, I gathered that his practice was a lucrative one. Before we left the ship, he had made me promise that I would stop two or three days in Washington before going on south.

We put up at a hotel in Boston for a couple of days and visited several of my new friend's acquaintances; they were all people of education and culture and, apparently, of means. I could not help being struck by the great difference between them and the same class of coloured people in the South. In speech and thought they were genuine Yankees. The difference was especially noticeable in their speech. There was none of that heavy-tongued enunciation which characterizes even the best-educated coloured people of the South. It is remarkable, after all, what an adaptable creature the Negro is. I have seen the black West Indian gentleman in London, and he is in speech and manners a perfect Englishman. I have seen natives of Haiti and Martinique in Paris, and they are more Frenchy than a Frenchman. I have no doubt that the Negro would make a good Chinaman, with exception of the pigtail.

My stay in Washington, instead of being two or three

days, was two or three weeks. This was my first visit to the national capital, and I was, of course, interested in seeing the public buildings and something of the working of the government; but most of my time I spent with the doctor among his friends and acquaintances. The social phase of life among coloured people is more developed in Washington than in any other city in the country. This is on account of the large number of individuals earning good salaries and having a reasonable amount of leisure time to draw from. There are dozens of physicians and lawyers, scores of school-teachers, and hundreds of clerks in the departments. As to the coloured department clerks, I think it fair to say that in educational equipment they average above the white clerks of the same grade; for, whereas a coloured college-graduate will seek such a job, the white university-man goes into one of the many higher vocations which are open to him.

In a previous chapter I spoke of social life among coloured people; so there is no need to take it up again here. But there is one thing I did not mention: among Negroes themselves there is the peculiar inconsistency of a colour question. Its existence is rarely admitted and hardly ever mentioned; it may not be too strong a statement to say that the greater portion of the race is unconscious of its influence; yet this influence, though silent, is constant. It is evidenced most plainly in marriage selection; thus the black men generally marry women fairer than themselves; while, on the other hand, the dark women of stronger mental endowment are very often married to light-complexioned men; the effect is a tendency toward lighter complexions, especially among the more active elements in the race. Some might claim that this is a tacit admission of coloured people among themselves of their own inferiority judged by the colour line. I do not think so. What I have termed an inconsistency is, after all, most natural; it is, in fact, a tendency in accordance with what might be called an economic necessity. So far as racial differences go, the United States puts a greater premium on colour, or, better, lack of colour, than upon anything else in the world. To paraphrase, "Have a white skin, and all things else may be added unto you." I have seen advertisements in newspapers for waiters, bell-boys, or elevator men, which read: "Light coloured man wanted." It is this tremendous pressure which the sentiment of the country exerts that is operating on the race. There is involved not only the question of higher opportunity, but often the question of earning a livelihood; and so I say it is not strange, but a natural tendency. Nor is it any more a sacrifice of self-respect that a black man should give to his children every advantage he can which complexion of the skin carries than that the new or vulgar rich should purchase for their children the advantages which ancestry, aristocracy, and social position carry. I once heard a coloured man sum it up in these words: "It's no disgrace to be black, but it's often very inconvenient."

Washington shows the Negro not only at his best, but also at his worse. As I drove round with the doctor, he commented rather harshly on those of the latter class which we saw. He remarked: "You see those lazy, loafing, good-for-nothing darkies; they're not worth digging graves for; yet they are the ones who create impressions of the race for the casual observer. It's because they are always in evidence on the street corners, while the rest of us are hard at work, and you know a dozen loafing darkies make a bigger crowd and a worse impression in this country than fifty white men of the same class. But they ought not to represent the race. We are the race, and the race ought to be judged by us, not by them. Every race and every nation should be judged by the best it has been able to produce, not by the worst."

The recollection of my stay in Washington is a pleasure to me now. In company with the doctor I visited Howard University, the public schools, the excellent coloured hospital, with which he was in some way connected, if I remember correctly, and many comfortable and even elegant homes. It was with some reluctance that I continued my journey south. The doctor was very kind in giving me letters to people in Richmond and Nashville when I told him that I intended to stop in both of these cities. In Richmond a man who was then editing a very creditable coloured newspaper gave me a great deal of his time and made my stay there of three or four days very pleasant. In Nashville I spent a whole day at Fisk University, the home of the "Jubilee Singers," and was more than repaid for my time. Among my letters of introduction was one to a very prosperous physician. He drove me about the city and introduced me to a number of people. From Nashville I went to Atlanta, where I stayed long enough to gratify an old desire to see Atlanta University again. I then continued my journey to Macon.

During the trip from Nashville to Atlanta I went into the smoking-compartment of the car to smoke a cigar. I was travelling in a Pullman, not because of an abundance of funds, but because through my experience with my millionaire a certain amount of comfort and luxury had become a necessity to me whenever it was obtainable. When I entered the car, I found only a couple of men there; but in a half-hour there were half a dozen or more. From the general conversation I learned that a fat Jewish-looking man was a cigar-manufacturer, and was experimenting in growing Havana tobacco in Florida; that a slender bespectacled young man was from Ohio and a professor in some State institution in Alabama; that a white-moustached, well-dressed man was an old Union soldier who had fought through the Civil War; and that a tall, raw-boned, red-faced man, who seemed bent on leaving nobody in ignorance of the fact that he was from Texas, was a cotton-planter.

In the North men may ride together for hours in a "smoker" and unless they are acquainted with each other never exchange a word; in the South men thrown together in such manner are friends in fifteen minutes. There is always present a warm-hearted cordiality which will melt down the most frigid reserve. It may be because Southerners are very much like Frenchmen in that they must talk; and not only must they talk, but they must express their opinions.

The talk in the car was for a while miscellaneous—on the weather, crops, business prospects; the old Union soldier had invested capital in Atlanta, and he predicted that that city would soon be one of the greatest in the country. Finally the conversation drifted to politics; then, as a natural sequence, turned upon the Negro question.

In the discussion of the race question the diplomacy of the Jew was something to be admired; he had the faculty of agreeing with everybody without losing his allegiance to any side. He knew that to sanction Negro oppression would be to sanction Jewish oppression and would expose him to a shot along that line from the old soldier, who stood firmly on the ground of equal rights and opportunity to all men; long traditions and business instincts told him when in Rome to act as a Roman. Altogether his position was a delicate one, and I gave him credit for the skill he displayed in maintaining it. The young professor was apologetic. He had had the same views as the G. A. R. man; but a year in the South had opened his eyes, and he had to confess that the problem could hardly be handled any better than it was being handled by the Southern whites. To which the G. A. R. man responded somewhat rudely that he had spent ten times as many years in the South as his young friend and that he could easily understand how holding a position in a State institution in Alabama would bring about a change of views. The professor turned very red and had very little more to say. The Texan was fierce, eloquent, and profane in his argument, and, in a lower sense, there was a direct logic in what he said, which was convincing; it was only by taking higher ground, by dealing in what Southerners call "theories," that he could be combated. Occasionally some one of the several other men in the "smoker" would throw in a remark to reinforce what he said, but he really didn't need any help; he was sufficient in himself.

In the course of a short time the controversy narrowed itself down to an argument between the old soldier and the Texan. The latter maintained hotly that the Civil War was a criminal mistake on the part of the North and that the humiliation which the South suffered during Reconstruction could never be forgotten. The Union man retorted just as hotly that the South was responsible for the war and that the spirit of unforgetfulness on its part was the greatest cause of present friction; that it seemed to be the one great aim of the South to convince the North that the latter made a mistake in fighting to preserve the Union and liberate the slaves. "Can you imagine," he went on to say, "what would have been the condition of things eventually if there had been no war, and the South had been allowed to follow its course? Instead of one great, prosperous country with nothing before it but the conquests of peace, a score of petty republics, as in Central and South America, wasting their energies in war with each other or in revolutions."

"Well," replied the Texan, "anything—no country at all—is better than having niggers over you. But anyhow, the war was fought and the niggers were freed; for it's no use beating around the bush, the niggers, and not the Union, was the cause of it; and now do you believe that all the niggers on earth are worth the good white blood that was spilt? You freed the nigger and you gave him the ball but you couldn't make a citizen out of him. He don't know what he's voting for, and we buy 'em like so many hogs. You're giving 'em education, but that only makes slick rascals out of 'em."

"Don't fancy for a moment," said the Northern man, "that you have any monopoly in buying ignorant votes. The same thing is done on a larger scale in New York and Boston, and in Chicago and San Francisco; and they are not black votes either. As to education's making the Negro worse, you might just as well tell me that religion does the same thing. And, by the way, how many educated coloured men do you know personally?"

The Texan admitted that he knew only one, and added that he was in the penitentiary. "But," he said, "do you mean to claim, ballot or no ballot, education or no education, that niggers are the equals of white men?"

"That's not the question," answered the other, "but if the Negro is so distinctly inferior, it is a strange thing to me that it takes such tremendous effort on the part of the white man to make him realize it, and to keep him in the same place into which inferior men naturally fall. However, let us grant for sake of argument that the Negro is inferior in every respect to the white man; that fact only increases our moral responsibility in regard to our actions toward him. Inequalities of numbers, wealth, and power, even of intelligence and morals, should make no difference in the essential rights of men."

"If he's inferior and weaker, and is shoved to the wall, that's his own look-out," said the Texan. "That's the law of nature; and he's bound to go to the wall; for no race in the world has ever been able to stand competition with the Anglo-Saxon. The Anglo-Saxon race has always been and always will be the masters of the world, and the niggers in the South ain't going to change all the records of history."

"My friend," said the old soldier slowly, "if you have studied history, will you tell me, as confidentially between white men, what the Anglo-Saxon has ever done?"

The Texan was too much astonished by the question to venture any reply.

His opponent continued: "Can you name a single one of the great fundamental and original intellectual achievements which have raised man in the scale of civilization that may be credited to the Anglo-Saxon? The art of letters, of poetry, of music, of sculpture, of painting, of the drama, of architecture; the science of mathematics, of astronomy of philosophy, of logic, of physics, of chemistry, the use of the metals, and the principles of mechanics, were all invented or discovered by darker and what we now call inferior races and nations. We have carried many of these to their highest point of perfection, but the foundation was laid by others. Do you know the only original contribution to civilization we can claim is what we have done in steam and electricity and in making implements of war more deadly? And there we worked largely on principles which we did not discover. Why, we didn't even originate the religion we use. We are a great race, the greatest in the world today, but we ought to remember that we are standing on a pile of past races, and enjoy our position with a little less show of arrogance. We are simply having our turn at the game, and we were a long time getting to it. After all, racial supremacy is merely a matter of dates in history. The man here who belongs to what is, all in all, the greatest race the world ever produced, is almost ashamed to own it. If the Anglo-Saxton is the source of everything good and great in the human race from the beginning, why wasn't the German forest the birthplace of civilization, rather than the valley of the Nile?"

The Texan was somewhat disconcerted, for the argument had passed a little beyond his limits, but he swung it back to where he was sure of his ground by saying: "All that may be true, but it hasn't got much to do with us and the niggers here in the South. We've got 'em here, and we got 'em to live with, and it's a question of white man or nigger, no middle ground. You want us to treat niggers as equals. Do you want to see 'em sitting around in our parlours? Do you want to see a mulatto South? To bring it right home to you, would you let your daughter marry a nigger?"

"No, I wouldn't consent to my daughter's marrying a nigger, but that doesn't prevent my treating a black man fairly. And I don't see what fair treatment has to do with niggers sitting round in your parlours; they can't come there unless they're invited. Out of all the white men I know, only a hundred or so have the privilege of sitting

round in my parlour. As to the mulatto South, if you
Southerners have one boast that is stronger than another,
it is your women; you put them on a pinnacle of purity and
virtue and bow down in a chivalric worship before them;
yet you talk and act as though, should you treat the Negro
fairly and take the anti-intermarriage laws off your statute
books, these same women would rush into the arms of
black lovers and husbands. It's a wonder to me that they
don't rise up and resent the insult."

"Colonel," said the Texan, as he reached into his hand-
bag and brought out a large flask of whisky, "you might
argue from now until hell freezes over, and you might convince me that you're right, but you'll never convince me
that I'm wrong. All you say sounds very good, but it's got
nothing to do with facts. You can say what men ought to
be, but they ain't that; so there you are. Down here in the
South we're up against facts, and we're meeting 'em like
facts. We don't believe the nigger is or ever will be the
equal of the white man, and we ain't going to treat him
as an equal; I'll be damned if we will. Have a drink." Everybody except the professor partook of the generous Texan's
flask, and the argument closed in a general laugh and good
feeling.

I went back into the main part of the car with the conversation on my mind. Here I had before me the bald, raw,
naked aspects of the race question in the South; and, in
consideration of the step I was just taking, it was far from
encouraging. The sentiments of the Texan—and he expressed the sentiments of the South—fell upon me like a
chill. I was sick at heart. Yet I must confess that underneath it all I felt a certain sort of admiration for the man
who could not be swayed from what he held as his principles. Contrasted with him, the young Ohio professor
was indeed a pitiable character. And all along, in spite of
myself, I have been compelled to accord the same kind
of admiration to the Southern white man for the manner
in which he defends not only his virtues, but his vices.
He knows that, judged by a high standard, he is narrow
and prejudiced, that he is guilty of unfairness, oppression,
and cruelty, but this he defends as stoutly as he would his
better qualities. This same spirit obtains in a great degree
among the blacks; they, too, defend their faults and fail-
ings. This they generally do whenever white people are
concerned. And yet among themselves they are their own

most merciless critics. I have never heard the race so terribly arraigned as I have by coloured speakers to strictly coloured audiences. It is the spirit of the South to defend everything belonging to it. The North is too cosmopolitan and tolerant for such a spirit. If you should say to an Easterner that Paris is a gayer city than New York, he would be likely to agree with you, or at least to let you have your own way; but to suggest to a South Carolinian that Boston is a nicer city to live in than Charleston would be to stir his greatest depths of argument and eloquence.

But today, as I think over that smoking-car argument, I can see it in a different light. The Texan's position does not render things so hopeless, for it indicates that the main difficulty of the race question does not lie so much in the actual condition of the blacks as it does in the mental attitude of the whites; and a mental attitude, especially one not based on truth, can be changed more easily than actual conditions. That is to say, the burden of the question is not that the whites are struggling to save ten million despondent and moribund people from sinking into a hopeless slough of ignorance, poverty, and barbarity in their very midst, but that they are unwilling to open certain doors of opportunity and to accord certain treatment to ten million aspiring, education-and-property-acquiring people. In a word, the difficulty of the problem is not so much due to the facts presented as to the hypothesis assumed for its solution. In this it is similar to the problem of the solar system. By a complex, confusing, and almost contradictory mathematical process, by the use of zigzags instead of straight lines, the earth can be proved to be the centre of things celestial; but by an operation so simple that it can be comprehended by a schoolboy, its position can be verified among the other worlds which revolve about the sun, and its movements harmonized with the laws of the universe. So, when the white race assumes as a hypothesis that it is the main object of creation and that all things else are merely subsidiary to its well-being, sophism, subterfuge, perversion of conscience, arrogance, injustice, oppression, cruelty, sacrifice of human blood all are required to maintain the position and its dealings with other races become indeed a problem, a problem which, if based on a hypothesis of common humanity, could be solved by the simple rules of justice.

When I reached Macon, I decided to leave my trunk and

all my surplus belongings, to pack my bag, and strike out
into the interior. This I did; and by train, by mule and ox-
cart, I travelled through many counties. This was my first
real experience among rural coloured people, and all
that I saw was interesting to me; but there was a great deal
which does not require description at my hands; for log-
cabins and plantations and dialect-speaking "darkies" are
perhaps better known in American literature than any other
single picture of our national life. Indeed, they form an
ideal and exclusive literary concept of the American Negro
to such an extent that it is almost impossible to get the
reading public to recognize him in any other setting; so I
shall endeavour to avoid giving the reader any already over-
worked and hackneyed descriptions. This generally accepted literary ideal of the American Negro constitutes
what is really an obstacle in the way of the thoughtful and
progressive element of the race. His character has been
established as a happy-go-lucky, laughing, shuffling, banjo-
picking being, and the reading public has not yet been
prevailed upon to take him seriously. His efforts to elevate
himself socially are looked upon as a sort of absurd
caricature of "white civilization." A novel dealing with
coloured people who lived in respectable homes and amidst
a fair degree of culture and who naturally acted "just
like white folks" would be taken in a comic-opera sense. In
this respect the Negro is much in the position of a great
comedian who gives up the lighter roles to play tragedy.
No matter how well he may portray the deeper passions,
the public is loath to give him up in his old character; they
even conspire to make him a failure in serious work, in
order to force him back into comedy. In the same
respect, the public is not too much to be blamed, for great
comedians are far more scarce than mediocre tragedians;
every amateur actor is a tragedian. However, this very fact
constitutes the opportunity of the future Negro novelist
and poet to give the country something new and unknown, in depicting the life, the ambitions, the struggles, and the passions of those of their race who are striving to break the narrow limits of traditions. A beginning
has already been made in that remarkable book by Dr.
DuBois, *The Souls of Black Folk*.

Much, too, that I saw while on this trip, in spite of my
enthusiasm, was disheartening. Often I thought of what my
millionaire had said to me, and wished myself back in

Europe. The houses in which I had to stay were generally uncomfortable, sometimes worse. I often had to sleep in a division or compartment with several other people. Once or twice I was not so fortunate as to find divisions; everybody slept on pallets on the floor. Frequently I was able to lie down and contemplate the stars which were in their zenith. The food was at times so distasteful and poorly cooked that I could not eat it. I remember that once I lived for a week or more on buttermilk, on account of not being able to stomach the fat bacon, the rank turnip-tops, and the heavy damp mixture of meal, salt, and water which was called corn bread. It was only my ambition to do the work which I had planned that kept me steadfast to my purpose. Occasionally I would meet with some signs of progress and uplift in even one of these backwood settlements—houses built of boards, with windows, and divided into rooms; decent food, and a fair standard of living. This condition was due to the fact that there was in the community some exceptionally capable Negro farmer whose thrift served as an example. As I went about among these dull, simple people—the great majority of them hard working, in their relations with the whites submissive, faithful, and often affectionate, negatively content with their lot—and contrasted them with those of the race who had been quickened by the forces of thought, I could not but appreciate the logic of the position held by those Southern leaders who have been bold enough to proclaim against the education of the Negro. They are consistent in their public speech with Southern sentiment and desires. Those public men of the South who have not been daring or heedless enough to defy the ideals of twentieth-century civilization and of modern humanitarianism and philanthropy, find themselves in the embarrassing situation of preaching one thing and praying for another. They are in the position of the fashionable woman who is compelled by the laws of polite society to say to her dearest enemy: "How happy I am to see you!"

And yet in this respect how perplexing is Southern character; for, in opposition to the above, it may be said that the claim of the Southern whites that they love the Negro better than the Northern whites do is in a manner true. Northern white people love the Negro in a sort of abstract way, as a race; through a sense of justice, charity, and philanthropy, they will liberally assist in his elevation. A number of them have heroically spent their lives in this effort (and just here I wish to say that when the coloured people reach the monument-building stage, they should not forget the men and women who went South after the war and founded schools for them). Yet, generally speaking, they have no particular liking for individuals of the race. Southern white people despise the Negro as a race, and will do nothing to aid in his elevation as such; but for certain individuals they have a strong affection, and are helpful to them in many ways. With these individual members of the race they live on terms of the greatest intimacy; they entrust to them their children, their family treasures, and their family secrets; in trouble they often go to them for comfort and counsel; in sickness they often rely upon their care. This affectionate relation between the Southern whites and those blacks who come into close touch with them has not been overdrawn even in fiction.

This perplexity of Southern character extends even to the intermixture of the races. That is spoken of as though it were dreaded worse than smallpox, leprosy, or the plague. Yet, when I was in Jacksonville, I knew several prominent families there with large coloured branches, which went by the same name and were known and acknowledged as blood relatives. And what is more, there seemed to exist between these black brothers and sisters and uncles and aunts a decidedly friendly feeling.

I said above that Southern whites would do nothing for the Negro as a race. I know the South claims that it has spent millions for the education of the blacks, and that it has of its own free will shouldered this awful burden. It seems to be forgetful of the fact that these millions have been taken from the public tax funds for education, and that the law of political economy which recognizes the land-owner as the one who really pays the taxes is not tenable. It would be just as reasonable for the relatively few land-owners of Manhattan to complain that they had to stand the financial burden of the education of the thousands and thousands of children whose parents pay rent for tenements and flats. Let the millions of producing and consuming Negroes be taken out of the South, and it would be quickly seen how much less of public funds there would be to appropriate for education or any other purpose.

In thus travelling about through the country I was sometimes amused on arriving at some little railroad-station town to be taken for and treated as a white man, and six hours later, when it was learned that I was stopping at the house of the coloured preacher or school-teacher, to note the attitude of the whole town change. At times this led even to embarrassment. Yet it cannot be so embarrassing for a coloured man to be taken for white as for a white man to be taken for coloured; and I have heard of several cases of the latter kind.

All this while I was gathering material for work, jotting down in my note-book themes and melodies, and trying to catch the spirit of the Negro in his relatively primitive state. I began to feel the necessity of hurrying so that I might get back to some city like Nashville to begin my compositions and at the same time earn at least a living by teaching and performing before my funds gave out. At the last settlement in which I stopped I found a mine of material. This was due to the fact that "big meeting" was in progress. "Big meeting" is an institution something like camp-meeting, the difference being that it is held in a permanent church, and not in a temporary structure. All the churches of some one denomination—of course, either Methodist or Baptist—in a county, or, perhaps, in several adjoining counties, are closed, and the congregations unite at some centrally located church for a series of meetings lasting a week. It is really a social as well as a religious function. The people come in great numbers, making the trip, according to their financial status, in buggies drawn by sleek, fleet-footed mules, in ox-carts, or on foot. It was amusing to see some of the latter class trudging down the hot and dusty road, with their shoes, which were brand-new, strung across their shoulders. When they got near the church, they sat on the side of the road and, with many grimaces, tenderly packed their feet into those instruments of torture. This furnished, indeed, a trying test of their religion. The famous preachers come from near and far and take turns in warning sinners of the day of wrath. Food, in the form of those two Southern luxuries, fried chicken and roast pork, is plentiful, and no one need go hungry. On the opening Sunday the women are immaculate in starched stiff white dresses adorned with ribbons, either red or blue. Even a great many of the men wear streamers of vari-coloured ribbons in the buttonholes of their coats. A few of them carefully cultivate a forelock of hair by wrapping it in twine, and on such festive occasions decorate it with a narrow ribbon streamer. Big meetings afford a fine opportunity to the younger people to meet each other dressed in their Sunday clothes, and much rustic court-ing, which is as enjoyable as any other kind, is indulged in.

This big meeting which I was lucky enough to catch was particularly well attended; the extra large attendance was due principally to two attractions, a man by the name of John Brown, who was renowned as the most powerful preacher for miles around; and a wonderful leader of sing-ing, who was known as "Singing Johnson." These two men were a study and a revelation to me. They caused me to reflect upon how great an influence their types have been in the development of the Negro in America. Both these types are now looked upon generally with condescension or contempt by the progressive element among the coloured people; but it should never be forgotten that it was they who led the race from paganism and kept it steadfast to Christianity through all the long, dark years of slavery.

John Brown was a jet-black man of medium size, with a strikingly intelligent head and face, and a voice like an organ peal. He preached each night after several lesser lights had successively held the pulpit during an hour or so. As far as subject-matter is concerned, all of the sermons were alike: each began with the fall of man, ran through various trials and tribulations of the Hebrew children, on to the redemption by Christ, and ended with a fervid pic-ture of the judgment-day and the fate of the damned. But John Brown possessed magnetism and an imagination so free and daring that he was able to carry through what the other preachers would not attempt. He knew all the arts and tricks of oratory, the modulation of the voice to al-most a whisper, the pause for effect, the rise through light, rapid-fire sentences to the terrific, thundering outburst of an electrifying climax. In addition, he had the intuition of a born theatrical manager. Night after night this man held me fascinated. He convinced me that, after all, elo-quence consists more in the manner of saying than in what is said. It is largely a matter of tone pictures.

The most striking example of John Brown's magnetism and imagination was his "heavenly march"; I shall never forget how it impressed me when I heard it. He opened his sermon in the usual way; then, proclaiming to his listeners that he was going to take them on the heavenly march, he seized the Bible under his arm and began to pace up and down the pulpit platform. The congregation immediately began with their feet a tramp, tramp, tramp, in time with the preacher's march in the pulpit, all the while singing in an undertone a hymn about marching to Zion. Suddenly he cried: "Halt!" Every foot stopped with the precision of a company of well-drilled soldiers, and the singing ceased. The morning star had been reached. Here the preacher described the beauties of that celestial body. Then the march, the tramp, tramp, tramp, and the singing were again taken up. Another "Halt!" They had reached the evening star. And so on, past the sun and moon—the intensity of religious emotion all the time increasing—along the milky way, on up to the gates of heaven. Here the halt was longer, and the preacher described at length the gates and walls of the New Jerusalem. Then he took his hearers through the pearly gates, along the golden streets, pointing out the glories of the city, pausing occasionally to greet some patriarchal members of the church, well-known to most of his listeners in life, who had had "the tears wiped from their eyes, were clad in robes of spotless white, with crowns of gold upon their heads and harps within their hands," and ended his march before the great white throne. To the reader this may sound ridiculous, but listened to under the circumstances, it was highly and effectively dramatic. I was a more or less sophisticated and non-religious man of the world, but the torrent of the preacher's words, moving with the rhythm and glowing with the eloquence of primitive poetry, swept me along, and I, too, felt like joining in the shouts of "Amen! Hallelujah!"

John Brown's powers in describing the delights of heaven were no greater than those in depicting the horrors of hell. I saw great, strapping fellows trembling and weeping like children at the "mourners' bench." His warnings to sinners were truly terrible. I shall never forget one expression that he used, which for originality and aptness could not be excelled. In my opinion, it is more graphic and, for us, far more expressive than St. Paul's "It is hard to kick against the pricks." He struck the attitude of a pugilist and thundered out: "Young man, your arm's too short to box with God!"

Interesting as was John Brown to me, the other man,

"Singing Johnson," was more so. He was a small, dark-brown, one-eyed man, with a clear, strong, high-pitched voice, a leader of singing, a maker of songs, a man who could improvise at the moment lines to fit the occasion. Not so striking a figure as John Brown, but, at "big meetings," equally important. It is indispensable to the success of the singing, when the congregation is a large one made up of people from different communities, to have someone with a strong voice who knows just what hymn to sing and when to sing it, who can pitch it in the right key, and who has all the leading lines committed to memory. Sometimes it devolves upon the leader to "sing down" a long-winded or uninteresting speaker. Committing to memory the leading lines of all the Negro spiritual songs is no easy task, for they run up into the hundreds. But the accomplished leader must know them all, because the congregation sings only the refrains and repeats; every ear in the church is fixed upon him, and if he becomes mixed in his lines or forgets them, the responsibility falls directly on his shoulders.

For example, most of these hymns are constructed to be sung in the following manner:

Leader. Swing low, sweet chariot.
Congregation. Coming for to carry me home.
Leader. Swing low, sweet chariot.
Congregation. Coming for to carry me home.
Leader. I look over yonder, what do I see?
Congregation. Coming for to carry me home.
Leader. Two little angels coming after me.
Congregation. Coming for to carry me home. . . .

The solitary and plaintive voice of the leader is answered by a sound like the roll of the sea, producing a most curious effect.

In only a few of these songs do the leader and the congregation start off together. Such a song is the well-known "Steal away to Jesus."

The leader and the congregation begin with part-singing:

Steal away, steal away,
Steal away to Jesus;

> Steal away, steal away home,
> I ain't got long to stay here.

Then the leader alone or the congregation in unison:

> My Lord he calls me,
> He calls me by the thunder,
> The trumpet sounds within-a my soul.

Then all together:

> I ain't got long to stay here.

The leader and the congregation again take up the opening refrain; then the leader sings three more leading lines alone, and so on almost *ad infinitum*. It will be seen that even here most of the work falls upon the leader, for the congregation sings the same lines over and over, while his memory and ingenuity are taxed to keep the songs going.

Generally the parts taken up by the congregation are sung in a three-part harmony, the women singing the soprano and a transposed tenor, the men with high voices singing the melody, and those with low voices a thundering bass. In a few of these songs, however, the leading part is sung in unison by the whole congregation, down to the last line, which is harmonized. The effect of this is intensely thrilling. Such a hymn is "Go down, Moses." It stirs the heart like a trumpet-call.

"Singing Johnson" was an ideal leader, and his services were in great demand. He spent his time going about the country from one church to another. He received his support in much the same way as the preachers—part of a collection, food and lodging. All of his leisure time he devoted to originating new words and melodies and new lines for old songs. He always sang with his eyes—or, to be more exact, his eye—closed, indicating the *tempo* by swinging his head to and fro. He was a great judge of the proper hymn to sing at a particular moment; and I noticed several times, when the preacher reached a certain climax, or expressed a certain sentiment, that Johnson broke in with a line or two of some appropriate hymn. The speaker understood and would pause until the singing ceased.

As I listened to the singing of these songs, the wonder of their production grew upon me more and more. How

did the men who originated them manage to do it? The
sentiments are easily accounted for; they are mostly taken
from the Bible; but the melodies, where did they come
from? Some of them so weirdly sweet, and others so wonderfully strong. Take, for instance, "Go down, Moses."
I doubt that there is a stronger theme in the whole musical
literature of the world. And so many of these songs
contain more than mere melody; there is sounded in them
that elusive undertone, the note in music which is not heard
with the ears. I sat often with the tears rolling down my
cheeks and my heart melted within me. Any musical person who had never heard a Negro congregation under the
spell of religious fervour sing these old songs has missed
one of the most thrilling emotions which the human heart
may experience. Anyone who without shedding tears can
listen to Negroes sing "Nobody knows de trouble I see,
Nobody knows but Jesus" must indeed have a heart of
stone.

As yet, the Negroes themselves do not fully appreciate
these old slave songs. The educated classes are rather
ashamed of them and prefer to sing hymns from books.
This feeling is natural; they are still too close to the conditions under which the songs were produced; but the
day will come when this slave music will be the most treas-
ured heritage of the American Negro.

At the close of the "big meeting" I left the settlement
where it was being held, full of enthusiasm. I was in that
frame of mind which, in the artistic temperament,
amounts to inspiration. I was now ready and anxious to get
to some place where I might settle down to work, and give
expression to the ideas which were teeming in my head;
but I strayed into another deviation from my path of life
as I had it marked out, which led me upon an entirely
different road. Instead of going to the nearest and most
convenient railroad station, I accepted the invitation of a
young man who had been present the closing Sunday at
the meeting to drive with him some miles farther to the
town in which he taught school, and there take the train.
My conversation with this young man as we drove along
through the country was extremely interesting. He had been
a student in one of the Negro colleges—strange coincidence
in the very college, as I learned through him, in which
"Shiny" was now a professor. I was, of course, curious to
hear about my boyhood friend; and had it not been

vacation time, and that I was not sure that I should find him, I should have gone out of my way to pay him a visit; but I determined to write to him as soon as the school opened. My companion talked to me about his work among the people, of his hopes and his discouragements. He was tremendously in earnest; I might say, too much so. In fact, it may be said that the majority of intelligent coloured people are, in some degree, too much in earnest over the race question. They assume and carry so much that their progress is at times impeded and they are unable to see things in their proper proportions. In many instances a slight exercise of the sense of humour would save much anxiety of soul. Anyone who marks the general tone of editorials in coloured newspapers is apt to be impressed with this idea. If the mass of Negroes took their present and future as seriously as do the most of their leaders, the race would be in no mental condition to sustain the terrible pressure which it undergoes; it would sink of its own weight. Yet it must be acknowledged that in the making of a race over-seriousness is a far lesser failing than its reverse, and even the faults resulting from it lean toward the right.

We drove into the town just before dark. As we passed a large, unpainted church, my companion pointed it out as the place where he held his school. I promised that I would go there with him the next morning and visit awhile. The town was of that kind which hardly requires or deserves description; a straggling line of brick and wooden stores on one side of the railroad track and some cottages of various sizes on the other side constituted about the whole of it. The young school-teacher boarded at the best house in the place owned by a coloured man. It was painted, had glass windows, contained "store bought" furniture, an organ, and lamps with chimneys. The owner held a job of some kind on the railroad. After supper it was not long before everybody was sleepy. I occupied the room with the school-teacher. In a few minutes after we got into the room he was in bed and asleep; but I took advantage of the unusual luxury of a lamp which gave light, and sat looking over my notes and jotting down some ideas which were still fresh in my mind. Suddenly I became conscious of that sense of alarm which is always aroused by the sound of hurrying footsteps on the silence of the night. I stopped work and looked at my watch. It was after eleven. I

listened, straining every nerve to hear above the tumult of
my quickening pulse. I caught the murmur of voices, then
the gallop of a horse, then of another and another. Now
thoroughly alarmed, I woke my companion, and together
we both listened. After a moment he put out the light
and softly opened the window-blind, and we cautiously
peeped out. We saw men moving in one direction, and
from the mutterings we vaguely caught the rumour that
some terrible crime had been committed. I put on my coat
and hat. My friend did all in his power to dissuade me
from venturing out, but it was impossible for me to remain
in the house under such tense excitement. My nerves
would not have stood it. Perhaps what bravery I exercised
in going out was due to the fact that I felt sure my
identity as a coloured man had not yet become known in
the town.

I went out and, following the drift, reached the railroad station. There was gathered there a crowd of men,
all white, and others were steadily arriving, seemingly
from all the surrounding country. How did the news
spread so quickly? I watched these men moving under the
yellow glare of the kerosene lamps about the station,
stern, comparatively silent, all of them armed, some
of them in boots and spurs; fierce, determined men. I
had come to know the type well, blond, tall, and lean,
with ragged moustache and beard, and glittering grey eyes.
At the first suggestion of daylight they began to disperse
in groups, going in several directions. There was no extra
noise or excitement, no loud talking, only swift, sharp
words of command given by those who seemed to be
accepted as leaders by mutual understanding. In fact,
the impression made upon me was that everything was
being done in quite an orderly manner. In spite of so many
leaving, the crowd around the station continued to grow;
at sunrise there were a great many women and children.
By this time I also noticed some coloured people; a few
seemed to be going about customary tasks; several were
standing on the outskirts of the crowd; but the gathering of
Negroes usually seen in such towns was missing.

Before noon they brought him in. Two horsemen rode
abreast; between them, half dragged, the poor wretch
made his way through the dust. His hands were tied behind him, and ropes round his body were fastened to the
saddle horns of his double guard. The men who at mid-

night had been stern and silent were now emitting that
terror-instilling sound known as the "rebel yell." A space
was quickly cleared in the crowd, and a rope placed about
his neck, when from somewhere came the suggestion,
"Burn him!" It ran like an electric current. Have you
ever witnessed the transformation of human beings into
savage beasts? Nothing can be more terrible. A railroad
tie was sunk into the ground, the rope was removed, and
a chain brought and securely coiled round the victim and
the stake. There he stood, a man only in form and stature,
every sign of degeneracy stamped upon his countenance.
His eyes were dull and vacant, indicating not a single
ray of thought. Evidently the realization of his fearful
fate had robbed him of whatever reasoning power he had
ever possessed. He was too stunned and stupefied even to
tremble. Fuel was brought from everywhere, oil, the
torch; the flames crouched for an instant as though to
gather strength, then leaped up as high as their victim's
head. He squirmed, he writhed, strained at his chains,
then gave out cries and groans that I shall always hear.
The cries and groans were choked off by the fire and
smoke; but his eyes, bulging from their sockets, rolled
from side to side, appealing in vain for help. Some of
the crowd yelled and cheered, others seemed appalled at
what they had done, and there were those who turned
away sickened at the sight. I was fixed to the spot where
I stood, powerless to take my eyes from what I did not
want to see.

It was over before I realized that time had elapsed.
Before I could make myself believe that what I saw was
really happening, I was looking at a scorched post, a
smouldering fire, blackened bones, charred fragments
sifting down through coils of chain; and the smell of
burnt flesh—human flesh—was in my nostrils.

I walked a short distance away and sat down in order
to clear my dazed mind. A great wave of humiliation and
shame swept over me. Shame that I belonged to a race
that could be so dealt with; and shame for my country,
that it, the great example of democracy to the world,
should be the only civilized, if not the only state on
earth, where a human being would be burned alive. My
heart turned bitter within me. I could understand why
Negroes are led to sympathize with even their worst
criminals and to protect them when possible. By all the

impulses of normal human nature they can and should do nothing less.

Whenever I hear protests from the South that it should be left alone to deal with the Negro question, my thoughts go back to that scene of brutality and savagery. I do not see how a people that can find in its conscience any excuse whatever for slowly burning to death a human being, or for tolerating such an act, can be entrusted with the salvation of a race. Of course, there are in the South men of liberal thought who do not approve lynching, but I wonder how long they will endure the limits which are placed upon free speech. They still cower and tremble before "Southern opinion." Even so late as the recent Atlanta riot those men who were brave enough to speak a word in behalf of justice and humanity felt called upon, by way of apology, to preface what they said with a glowing rhetorical tribute to the Anglo-Saxon's superiority and to refer to the "great and impassable gulf" between the races "fixed by the Creator at the foundation of the world." The question of the relative qualities of the two races is still an open one. The reference to the "great gulf" loses force in face of the fact that there are in this country perhaps three or four million people with the blood of both races in their veins; but I fail to see the pertinency of either statement subsequent to the beating and murdering of scores of innocent people in the streets of a civilized and Christian city.

The Southern whites are in many respects a great people. Looked at from a certain point of view, they are picturesque. If one will put oneself in a romantic frame of mind, one can admire their notions of chivalry and bravery and justice. In this same frame of mind an intelligent man can go to the threatre and applaud the impossible hero, who with his single sword slays everybody in the play except the equally impossible heroine. So can an ordinary peace-loving citizen sit by a comfortable fire and read with enjoyment of the bloody deeds of pirates and fierce brutality of vikings. This is the way in which we gratify the old, underlying animal instincts and passions; but we should shudder with horror at the mere idea of such practices being realities in this day of enlightened and humanitarianized thought. The Southern whites are not yet living quite in the present age; many of their general ideas hark back to a former century, some of them to the Dark Ages. In the light of other days they are sometimes magnificent. Today they are often cruel and ludicrous.

How long I sat with bitter thoughts running through my mind I do not know; perhaps an hour or more. When I decided to get up and go back to the house, I found that I could hardly stand on my feet. I was as weak as a man who had lost blood. However, I dragged myself along, with the central idea of a general plan well fixed in my mind. I did not find my school-teacher friend at home, so I did not see him again. I swallowed a few mouthfuls of food, packed my bag, and caught the afternoon train.

When I reached Macon, I stopped only long enough to get the main part of my luggage and to buy a ticket for New York. All along the journey I was occupied in debating with myself the step which I had decided to take. I argued that to forsake one's race to better one's condition was no less worthy an action than to forsake one's country for the same purpose. I finally made up my mind that I would neither disclaim the black race nor claim the white race; but that I would change my name, raise a moustache, and let the world take me for what it would; that it was not necessary for me to go about with a label of inferiority pasted across my forehead. All the while I understood that it was not discouragement or fear or search for a larger field of action and opportunity that was driving me out of the Negro race. I knew that it was shame, unbearable shame. Shame at being identified with a people that could with impunity be treated worse than animals. For certainly the law would restrain and punish the malicious burning alive of animals.

So once again I found myself gazing at the towers of New York and wondering what future that city held in store for me.

CHAPTER XI

I have now reached that part of my narrative where I must be brief and touch only on important facts; therefore the reader must make up his mind to pardon skips and jumps and meagre details.

When I reached New York, I was completely lost. I could not have felt more a stranger had I been suddenly dropped into Constantinople. I knew not where to turn or how to strike out. I was so oppressed by a feeling of loneliness that the temptation to visit my old home in Connecticut was well-nigh irresistible. I reasoned, however, that unless I found my old music teacher, I should be, after so many years of absence, as much of a stranger there as in New York; and, furthermore, that in view of the step which I had decided to take, such a visit would be injudicious. I remembered, too, that I had some property there in the shape of a piano and a few books, but decided that it would not be worth what it might cost me to take possession.

By reason of the fact that my living-expenses in the South had been very small, I still had nearly four hundred dollars of my capital left. In contemplation of this, my natural and acquired Bohemian tastes asserted themselves, and I decided to have a couple of weeks' good time before worrying seriously about the future. I went to Coney Island and the other resorts, took in the pre-season shows along Broadway, and ate at first-class restaurants; but I shunned the old Sixth Avenue district as though it were pest-infected. My few days of pleasure made appalling inroads upon what cash I had, and caused me to see that it required a good deal of money to live in New York as I wished to live and that I should have to find, very soon, some more or less profitable employment. I was sure that unknown, without friends or prestige, it would be useless to try to establish myself as a teacher of music; so I gave that means of earning a livelihood scarcely any consideration. And even had I considered it possible to secure pupils, as I then felt, I should have hesitated about taking up a work in which the chances for any considerable financial success are necessarily so small. I had made up my mind that since I was not going to be a Negro, I would avail myself of every possible opportunity to make a white man's success; and that, if it can be summed up in any one word, means "money."

I watched the "want" columns in the newspapers and answered a number of advertisements, but in each case found the positions were such as I could not fill or did not want. I also spent several dollars for "ads" which brought me no replies. In this way I came to know the hopes and disappointments of a large and pitiable class of humanity in this great city, the people who look for work through the newspapers. After some days of this sort of experience I concluded that the main difficulty with me was that I was not prepared for what I wanted to do. I then decided upon a course which, for an artist, showed an uncommon amount of practical sense and judgment. I made up my mind to enter a business college. I took a small room, ate at lunch counters, in order to economize, and pursued my studies with the zeal that I have always been able to put into my work upon which I set my heart. Yet, in spite of all my economy, when I had been at the school for several months, my funds gave out completely. I reached the point where I could not afford sufficient food for each day. In this plight I was glad to get, through one of the teachers, a job as an ordinary clerk in a downtown wholesale house. I did my work faithfully, and received a raise of salary before I expected it. I even managed to save a little money out of my modest earnings. In fact, I began then to contract the money fever, which later took strong possession of me. I kept my eyes open, watching for a chance to better my condition. It finally came in the form of a position with a house which was at the time establishing a South American department. My knowledge of Spanish was, of course, the principal cause of my good luck; and it did more for me: it placed me where the other clerks were practically put out of competition with me. I was not slow in taking advantage of the opportunity to make myself indispensable to the firm.

What an interesting and absorbing game is money-making! After each deposit at my savings-bank I used to sit and figure out, all over again, my principal and interests, and make calculations on what the increase would be in such and such time. Out of this I derived a great deal of pleasure. I denied myself as much as possible in order to swell my savings. As much as I enjoyed smoking, I limited myself to an occasional cigar, and that was generally of a variety which in my old days at the "Club" was known as a "Henry Mud." Drinking I cut out altogether, but that was no great sacrifice.

The day on which I was able to figure up a thousand

dollars marked an epoch in my life. And this was not because I had never before had money. In my gambling days and while I was with my millionaire I handled sums running high up into the hundreds; but they had come to me like fairy godmother's gifts, and at a time when my conception of money was that it was made only to spend. Here, on the other hand, was a thousand dollars which I had earned by days of honest and patient work, a thousand dollars which I had carefully watched grow from the first dollar; and I experienced, in owning them, a pride and satisfaction which to me was an entirely new sensation. As my capital went over the thousand-dollar mark, I was puzzled to know what to do with it, how to put it to the most advantageous use. I turned down first one scheme and then another, as though they had been devised for the sole purpose of gobbling up my money. I finally listened to a friend who advised me to put all I had in New York real estate; and under his guidance I took equity in a piece of property on which stood a rickety old tenement-house. I did not regret following this friend's advice, for in something like six months I disposed of my equity for more than double my investment. From that time on I devoted myself to the study of New York real estate and watched for opportunities to make similiar investments. In spite of two or three speculations which did not turn out well, I have been remarkably successful. Today I am the owner and part-owner of several flat-houses. I have changed my place of employment four times since returning to New York, and each change has been a decided advancement. Concerning the position which I now hold I shall say nothing except that it pays extremely well.

As my outlook on the world grew brighter, I began to mingle in the social circles of the men with whom I came in contact; and gradually, by a process of elimination, I reached a grade of society of no small degree of culture. My appearance was always good and my ability to play on the piano, especially rag-time, which was then at the height of its vogue, made me a welcome guest. The anomaly of my social position often appealed strongly to my sense of humour. I frequently smiled inwardly at some remark not altogether complimentary to people of colour; and more than once I felt like declaiming: "I am a coloured man. Do I not disprove the theory that one drop of Negro blood renders a man unfit?" Many a night when I returned to my room after an enjoyable evening, I laughed heartily over what struck me as the capital joke I was playing.

Then I met her, and what I had regarded as a joke was gradually changed into the most serious question of my life. I first saw her at a musical which was given one evening at a house to which I was frequently invited. I did not notice her among the other guests before she came forward and sang two sad little songs. When she began, I was out in the hallway, where many of the men were gathered; but with the first few notes I crowded with others into the doorway to see who the singer was. When I saw the girl, the surprise which I had felt at the first sound of her voice was heightened; she was almost tall and quite slender, with lustrous yellow hair and eyes so blue as to appear almost black. She was as white as a lily, and she was dressed in white. Indeed, she seemed to me the most dazzlingly white thing I had ever seen. But it was not her delicate beauty which attracted me most; it was her voice, a voice which made one wonder how tones of such passionate colour could come from so fragile a body.

I determined that when the program was over, I would seek an introduction to her; but at the moment, instead of being the easy man of the world, I became again the bashful boy of fourteen, and my courage failed me. I contended myself with hovering as near her as politeness would permit; near enough to hear her voice, which in conversation was low, yet thrilling, like the deeper middle tones of a flute. I watched the men gather round her talking and laughing in an easy manner, and wondered how it was possible for them to do it. But destiny, my special destiny, was at work. I was standing near, talking with affected gaiety to several young ladies, who, however, must have remarked my preoccupation; for my second sense of hearing was alert to what was being said by the group of which the girl in white was the centre, when I heard her say: "I think his playing of Chopin is exquisite." And one of my friends in the group replied: "You haven't met him? Allow me——" Then turning to me, "Old man, when you have a moment I wish you to meet Miss——." I don't know what she said to me or what I said to her. I can remember that I tried to be clever, and experienced a growing conviction that I was making myself appear more and more idiotic. I am certain, too, that, in spite of my Italian-like complexion, I was as red as a beet.

Instead of taking the car, I walked home. I needed the air and exercise as a sort of sedative. I am not sure whether my troubled condition of mind was due to the fact that I had been struck by love or to the feeling that I had made a bad impression upon her.

As the weeks went by, and when I had met her several more times, I came to know that I was seriously in love; and then began for me days of worry, for I had more than the usual doubts and fears of a young man in love to contend with.

Up to this time I had assumed and played my role as a white man with a certain degree of nonchalance, a carelessness as to the outcome, which made the whole thing more amusing to me than serious; but now I ceased to regard "being a white man" as a sort of practical joke. My acting had called for mere external effects. Now I began to doubt my ability to play the part. I watched her to see if she was scrutinizing me, to see if she was looking for anything in me which made me differ from the other men she knew. In place of an old inward feeling of superiority over many of my friends I began to doubt myself. I began even to wonder if I really was like the men I associated with; if there was not, after all, an indefinable something which marked a difference.

But, in spite of my doubts and timidity, my affair progressed, and I finally felt sufficiently encouraged to decide to ask her to marry me. Then began the hardest struggle of my life, whether to ask her to marry me under false colours or to tell her the whole truth. My sense of what was exigent made me feel there was no necessity of saying anything; but my inborn sense of honour rebelled at even indirect deception in this case. But however much I moralized on the question, I found it more and more difficult to reach the point of confession. The dread that I might lose her took possession of me each time I sought to speak, and rendered it impossible for me to do so. That moral courage requires more than physical courage is no mere poetic fancy. I am sure I should have found it easier to take the place of a gladiator, no matter how fierce the Numidian lion, than to tell that slender girl that I had

Negro blood in my veins. The fact which I had at times wished to cry out, I now wished to hide for ever.

During this time we were drawn together a great deal by the mutual bond of music. She loved to hear me play Chopin and was herself far from being a poor performer of his compositions. I think I carried her every new song that was published which I thought suitable to her voice, and played the accompaniment for her. Over these songs we were like two innocent children with new toys. She had never been anything but innocent; but my innocence was a transformation wrought by my love for her, love which melted away my cynicism and whitened my sullied soul and gave me back the wholesome dreams of my boyhood.

My artistic temperament also underwent an awakening. I spent many hours at my piano, playing over old and new composers. I also wrote several little pieces in a more or less Chopinesque style, which I dedicated to her. And so the weeks and months went by. Often words of love trembled on my lips, but I dared not utter them, because I knew they would have to be followed by other words which I had not the courage to frame. There might have been some other woman in my set whom I could have fallen in love with and asked to marry me without a word of explanation; but the more I knew this girl, the less could I find in my heart to deceive her. And yet, in spite of this spectre that was constantly looming up before me, I could never have believed that life held such happiness as was contained in those dream days of love.

One Saturday afternoon, in early June, I was coming up Fifth Avenue, and at the corner of Twenty-third Street I met her. She had been shopping. We stopped to chat for a moment, and I suggested that we spend half an hour at the Eden Musée. We were standing leaning on the rail in front of a group of figures, more interested in what we had to say to each other than in the group, when my attention became fixed upon a man who stood at my side studying his catalogue. It took me only an instant to recognize in him my old friend "Shiny." My first impulse was to change my position at once. As quick as a flash I considered all the risks I might run in speaking to him, and most especially the delicate question of introducing him to her. I confess that in my embarrassment and confusion I felt small and mean. But before I could decide what to do, he looked round at me and, after an instant, quietly asked: "Pardon me; but isn't this——?" The nobler part in me responded to the sound of his voice and I took his hand in a hearty clasp. Whatever fears I had felt were quickly banished, for he seemed, at a glance, to divine my situation, and let drop no word that would have aroused suspicion as to the truth. With a slight misgiving I presented him to her and was again relieved of fear. She received the introduction in her usual gracious manner, and wihout the least hesitancy or embarrassment joined in the conversation. An amusing part about the introduction was that I was upon the point of introducing him as "Shiny," and stammered a second or two before I could recall his name. We chatted for some fifteen minutes. He was spending his vacation north, with the intention of doing four or six weeks' work in one of the summer schools; he was also going to take a bride back with him in the fall. He asked me about myself, but in so diplomatic a way that I found no difficulty in answering him. The polish of his language and the unpedantic manner in which he revealed his culture greatly impressed her; and after we had left the Musée she showed it by questioning me about him. I was surprised at the amount of interest a refined black man could arouse. Even after changes in the conversation she reverted several times to the subject of "Shiny." Whether it was, more than mere curiosity I could not tell, but I was convinced that she herself knew very little about prejudice.

Just why it should have done so I do not know, but somehow the "Shiny" incident gave me encouragement and confidence to cast the die of my fate. I reasoned, however, that since I wanted to marry her only, and since it concerned her alone, I would divulge my secret to no one else, not even her parents.

One evening, a few days afterwards, at her home we were going over some new songs and compositions when she asked me, as she often did, to play the Thirteenth Nocturne. When I began, she drew a chair near to my right and sat leaning with her elbow on the end of the piano, her chin resting on her hand, and her eyes reflecting the emotions which the music awoke in her. An impulse which I could not control rushed over me, a wave of exultation, the music under my fingers sank almost to a whisper, and calling her for the first time by her Christian name, but without daring to look at her, I said: "I love you, I love you, I love you." My fingers were trembling so that I ceased playing. I felt her hand creep to mine, and when I looked at her, her eyes were glistening with tears. I understood, and could scarcely resist the longing to take her in my arms; but I remembered, remembered that which has been the sacrificial altar of so much happiness—Duty; and bending over her hand in mine, I said: "Yes, I love you; but there is something more, too, that I must tell you." Then I told her, in what words I do not know, the truth. I felt her hand grow cold, and when I looked up, she was gazing at me with a wild, fixed stare as though I was some object she had never seen. Under the strange light in her eyes I felt that I was growing black and thick-featured and crimp-haired. She appeared not to have comprehended what I had said. Her lips trembled and she attempted to say something to me, but the words stuck in her throat. Then, dropping her head on the piano, she began to weep with great sobs that shook her frail body. I tried to console her, and blurted out incoherent words of love, but this seemed only to increase her distress, and when I left her, she was still weeping.

When I got into the street, I felt very much as I did the night after meeting my father and sister at the opera in Paris, even a similar desperate inclination to get drunk; but my self-control was stronger. This was the only time in my life that I ever felt absolute regret at being coloured, that I cursed the drops of African blood in my veins and wished that I were really white. When I reached my rooms, I sat and smoked several cigars while I tried to think out the significance of what had occurred. I reviewed the whole history of our acquaintance, recalled each smile she had given me, each word she had said to me that nourished my hope. I went over the scene we had just gone through, trying to draw from it what was in my favour and what was against me. I was rewarded by feeling confident that she loved me, but I could not estimate what was the effect upon her of my confession. At last, nervous and unhappy, I wrote her a letter, which I dropped into the mail-box before going to bed, in which I said:

I understand, understand even better than you, and so I suffer even more than you. But why should either of

us suffer for what neither of us is to blame for? If there
is any blame, it belongs to me and I can only make the
old, yet strongest plea that can be offered, I love you; and
I know that my love, my great love, infinitely over-
balances that blame and blots it out. What is it that
stands in the way of our happiness? It is not what you
feel or what I feel; it is not what you are or what I am.
It is what others feel and are. But, oh! is that a fair
price? In all the endeavours and struggles of life, in all
our strivings and longings, there is only one thing worth
seeking, only one thing worth winning, and that is
love. It is not always found; but when it is, there is
nothing in all the world for which it can be profitably
exchanged.

The second morning after, I received a note from her
which stated briefly that she was going up into New Hampshire to spend the summer with relatives there. She made
no reference to what had passed between us; nor did she
say exactly when she would leave the city. The note contained no single word that gave me any clue to her feelings.
I could gather hope only from the fact that she had written
at all. On the same evening, with a degree of trepidation
which rendered me almost frightened, I went to her house.
I met her mother, who told me that she had left for
the country that very afternoon. Her mother treated me in
her usual pleasant manner, which fact greatly reassured
me; and I left the house with a vague sense of hope stirring
in my breast, which sprang from the conviction that she
had not yet divulged my secret. But that hope did not
remain with me long. I waited one, two, three weeks,
nervously examining my mail every day, looking for some
word from her. All of the letters received by me seemed
so insignificant, so worthless, because there was none from
her. The slight buoyancy of spirit which I had felt gradually
dissolved into gloomy heart-sickness. I became preoccupied;
I lost appetite, lost sleep, and lost ambition. Several of my
friends intimated to me that perhaps I was working too
hard.
She stayed away the whole summer. I did not go to the
house, but saw her father at various times, and he was as
friendly as ever. Even after I knew that she was back in
town, I did not go to see her. I determined to wait for some
word or sign. I had finally taken refuge and comfort in my

pride, pride which, I suppose, I came by naturally enough.

The first time I saw her after her return was one night at the theatre. She and her mother sat in company with a young man whom I knew slightly, not many seats away from me. Never did she appear more beautiful; and yet, it may have been my fancy, she seemed a trifle paler, and there was a suggestion of haggardness in her countenance. But that only heightened her beauty; the very delicacy of her charm melted down the strength of my pride. My situation made me feel weak and powerless, like a man trying with his bare hands to break the iron bars of his prison cell. When performance was over, I hurried out and placed myself where, unobserved, I could see her as she passed out. The haughtiness of spirit in which I had sought relief was all gone, and I was willing and ready to undergo any humiliation.

Shortly afterward we met at a progressive card party, and during the evening we were thrown together at one of the tables as partners. This was really our first meeting since the eventful night at her house. Strangely enough, in spite of our mutual nervousness, we won every trick of the game, and one of our opponents jokingly quoted the old saw: "Lucky at cards, unlucky in love." Our eyes met and I am sure that in the momentary glance my whole soul went out to her in one great plea. She lowered her eyes and uttered a nervous little laugh. During the rest of the game I fully merited the unexpressed and expressed abuse of my various partners; for my eyes followed her wherever she was and I played whatever card my fingers happened to touch.

Later in the evening she went to the piano and began to play very softly, as if to herself, the opening bars of the Thirteenth Nocturne. I felt that the psychic moment of my life had come, a moment which, if lost, could never be called back; and, in as careless a manner as I could assume, I sauntered over to the piano and stood almost bending over her. She continued playing, but, in a voice that was almost a whisper, she called me by my Christian name and said: "I love you, I love you, I love you." I took her place at the piano and played the Nocturne in a manner that silenced the chatter of the company both in and out of the room, involuntarily closing it with the major triad. We were married the following spring, and went to Europe for several months. It was a double joy for me to be in France again under such conditions.

First there came to us a little girl, with hair and eyes dark like mine, but who is growing to have ways like her mother. Two years later there came a boy, who has my temperament, but is fair like his mother, a little golden-headed god, with a face and head that would have delighted the heart of an old Italian master. And this boy, with his mother's eyes and features, occupies an inner sanctuary of my heart; for it was for him that she gave all; and that is the second sacred sorrow of my life.

The few years of our married life were supremely happy, and perhaps she was even happier than I; for after our marriage, in spite of all the wealth of her love which she lavished upon me, there came a new dread to haunt me, a dread which I cannot explain and which was unfounded, but one that never left me. I was in constant fear that she would discover in me some shortcoming which she would unconsciously attribute to my blood rather than to a failing of human nature. But no cloud ever came to mar our life together; her loss to me is irreparable. My children need a mother's care, but I shall never marry again. It is to my children that I have devoted my life. I no longer have the same fear for myself of my secret's being found out, for since my wife's death I have gradually dropped out of social life; but there is nothing I would not suffer to keep the brand from being placed upon them.

It is difficult for me to analyse my feelings concerning my present position in the world. Sometimes it seems to me that I have never really been a Negro, that I have been only a privileged spectator of their inner life; at other times I feel that I have been a coward, a deserter, and I am possessed by a strange longing for my mother's people.

Several years ago I attended a great meeting in the interest of Hampton Institute at Carnegie Hall. The Hampton students sang the old songs and awoke memories that left me sad. Among the speakers were R. C. Ogden, ex-Ambassador Choate, and Mark Twain; but the greatest interest of the audience was centred in Booker T. Washington, and not because he so much surpassed the others in eloquence, but because of what he represented with so much earnestness and faith. And it is this that all of that small but gallant band of coloured men who are publicly fighting the cause of their race have behind them. Even those who oppose them know that these men have the eternal principles of right on their side, and they will be victors even though they should go down in defeat. Beside them I feel small and selfish. I am an ordinarily successful white man who has made a little money. They are men who are making history and a race. I, too, might have taken part in a work so glorious.

My love for my children makes me glad that I am what I am and keeps me from desiring to be otherwise; and yet, when I sometimes open a little box in which I still keep my fast yellowing manuscripts, the only tangible remnants of a vanished dream, a dead ambition, a sacrificed talent, I cannot repress the thought that, after all, I have chosen the lesser part, that I have sold my birthright for a mess of pottage.

THE AFRICAN AMERICAN EXPERIENCE

BULLWHIP DAYS
edited by James Mellon
0-380-70884-1/ $14.00 US/ $19.00 CAN
In their own voices, an oral history of the personal memories of the last survivors of American slavery.

CELIA: A SLAVE
by Melton A. McLaurin
0-380-71935-3/ $12.50 US/ $18.50 CAN
An account of a landmark courtroom battle that threatened to undermine the very foundation of the old South's most cherished institution.

SUCCESS RUNS IN OUR RACE
by George Fraser
0-380-72622-X/ $13.00 US/ $19.95 CAN

GROWING UP BLACK
edited by Jay David
0-380-76632-9/ $12.50 US/ $16.00 CAN
From slave days to the present—25 African-Americans reveal the trials and triumphs of their childhoods.